Polymer Science Overview

A Tribute to Herman F. Mark

Polymer Science Overview

A Tribute to Herman F. Mark

G. Allan Stahl, EDITOR
Phillips Research Center

Based on a symposium
sponsored by the Organic Coatings
and Plastics Chemistry Division
at the Second Chemical Congress
of the North American Continent
(180th ACS National Meeting),
Las Vegas, Nevada,
August 25–28, 1980.

ACS SYMPOSIUM SERIES **175**

AMERICAN CHEMICAL SOCIETY
WASHINGTON, D. C. 1981

Library of Congress CIP Data

Polymer science overview.
(ACS symposium series, ISSN 0097-6156; 175)

"Based on a symposium sponsored by the Organic Coatings and Plastics Chemistry Division at the Second Chemical Congress of the North American Continent (180th ACS National Meeting), Las Vegas, Nevada, August 27–28, 1980."

Bibliography and index.

1. Polymers and polymerization—Congresses. 2. Mark, Herman Francis, 1895– . 3. Chemists—United States—Biography.
I. Stahl, G. Allan. II. American Chemical Society. Division of Organic Coatings and Plastics Chemistry. III. Chemical Congress of the North American Continent (2nd: 1980: Las Vegas, Nev.) IV. Series.

QD380.P645 547.7 81–17647
ISBN 0–8412–0668–6 AACR2 ACSMC8 175 1–356
ISBN 0-8412-0703-8 (pbk.) 1981

PRINTED IN THE UNITED STATES OF AMERICA

ACS Symposium Series

M. Joan Comstock, *Series Editor*

FOREWORD

The ACS SYMPOSIUM SERIES was founded in 1974 to provide a medium for publishing symposia quickly in book form. The format of the Series parallels that of the continuing ADVANCES IN CHEMISTRY SERIES except that in order to save time the papers are not typeset but are reproduced as they are submitted by the authors in camera-ready form. Papers are reviewed under the supervision of the Editors with the assistance of the Series Advisory Board and are selected to maintain the integrity of the symposia; however, verbatim reproductions of previously published papers are not accepted. Both reviews and reports of research are acceptable since symposia may embrace both types of presentation.

CONTENTS

EDITOR'S PREFACE

Polymer science is stimulating and exciting because it is a new and still rapidly growing discipline. Consequently, it seemed clear that a symposium on this field would be informative as well as popular. The symposium upon which this book is based was sponsored jointly by the Division of Organic Coatings and Plastics Chemistry and the Division of Polymer Chemistry of the American Chemical Society. It was planned as a tribute to one of the pioneers of the field, Herman F. Mark, on his 85th birthday and brought together many notables onto a common platform.

This volume is a collection of several papers presented in Las Vegas and a number of special contributions submitted in honor of Professor Mark. The collection is as diverse as the field of polymers, and is intended as a general overview of polymer science. The book will be interesting reading for anyone working in the field of polymers, interested in the history of science, or curious about specific developments of science.

Professor Mark's story is told in three chapters by the Editor and four reminiscences by Rudolf Brill (whose association with Mark dates back to 1922), Hans Mark (his son), Linus Pauling, and Maurice Morton. The history of polymer science is given in separate chapters by the Editor, Robert Simha (who has worked with Professor Mark in two countries), and Carl "Speed" Marvel. One chapter by Charles Carraher gives an up to the minute report on the status of polymer education. The remainder of the book is a collection of reviews and previews of specific, timely topics in polymer science. Despite the diversity of topics, each area covered has contributions from Herman Mark.

It has been said that Herman Mark's story is the story of polymer science. Considering the importance of polymers in today's and tomorrow's world, Mark's life and contributions are important to us all. This volume is respectfully submitted with this importance in mind as a tribute to Herman F. Mark.

G. A. STAHL
Bartlesville, Oklahoma

July 1981

Editor Stahl, Carl "Speed" Marvel, Herman Mark

PREFACE

Born in Vienna, Austria, on May 3, 1895, Professor Herman Mark is as vital today as his almost innumerable contributions to polymer science. He continues to maintain an active schedule as a world traveler and lecturer at numerous national and international symposia and seminars. His home base, however, remains at Polytechnic Institute of New York, where he is Dean Emeritus and an emeritus member of the Board of Trustees.

Dr. Mark is acknowledged throughout the scientific community as the father of polymer science. He received the National Medal of Science in 1980, the 1978 Senior U.S. Scientist Award, 29 other medals and awards from various international organizations, and 17 honorary degrees. In 1977, he was appointed as an Honorable Member of the Japan Chemical Society. He has published over 500 original and review articles and some 20 books on topics related to polymer chemistry. Dr. Mark is founder and first editor of the *Journal of Polymer Science,* the definitive publication in its field, and is currently editor of the *Journal of Applied Science* and associate editor of the *Textile Research Journal.* In addition to these duties he serves as Chairman of the Editorial Board of the *Encyclopedia of Polymer Science and Technology.*

Dr. Mark's World War I exploits (including an escape from an enemy prisoner-of-war camp) read like a Hollywood film script. During World War II, he contributed to the Allied cause through development of plastic-coated tents, mildewproof fabrics, gasproof clothing, an amphibious truck, and the snow vehicle called "the Weasel."

Dr. Herman Mark is a teacher of others and possesses great warmth, sensitivity, and wit. He remains a scientist without comparison in his discipline, to which the greatness of his work testifies. He is a man for all seasons.

ELI M. PEARCE
Brooklyn, New York

July 1981

The Geheimrat

Program speakers of symposium presented in honor of Dr. Herman F. Mark. Second Chemical Congress of the North American Continent in Las Vegas on August 25, 1980 (l-r, E. M. Pearce, C. G. Overberger, M. Goodman, P. J. Flory, H. Ringsdorf, "Speed" Marvel, Hans Mark, the Geheimrat, W. O. Baker, Jr., and R. Simha).

Review of Polymer Science: Introductory Remarks

TURNER ALFREY

Dow Chemical Company, Central Research, 1702 Building, Midland, MI 48640

A review of Polymer Technology must begin with Prehistoric Man.

A great deal of our present economic activity has its roots either in prehistoric times or in the earliest historic period: for example, agriculture, including the domestication of plants and animals. The ostrich was apparently domesticated during historic times, but the other domestic animals are a heritage from our prehistoric ancestors.

Cyril Smith has reported how much modern materials technology owes to very early man. In the case of polymeric materials, this includes the utilization of many naturally occurring macromolecular substances: wood, leather from hides, various fibers, vegetable gums, and animal adhesives. This utilization was not merely of the nature of using tree branches as clubs. Complex and sophisticated technology was developed. One example was the utilization of fibers: spinning short fibers to make yarn; weaving yarn to make cloth; cutting and sewing cloth to form garments which fit surfaces of positive and negative Gaussian Curvature and non-Euclidean intrinsic geometry. Other early technologies included adhesives, composites, surface coatings, and leather.

The invention of paper (as distinct from Papyrus) occurred well into historic times; but apart from paper, the polymer technology of the year 1800 was essentially the legacy of prehistoric and earliest historic times.

The Industrial Revolution, of course, had been largely concerned with polymeric materials--but the new developments were principally of a mechanical nature, the mechanization of ancient hand-crafts, rather than fundamental innovations in polymeric materials per se.

The 19th Century saw the discovery of rubber vulcanization, and other chemical modifications of natural polymers. The science of Chemistry began to play some part in this technology, and around the beginning of the 20th Century, truly synthetic polymers began to emerge (for example, Bakelite).

0097-6156/81/0175-0001$5.00/0

At the time of World War I, here was the situation: The production, processing, and fabrication of polymeric materials constituted an important part of the Gross National Product. These materials were the subject of serious and in some respects highly astute scientific investigation.

There were two principle limitations to this science and technology: First, it was almost exclusively concerned with materials of natural origin, and derivatives thereof. Second, the scientific understanding did not include adequate concepts of molecular structure. (For example, natural rubber was recognized as being in some sense a polymer of isoprene units--but suggested specific structures such as a cyclic octamer could not serve as the basis for structure-property relationships).

During the following decade, the central structural concept--that of long chains of covalently bonded atoms--was developed and established.

Once this central structural concept was accepted, it rapidly became possible to apply to polymers that general fund of knowledge of synthetic organic chemistry, structural organic chemistry, and physical chemistry, which had accumulated during the past century.

Structure-property relationships, and rational polymer synthesis, facilitated the rapid growth of synthetic polymer industries: synthetic fibers; synthetic rubbers; synthetic plastics; coatings; adhesives; and others. This rapid industrial growth was accompanied by growth in scientific research, journals, books, and technical societies dealing with polymers. This section of the book will deal with the developments of this eventful half-century.

An important event in the history of American polymer science and technology occurred in 1940--the arrival of Dr. Herman Mark in this country and at the Polytechnic Institute of Brooklyn. By that time, Dr. Mark had established in Europe a whole set of distinguished careers: military, scientific; industrial; academic--any one of which would have marked him as a most outstanding person. In Brooklyn, and operating out of Brooklyn, he set about building a new set of careers, involving research, teaching, consulting, publishing, and other activities in support of polymer science. The Polymer Institute at Brooklyn; the broad array of polymer courses; the Saturday seminars; the Journal of Polymer Science; the Interscience High Polymer Monograph Series; Dr. Mark's extensive industrial consultation; his activities as an international lecturer and with international scientific societies; his assistance to foreign scientific visitors; his many services to the various ACS divisions concerned with polymers; and his deep, helpful interest in the careers of younger scientists. Such activities have spanned the years and continued until the present. Just recently, Dr. Mark has served as chairman of the largest single study of Fire Safety of Polymers.

This section will deal with the developments of polymer science and polymer technology which have taken place during the past half-century, and with Dr. Mark's extensive involvement in these developments. Two topics which I will not even begin on here--because they would require the entire time just to list--are Dr. Mark's numerous well-deserved honors and awards, and my own deep indebtedness to him.

I am sorry that I was unable to attend the symposium in person and extend congratulations and best wishes to Dr. Herman Mark. Only his youthfulness prevents me from calling him "the elder statesman of Polymer Science".

RECEIVED September 28, 1981.

1

Herman F. Mark: The Early Years, 1895–1926

G. ALLAN STAHL

Phillips Petroleum Company, Research and Development, Bartlesville, OK 74004

"In my boyhood, life in Vienna was like a wheel with equal parts of discipline, creativity, and sports, Oh yes, sports."

H. F. Mark

Herman Francis Mark was born in Vienna on May 3, 1895. The oldest of three children and son of a family physician, he was raised in that city's Fourth District, then a semi-commercial area inhabited by lower middle class families. Mark's father was a German Jew who embraced Lutheranism on marriage.

Mark's turn of the century Vienna was exciting yet genteel. Its atmosphere was one which fostered new things. An atmosphere which has been described as being at its best because it was a time of change tempered by things old. The model of the exposures which Mark was refering above was not inaccurate, for Mark was born into a special place in a special time.

The discipline he mentions was taught by example by intelligent, benevolent adults. A young middle-class Viennese was taught to seek knowledge and maintain an open mind in judging facts. He was taught cooperation and how to listen, yet intellectual freedom was also fostered. Freedom to question and analyze, and rethink. A freedom which, Mark cautions, should not be confused with undisciplined, boundless movement.

Creativity was all about him since Vienna was leading the world in the development of new music, art, and architecture. Viennese Arnold Schonberg, Johann Strauss, and Richard Strause were making their indelible marks on music of the era, and the expressionist painters Oskar Kokorshka and Egon Schiele were stirring the art world. There was also Sigmund Freud and his revolutionary ideas about the framework of the human mind, and Theodor Herzl who, with the scientist-author Chaim Weizmann, founded and lead the Zionist movement. Many of these individuals were visitors in the Mark home, and their presence, by the sheer

0097-6156/81/0175-0005$5.00/0

Figure 1. Herman Mark, age 12, near Vienna.

drive of their personalities, had a tremendous influence on Herman.

Completing the wheel was sports. "Oh yes, sports", rendered with a wistfulness which shows their importance in Mark's life. In his boyhood, he was an exceptional performer in soccer, tennis, and mountain climbing. A gifted athlete, Mark ran to and from school, often exhausting his friends with seeming unending energy. He played on a local football club, and became an accomplished skiier and mountain climber. Besides the obvious physical maturation sports nourished, there was an additional benefit. Through sports, an attitude of healthy competition was developed. Not as established as today, sports of that era stressed physical exercise and competition of skills. Winning was not as important as it is now.

> "My brother always had high grades except in behavior, and this was a great pain to our father. One day he came home early and told our mother that he had heard one gentleman tell another: 'The worse boys in the area are the Mark boys!'"
>
> Hans Mark, M.D.

The three segments in place, the mischievious Mark grew up in a wholesome world. On a typical day he would arise and breakfast, run two miles to school at the Theresianiche Akademie, attend classes lunching at school, and then run home. After school he would play tennis or soccer until dark, have dinner, and study until bedtime. This is, of course, unless the family attended a concert or opera, which was two or three times a month. On these occasions, he would study during lunch. Between 8 and 15 years of age, Mark has described himself as "ambitious". From 16 to 18 years, he says he "worked less" and was "less concerned with excellence". His extraordinary energy was evident at all times.

At age twelve, Mark experienced one of those molding events which most scientists appear to have had, and almost all seem to enjoy relating. Mark and his close friend, Gerhardt Kirsch, toured the laboratories of the University of Vienna at the invitation of Gerhardt's father, a professor of Material Science. The boys returned home excited by "the sight of flasks and beakers and retorts, of blue flamed Bunsen burners, of bubbling liquids and the lengths of rubber tubing that carried off their vapors". The cliche' is probably accurate in so far as "nothing was ever the same again".

Before long both Herman and Gerhardt had made laboratories of their bedrooms. Their access to chemicals, through their fathers, was broad. They performed color changing reactions, burned sulfur, and eventually graduated into more sophisticated, qualitative analysis of metal salts and the generation of gases. They even once obtained a spectroscope to experiment with, through the avuncular generosity of a Mr. Waldstein.

It should also be noted, that Herman and Gerhardt's interest in science was furthered by one of their teachers, a priest named Hlavati. Fr. Hlavati taught mathematics, physics, and chemistry. With his lectures and especially his demonstrations and experiments, he piqued their imaginations, and, according to Mark, was the one man who most interested them in the study of nature and, in particular, physical science. Gerhardt Kirsch, three years older than Herman, graduated from gymnasium, attended the University of Vienna, and eventually became a professor of physics.

> "Oh, they always give big medals to the high ranks, and lesser ones to the others. I was just fortunate enough to be a Lieutenant."
>
> H. F. Mark

> "I first met Mark in 1918 at the University of Vienna where we both had a summer study leave from the armed forces. I quickly recognized that he was a gifted student in addition to having gained exceptional merit in the war. That summer he was presented the Leopold Orden which was unusual for a junior officer."
>
> E. Schmid

Herman took his final examination in July, 1913. At that time military service was mandatory for all young men, thus his only decision was whether to go directly into the army for his one year's service, or to take his studies at the University and then serve. He, after consultation with his father, enlisted as a private in the elite Alpine infantry unit, the Kaiserschutzen regiment. He was immediately stationed in the mountains of South Tirol.

Pre-World War I army life was far from disagreeable to Mark. It satisfied both his need for physical activity and thirst for knowledge. After a few months he was an acknowledged mountaineer, spoke Italian, and even, he claimed, was growing on the "good" army food.

Figure 2. The 1913 externe class of the Theresianiche Akademie.

Figure 3. Lt. Herman Mark and comrades in the South Tirol, 1917.

On June 23, 1914, as Mark neared the end of his service, the events of Serajevo changed everything. The Kaiserschutzen regiment was immediately reassigned to combat, not in the mountains, but on the plains of Galicia on the Russian front.

Ill trained for defensive warfare on a massive front the Austrians fell back before a coordinated Russian attack. In the ensuing two months they retreated nearly 300 miles. Corporal Mark, however, frequently stood out. In the bitter battles of Lvov and Przemysl, in which the regiment reportedly lost eighty-five percent of its men, Mark was decorated for bravery. Once in a rear guard action he single handedly held ten Russians at bay until his unit was safely withdrawn. His stamina kept him in good stead, just out of harm, until in September, 1914, he was shot in the right ankle.

Wounded, he was evacuated to Hungary, where it was ascertained that the bullet had broken a bone, and he would require additional hospitalization. He was then sent to Vienna.

Back in Vienna, Mark spent the Christmas of 1914 healing at home. But rest and relaxation for this soldier took an unusual form, for during this break he began his study of chemistry at the University of Vienna. He completed one of the eight required semesters of study, and then returned with his regiment to the front. This time they were sent to the more familiar but no longer hospitable mountains along the Austrian-Italian border.

During the next three years Austria and Italy fought a stalemate, neither side able to win a decisive victory. Slowly, however, the tide began to rise in favor of the Italians. In those days, the desperate Austrians used the Kaiserschutzen regiment as an attack force to regain lost terrain. On one such occasion Lieutenant Mark, now decorated for valor on numerous occasions, won one of Austria's highest awards for bravery, the Leopolds-Orden, at Zugna Torta.

In the summer of 1918, the Italians captured Zugna Torta, a peak, and threatened the city of Trento. Tossed into the conflict, Mark's regiment suffered greatly trying to recapture the heights. In a final and desperate drive, Mark led several hundred men over the ridge, and, thus, saved the city. During the same summer, he returned to the campus and completed another semester of study. The semester in which, as Schmid points out, his exceptionally high academic abilities were recognized.

Mark returned to the front during the fall, but his sacrifices were in vain. The Austrian front collasped in November, 1918, and Mark, now the most decorated company-grade officer in the Austrian Army, was imprisoned by the Italians.

Prison as it turned out was in an old convent near Monopoli in Southern Italy. Mark and his comrades had a relatively good life. Certainly they would have prefered to have been home, but most of them, profitably, used the time to study. Italian was the most popular subject, but a few, like Mark, included French and English. Very few, however, joined Mark in extending their studies to include chemistry.

Good hosts, the Italians purchased books for their prisoners. These were different times in this first of the modern wars. Many of the old ways still prevailed. The prisoners, just as an example, received their pay and mail.

Mark would have passed these times patiently had he not received a letter from Vienna in May, 1919, informing him that his father was seriously ill. In a made for Hollywood script, Mark prearranged a fist fight with another prisoner in order to be sent to a disciplinary camp. In route, he bribed an Italian sergeant to look the other way as he escaped in the washroom of a railway station. Next, he sewed on a few brass buttons to make his green Austrian uniform look more like that of a British private, and boarded a train headed north.

Enroute, a British Colonel sat near him. For miles he avoided the Colonel's stare until, finally, Mark decided that candor was his best course. "I feel obliged to explain the situation," he said in his recently polished English. He did, and continued home without further incident.

On arrival in Vienna in August, 1919, he found his father much improved and himself a national hero. Unimpressed, Mark enrolled in the University in September, 1919, to continue his studies.

> "I have especially high expectations of young Mark."
>
> E. Schroedinger

> "The concept of 'Free Radicals' was not known in 1920, well perhaps in politics but not in chemistry."
>
> H. F. Mark

Mark took advantage of every opportunity to recover the time lost during the war; completing three academic semesters a year he graduated suma cum laude with a Ph.D. in chemistry in July, 1921. His major professor was Wilhelm Schlenk, a highly respected organic chemist. His work, in which his experimental dexterity was especially notable, was both timely and well received. Mark, in his dissertation titled "The Synthesis of Pentaphenylethyl", discussed the presence of the recently proposed free radical. The findings were published in two papers with Schlenk (1, 2).

At the time of Mark's graduation, Schlenk accepted a position as successor to Emil Fischer at the University of Berlin. Mark and three other graduates with great potential went with Schlenk. In Berlin, he was exposed to scientists who were world renown experts in their respective research fields. Among the chemists he met Karl Fruedenberg (polysaccharides), Herman Leuchs (alkaloids), Max Bergmaun (polypeptides), and Carl Harries (rubber), and the great Walther Nernst. He also met some celebrated physicists including Albert Einstein, Erwin Schrodinger, Max Planck, and Max von Laue. Mark's research was an extension of his studies with Schlenk in Vienna.

A second and greater opportunity came his way in the spring of 1922. Professor Fritz Haber, discoverer of the Haber ammonia synthesis process and head of the Kaiser Wilhelm Institute for Physical Chemistry (now known as the Max Planck Institute), contacted Professor Schlenk.

Haber's Institute, although established to study pure research in 1911, was completely involved with the development of gas warfare in the First World War. After the war, as Germany's leading scientific organization, it was again diverted. This time to study the extraction of gold from sea water in an effort to pay Germany's war reparation. This effort would be unsuccessful and would torment Haber for the remainder of his life. However, while diverted, Haber recognized the need to stimulate the large German chemical industry, and thus he helped establish similar institutes for applied research. The Institute for Fiber Research, established in 1920 and directed by R. O. Herzog, was one such organization.

In 1922, Herzog had gathered a team of excellent physical chemists and physicists. Among them, Michael Polanyi, Karl Weissenberg, Erich Schmid, and Rudolf Brill. Since he intended to study organic fibers starting with cellulose, he felt it advisable to broaden this team with an organic chemist.

Haber asked his old friend Schlenk to recommend a young organic chemist with a wide range of interests. As a result of Schlenk's recommendation, Mark was called to meet with Haber. Mark recalls that Haber and Schlenk sat on either side of him, enveloped his head in a cloud of cigar smoke, and then, appropriately, recommended that a change in atmosphere would be to his advantage. That summer, Mark moved to the Berlin suburb of Dahlem, and joined Polanyi's research section.

> "After I finished my high school studies in Vienna, I joined the staff in Dahlem, and was told I would work in X-ray crystal structure. I told Mark I couldn't imagine the vaguest thing about it. He said, "Don't worry." You shouldn't let people intimidate you with their Hokus-Pokus."
>
> O. Kratky

Mark's first assignment was, with Brill, to set up two X-ray tubes in order to study the structure of simple compounds. Mark credits this approach to Herzog and Polanyi, and has said that it was a key to his later successes in the polymer areas. "If we had started", Mark summarized over a half century later, "with cellulose and rubber, and not zinc and tin , we would have failed." However, in 1922 their problems were even more basic. Neither Mark nor Brill had any experience in the fledging study of X-rays.

Although Herzog would eventually devote one-third of the Institute to X-ray structure determination, this particular early effort could not even be placed in the Institute's building. Thus Mark and Brill began hand construction of their equipment in the basement of Haber's villa. Haber did not mind the intrusion, and although she did not approve, Mrs. Haber accepted the workers. The Haber's dog had different views. Had the resourceful novice crystallographers not learned to bribe the pug with sausage, the course of X-ray structure determination might have been different.

When Mark and Brill began using their X-ray equipment to investigate the lattice structure of metals, and simple organic and inorganic compounds, equipment Mark has since described as an air filled X-ray tube, a ruler, and a log table, they were early students of an infant analytical technique. Only a few laboratories were, at that time, working extensively in the field. They included the laboratories of Debye and Scherrer (Gottingen), the Braggs (London), Nishikawa and Ono (Tokyo), and Herzog (Dahlem).

X-ray diffraction was in 1922 only recently born of the brilliant hypothesis and experimental verification by Max von Laue, and equally brilliant confirmation of von Laue's work by the British physicists, W. H. Bragg and his son W. L. Bragg.

Prior to von Laue, physicists had obtained information about crystal structure through chemical analysis, interfacial angle measurements, and the determination of physical properties. There was little knowledge of the internal structure, although it was believed that crystals were built up by periodic repetition of some unit, probably an atom or molecule, and that these units were about 1 or 2 Å apart. It was also suspected, but not confirmed, that X-rays were electromagnetic waves of about 1 or 2 Å in wavelength.

Von Laue, with these points in mind, hypothesized in 1912 that if crystals were composed of regularly spaced atoms of interatomic distance about equal to the wavelength of X-rays they might act as scattering centers for X-rays. As a model, he knew of the phenomenon of diffraction, and as an example he knew that visible light was scattered by a set of regularly spaced objects, such as a ruled grating. Under von Laue's direction, two young physicists Friedrich and Knipping tested this hypothesis by a simple experiment. They set a crystal of copper sulfate in the path of a narrow beam of X-rays and in front of a photographic plate to record the presence of diffracted beams. On development of the plate, they found that the first experiment was successful. It demonstrated without a doubt that X-rays were diffracted by the crystal out of the primary beam forming a pattern of spots on the photographic plate.

By the following year, the Braggs had confirmed von Laue's diffraction experiment, given direct proof of the existance of the space lattice, and provided a simple expression (the Bragg Law) for the relationship between the wave length of X-rays used and the lattice spacings in the crystal. The routine collection of diffraction patterns was later promoted by the invention of the X-ray goniometer by K. Weissenberg in Dahlem in 1924.

The technique received immediate acceptance by those studying crystal structure, and was quickly applied to the investigation of natural fibers by R. O. Herzog at the Institute for Fiber Research. By the summer of 1920, Herzog and his assistant Willie Jancke were obtaining powder patterns from crushed cellulosic fibers and had began to irradiate entire fibers in the hope that evidence of a regular structure along the length of the fiber might be found.

> "He (Mark) once showed me an X-ray burn on his finger that he got while adjusting a camera. Herzog characterized his manner of working as, 'Mark does not experiment with X-rays; he fights with them.'"
>
> O. Kratky

Figure 4. *Fiber Research Institute in Dahlem in 1923.*

Mark, in the five years he worked at the Institute, established himself as an expert crystallographer. Between 1923 and 1927 more than fifty papers were (co)authored on the lattice structure of metals, simple organic and inorganic compounds, and polymers. The laboratory became known as an efficient, smoothly functioning research organization which attracted all kinds of visitors. Max von Laue was a frequent visitor as was P. P. Ewald, the author of the first book on X-ray diffraction. Others who visited Dahlem were J. R. Katz from Holland, O. Hassel from Norway, C.T. Greenwood from Scotland, J. J. Trillat from France, S. N. Bose from India, and Albert Einstein from Berlin.

Working first with Polanyi, Weissenberg, and Brill, and later as the leader of the Textile Chemistry Section, Mark successively published papers on the crystal structures of hexamethylenetetramine, pentaerythritol, zinc salts, tin, urea, tin salts, triphenylmethane, bismuth, graphite, sulfur, oxalic acid, acetaldehyde, ammonia, ethane, diborane, carbon dioxide, and some aluminum silicates. Each paper showed his and the laboratory's increasing sophistication in the technique of X-ray diffraction. Their work over the period broadened to include contributions to the theories of atomic and molecular structure and X-ray scattering theory. A number of his papers were particularly notable including his work with Polanyi on the structure of white tin (3, 4), E. Wigner on the structure of rhombic sulfur (5), and E. Pohland on the low temperature crystal structure of ammonia and carbon dioxide (6, 7). The Mark-Szilard effect, a classical component of X-ray physics, was a result of his collaboration with Leo Szilard (8). And his work with E. A. Hauser (9, 10, 11) on rubber and J. R. Katz (12, 13, 14) on cellulose introduced him to the research field in which he gained his greatest fame.

Mark's success at this time can be attributed to his aggressive and tireless laboratory work complimented by acute powers of observation, a willingness to unselfishly publish, and, by no means last, to his remarkable ability to lead. The excellence of Mark and his associates was quickly recognized throughout the scientific circles of Europe. This recognition was demonstrated in many ways, but, as Mark adds, "one of the most pleasing acceptances" came in 1923 when Albert Einstein asked him to conduct some experiments of "special interest". The result of this collaboration was the repetition by Mark and his associate, Hartmut Kallmann, of A. H. Compton's famous confirmation of Einstein's light-quantum hypothesis (15,16).

Figure 5. X-Ray diffraction apparatus in the Fiber Research Institute.

Throughout his time at Dahlem, Mark worked closely on the development of more powerful X-ray tubes which made possible new experiments on the fundamental properties of X-rays such as polarization, coherence length, anomalous dispersion, and inelastic scattering. Basic investigations of these properties were carried out by Mark together with Weissenberg, Ewald, Ehrenberg, Kallman, Wigner, Bose, and Szilard. As the amount of results and the degree of experience progressed, Mark was encouraged to compile his findings. In 1927 he published an over 500 page treatise on the application of X-rays in chemistry and technology, "Die Verwendung der Roentgenstrahlen in Chemie und Technik" (17). The book was a truly imposing achievement for a scientist of just over thirty years of age. Otto Kratky, himself a highly respected scientist, recently said that the book, Mark's first, "is after fifty years most instructive and very readable".

Acknowledgements

In reviewing Herman Mark's early years and rise to prominence as a physical scientist the author wishes to make special note of the use of an earlier profile of Prof. Mark by Morton M. Hunt (18), a special issue of Chemie Kunststoffe Aktuell (19), and two taped interview sessions with Prof. Mark (20, 21, 22). A special thanks is given to Ms. Jeanne Strauss, Phillips Petroleum Co., for her patience in clarifying a great deal of German for the author.

Literature Cited

1. Schlenk, W. and Mark, H. Ber., 1922, 55B, 2285.
2. Schlenk, W. and Mark, H. Ber., 1922, 55B, 2299.
3. Mark, H. and Polanyi, M. Z. Physik, 1923, 18, 75.
4. Mark, H. and Polanyi, M. Z. Physik, 1924, 22, 200.
5. Mark, H. and Wigner, E. Z. Physik Chem., 1924, 111, 398.
6. Mark, H. and Pohland, E. Z. Kristi, 1925, 61, 532.
7. Mark, H. and Pohland, E. Z. Kristi, 1926, 64, 113.
8. Mark, H. and Szilard, L. Z. Physik, 1925, 33, 688.
9. Hauser, E. A. and Mark, H. Kautschuk, 1925, 12, 10.
10. Hauser, E. A. and Mark, H. Kolloidchem. Beihefte, 1926, 22, 63.
11. Hauser, E. A. and Mark, H. Kolloidchem. Beihefte, 1926, 23, 64.
12. Katz, J. R. and Mark, H. Verslag. Akod. Wetenschappen Ansterdam, 1924, 33, 294.

13. Katz, J. R. and Mark, H. Z. Elektrochem., 1925, 31, 105.
14. Katz, J. R. and Mark, H. Z. Physik Chem., 1925, 115, 385.
15. Kallmann, H. and Mark, H. Naturwissenschaften, 1925, 13, 1012.
16. Kallmann, H. and Mark, H. Z. Physik, 1926, 36, 120.
17. Mark, H. "Die Verwendung der Roentgenstrahlen in Chemie und Technik"; Johann Anbrosius Barth: Leipsig, 1926; 528 pp..
18. Hunt, M. M. New Yorker, 1958.
19. Chemie Kunststoffe Aktwell, 1980, 34(2), 1-18.
20. Stahl, G. A. "Taped Interview with H. Mark", October 11, 1976.
21. Stahl, G. A. J. Chem. Educ., 1979, 56, 83.
22. Stahl, G. A. "Taped Interview with H. Mark", June 2, 1979.

RECEIVED August 10, 1981.

2

Reminiscences of the Early Twenties

RUDOLF BRILL

8172 Lenggries, Arzbacher Strase 6, West Germany

Shortly after the I. World War scientific life began to grow again in Germany. One of the famous scientific enterprises in this country, the Kaiser-Wilhelm-Gesellschaft, had several institutes in Berlin and a new one, the Kaiser-Wilhelm-Institut für Faserstoffchemie (Institute for Chemistry of Fibrous Material), was just founded. In this institute a very interesting discovery was made: the x-ray investigation of cellulose fibers showed that these fibers were crystalline, i.e., x-rays were diffracted by them and the evaluation of the x-ray diagram resulted in a small elementary cell containing only four units of glucose. Since, by definition, the elementary cell must contain at least one molecule, the conclusion was drawn that cellulose has a low molecular weight of not more than four units of glucose. Consequently, an interscientific fight began and it might be that this interesting situation motivated Herman Mark to go from Vienna, where he had studied, to Berlin to join just that institute where the x-ray investigation of cellulose had been performed even though the situation there did not look too favorable: The just founded institute had no own building yet and was located as guest in the institute of Physical and Electrochemistry of the Kaiser-Wilhelm-Gesellschaft, headed by Fritz Haber the famous inventor of the technical synthesis of ammonia. However, there was not much space in Haber's institute and, therefore, the x-ray laboratory of the guest-institute was rather primitively located in the basement of Haber's residence. But this in no way impressed Mark. Obviously he was very much interested in the application of the new method of x-ray diffraction. Consequently, he joined the group of Polanyi and Weissenberg, the two authors of the work on cellulose. Polanyi was a theoretical physicist and Weissenberg a mathematician so that the addition of a chemist to this group was rather favorable.

For work with the method of x-ray diffraction two items were necessary: to study the theory and to learn to work with x-rays. So Mark studied together with Weissenberg Niggli's book on crystallography thoroughly. To work with soft x-rays, as it is necessary for diffraction work, was rather difficult at those times. X-Ray tubes for these purposes were not available com-

0097-6156/81/0175-0021$5.00/0

mercially. One had to construct the tubes oneself with the help of a mechanic and a glass blower. Tecause the x-ray tube was half glass and half metal these materials had to be compounded by means of a cement and, to keep up an excellent high vacuum, the tubes must be evacuated permanently during operation. Anyway, this was a job for a vacuum physicist and Mark was eager to learn this too.

In 1922 Mark had joined the institute in Berlin and in 1923 the first papers were published by him and his collaborators (mostly students performing their doctoral thesis). These papers concern the determination of the crystal structures of Sn, SnJ_4, urea ($CO(NH_2)_2$), pentaerythritol ($C(CH_2OH)_4$), hexamethylenetetramine ($C_6H_{12}N_4$), and triphenylmethane ($CH(C_6H_5)_3$). It is interesting to note that these papers had nothing to do with work on fibers or high polymers at all. This characterizes the situation at that time. Only a few crystal structures were known and to perform structural work one had the choice. On the other hand one has to keep in mind that the determination of a crystal structure, at these times, was a difficult task. Because x-ray intensities of the self-constructed x-ray tubes were rather low, the diffracted intensities had to be guessed from photographs. Hence, the only data which could be determined precisely from the location of the spots or rings on the photographs were the dimensions of the elementary cell. Missing reflections gave a hint on symmetry elements in the cell and, by comparing them with the symmetric structure of the molecule of the investigated substance, molecular and even atomic positions could be guessed. For the guessed structure x-ray intensities could be calculated and from the agreement between guessed and calculated intensities the degree of exactness of the determined structure could be obtained. But this method was complicated because electronic calculators were not available at that time. Consequently, many structure determinations must be finished by having an idea of the symmetry of a structure without being able to give exact atomic positions. So, of the above-mentioned five substances only the structures of tin, urea and hexamethylene-tetramine could be determined up to the exact atomic positions and that, at those times, was extremely good.

The situation in the institute changed favorably when a new building was finished and a much better x-ray laboratory was available. There was more space so that two x-ray tubes could be operated simultaneously and the time of operation was extended from the morning up to almost midnight. Here, Mark, who had married meanwhile, got a flat in the institute. Often late in the evening Mrs. Mark came down to the laboratory with some refreshments for those working in the x-ray laboratory.

Mark was very popular and well-liked. Consequently, he always had many students and was in the position to select the best ones to work with him. So between 1922 and 1927, when he left the institute in Berlin, 34 papers were published by him, almost all of them together with his students in the frame of a doctoral thesis. These papers concern simple substances like Sn, C (graphite), S, NH_3, CO_2, (the last two technically difficult) but also more complicated organic substances like hexamethylene-tetramine and many substances of the composition ABX_4 as sulfates of many metals, AR_4 as $C(CH_3)_4$, $C(CH_2OH)_4$ but also high polymers like cellulose and 'Kautschuk'. So it was already at these early times evident that Mark was an extremely good scientist and, when he left the institute, he became a lifelong member of the Kaiser-Wilhelm-Gesellschaft and its successor the Max-Planck-Gesellschaft.

RECEIVED March 16, 1981.

3

A Short History of Polymer Science

G. ALLAN STAHL

Phillips Petroleum Company, Research and Development, Bartlesville, OK 74004

The existence of large molecules with linear molecular weights of hundreds, thousands, and even millions is today a recognized fact. For today billions of dollars of technology and an army of scientists, technicians, and engineers (more than 60,000 in the United States alone) work with materials which often share only the common title, polymer. These workers and their predecessors have produced countless variations of polymers, each with properties designed to satisfy certain criterion. Research in the field of polymer research is so active that more than 60,000 pages of findings are published annually in several dozen journals.

The common chemical property, high molecular weight through repeating covalent bonds, is the single most important property of these materials. It is this feature which accounts for the characteristic physical properties which set polymers apart from other forms of matter. Useful physical properties such as high viscosity, long range elasticity, and high strength are all direct consequences of high molecular weight. Yet acceptance of the concept of high molecular weight in all scientific quarters is a recent event, only having occurred since 1930.

The word, polymer, was introduced a century before in 1833 by Jons Jacob Berzelius in his famous book, the "Jahres-Bericht". He recognized the fact that two compounds may have the same composition yet differ in molecular weight. Thus, he classified this polymerism as a special type of isomerism. In order to prevent confusion, it should be pointed out that Berzelius had in mind a series of compound related to each other as acetylene C_2H_2, benzene C_6H_6, and styrene C_8H_8 are related.

Although Berzelius had not considered high molecular weight substances, his definition contained the necessary elements to account for the isomerism that workers would find a few years later. The concept was, thus, not challenged through the middle Nineteenth Century, and with the establishment of

0097-6156/81/0175-0025$5.00/0

structure as an important part of organic chemistry it was, in fact, strengthened. In particular, the work of August Kekule (1) provided the solid foundation for polymerism.

Kekule' described the quadrivalency of carbon, and carbon atoms bound "with an equal part of the affinity of the other". He even proposed the existence of polyvalent atoms producing "sponge or net like" molecular mass in 1878. Towards the end of his career, he advanced the hypothesis that natural organic substances-those most closely associated with life, proteins, starch, and cellulose-may consist of very long chains, and derive their special properties from this structure.

As a result of the work of early pioneers like Berzelius and Kekule' reports on studies of materials we now recognize as high molecular weight, the natural polymers as well as those inadvertent tars from work performed in the pursuit of other goals, frequently used the term, polymer.

The early coordinated studies of polymeric substances were conducted by two essentially independent groups of investigators. On one hand there were those concerned with the chemical and physical constitution of natural materials. While on the other, there was the synthetic organic chemists. Special note should be made of those in Germany, those who stewarded the extraordinary advances of the second half of the Nineteenth Century. The path to the acceptance of the concept of high molecular weight might have been less torturous had the former group recognized the significance of the occasionally reported synthesis of polymeric products by the latter. Certainly the possibility of the existence of indefinitely large covalent structures is present in the basic concepts of structural chemistry.

An early example of a polymer preparation is found in the work of A.-V. Lourenco (2). In 1860 he reported the preparation of a series of adducts of ethylene glycol and ethylene dihalide with the general formula $HO{+}C_2H_4O{+}_nH$. He isolated and identified several members of the series of n=2 to n=6 by distillation, and noted that the boiling point and viscosity of each member increased with n. With great insite, he predicted that the highly viscous undistillable products obtained with more drastic reaction conditions were of a correspondingly greater complexity (or n must be greater than 6). In a remarkably accurate conclusion Lourenco reasoned that these materials "have the same apparent composition, present the same reactions, and have, however, entirely different degrees of condensation." Lourenco (3) prepared a viscous, undistillable material by heating ethylene glycol and succinic acid in 1863. He also concluded that this material was "highly condensed".

A short while afterwards K. Kraut (4) reported the preparation of dimeric and tetrameric chain structures by intermolecular esterification of acetylsalicyclic acid. In the same paper he assigned the analogous octameric chain formula to the "salicylide" prepared by C. Gerhardt (5) by the action of $POCl_3$ on sodium salicylate. Interestingly he noted the condensation of the intermediates in discrete steps each doubling the degree of condensation of its predecessor. Yet, he failed to recognize that preparation of the higher order structures could occur in single steps.

In a much later study, H. Schiff (6) and A. Klepl (7,8), although working independently, prepared polymers isomeric with the polysalicylides from hydroxybenzoic acids. Schiff, like Kraut, assigned dimeric, tetrameric, and octameric chain formulas. Klepl, however, obtained a product, $C_7H_4O_2$, which he concluded was high molecular weight. Working in Schiff's laboratory, A. Piutti (9) prepared an analogous material from m-aminobenzoic acid. This polymer is probably the first synthetic polyamide.

According to Flory (10), the concept that proteins and carbohydrates are polymeric goes back to at least the work of Hlasiwetz and Habermann (11). In 1871 they proposed that these substances were made up of a number of species differing from one another with respect to the degree of molecular condensation. Flory also noted that Hlasiwetz and Habermann differentiate "soluble and unorganized" members of these substances, for example dextrin and albumin, from "insoluble organized" members, such as cellulose or keratin. This distinction is the precursor of the present day differentiation between non-crystalline and crystalline polymers.

Musculus and Meyer (12) measured the diffusion rates of some starches and dextrins in 1881. The work was designed to determine the relationship of these "isomeric or polymeric" forms to the simple sugars from which they were formed. They concluded that dextrin molecules must be much larger than those of the sugars. This work, however, preceeded Raoult's (13) development of the cryoscopic technique for the determination of the molecular weights of dissolved substances, and van't Hoff's (14) formulation of the solution laws. Further, since the vapor density method was obviously inapplicable, it was not possible for them to actually determine the degree of polymerization.

It was Brown and Morris (15) in 1888 who employed Raoult's method. They reported a value of 30,000 for the molecular weight of amylodextrin, a degradation product of the hydrolysis of starch. Subsequently Lintner and Dull (16) also using cryoscopy reexamined amylodextrin, and reported the molecular weight as 17,500. In a third paper, Rodewald and Kattein (17) in 1900, measured the molecular weight of starch by osmotic pressure experiments carried out on aqueous solutions of starch iodide. They obtained somewhat higher molecular weights, 36,700 and 39,700.

As if to give final confirmation to the concept of high molecular weight, the eminent Emil Fischer in his first years in Berlin turned his attention to the study of polypeptides. With his characteristic thoroughness, he systematically prepared numerous polypeptides eventually obtaining a polypeptide with a molecular weight of 4200. During many years of intense studies Fischer never postulated any structure for these synthetic products, or for natural proteins, except linear chains consisting of covalently linked amino acids connected to each other by the -CO-NH- linkages now known to occur in all amides and peptides. On one occasion in 1906, he proposed that there was an uninterrupted continuous line between the simplest dimeric and trimeric amino acids and the native proteins (18). Later he reemphasized his beliefs stating that the ultimate proof of high molecular weight came from his synthetic products made by analogous, controlled chemical reactions (19).

Association Theory

The trial-by-fire methods of science, however, sidetracked the linear development of high polymer theory, for the theory was swept up by the development of the association theory of collodal particles at about the turn of the century. The peculiar and hard to understand chemical and physical behavior of polymers had, on occasions, lead to the suggestion that unusual or special forces were involved in their formation. In order to explain the forces, workers turned to the work of Thomas Graham.

In the most brilliant period of his career, Graham had demonstrated that certain polymers, including many gums, were unable to diffuse through certain gelatinous substances, membranes, and paper. He called the materials incapable of permeating a membrane "colloids", and reported that they could be obtained in a state free from diffusable "crystalloid materials" by this technique (20).

Graham's definitions were expanded, and the concept of a colloidal state of matter evolved. According to this view, a substance could occur in a colloidal state just as it could occur under various conditions as a gas, liquid, or solid. If a colloidal solution was, at that time, defined as a solution in which the dispersed particles were comprised of large molecules, the ascertion would have been more acceptable. Many workers, however, chose to ignore this possibility, and unfortunate and misleading misinterpretations occurred.

Thus in 1899, Johannes Thiele extended his valence theory of double bonds to include colloids. Thiele suggested that in such materials as polystyrene the molecules of styrene were bound together merely by association of the double bonds. He referred to this association as "partial valence" (21). In 1901, Rohm concluded that the transformation of acrylic esters into polymers was from an "allotropic alteration" and not a chemical reaction (22). Schroeter, working with salicylides just as Kraut, Schiff, and Klepl before him, concluded that the tetrameric salicylide was formed by "external forces about the monomeric molecules", and that the chemical structures of the monomers were unaltered (23). Thus the association theory rapidly grew in popularity.

In the light of what we know of polymer structure today, these suggestions seem ridiculous, but in the early Twentieth Century they were widely accepted as explaining the peculiarities of many substances. Pringsheim (24) and Hess (25) applied the association theory to explain the properties of cellulose just as Bergmann (26) and Abderhalden (27) did with proteins. Use of the theory to explain properties is demonstrated by the following example. It was reasoned that cellulose might be an anhydroglucose with the molecular formula $C_6H_{10}O_5$. Because of the unusual strain of this cyclic molecule (or for some other reason), it was supposed to exhibit exaggerated forces of association--or residual valence. Association of molecules with residual valence would thus produce behavior as though they were of very high molecular weight. Similarly it was believed that proteins might be comprised of associating diketopiperazine units (28).

In support of the association theory, colloid chemists cited non-reproduceable cryoscopic molecular weight determinations (which were eventually shown to be caused by errors in technique) and claimed that the ordinary laws of chemistry were not applicable to matter in the colloid state. The latter claim was based, not completely without merit, on the ascertation that the colloid particles are large aggregates of molecules, and thus not accessible to chemical reactants. After all many natural colloids were shown to form double electrical layers and adsorb ions, thus they were "autoregulative" by action of their "surface field" (29). Furthermore, colloidal solutions were known to have abnormally high solution viscosities and abnormally low osmotic pressures.

When challenged, supporters of the theory quickly rejected any possibility of high molecular weight. Sometimes they cited supportive experimental results, and at times as shown below, they evoked elaborate rationalization. It is only fair to point out that Crompton's following argument, although amusing from the vantage of nearly three-fourths of a century, was tendered with thought and sincerity.

> "No upper limit is usually assigned to molecular magnitude. E. Fischer has synthesized a polypeptide with the molecular weight 1212, and in the case of colloids, molecular weights of the order 10^4, and even 10^5, are commonly spoken of. A difficulty arises, however, in admitting that molecular weights can exceed a certain value, unless the density increases as the molecular weight increases.
>
> For suppose that a compound can exist, such as a protein, with a density at 0° not much greater than that of water, and with a molecular weight of rather more than 30,000, the grammolecule of such a compound at 0° would occupy about 30,000 cc. The grammolecule of a perfect gas under the standard conditions occupies only 22,400 cc, and we should therefore have a solid compound, at 0° and under a pressure that cannot be less than one atmosphere, occupying a greater molecular volume than that of any gas.
>
> That the molecules of liquids and solids should occupy greater volumes than those of gases under similar conditions, seems at first contrary to the usual conceptions of the gaseous, liquid, and solid states. It is true that at sufficiently low but a simple calculation shows that for the majority of chemical compounds it would only occur at temperatures not far removed from the absolute zero.
>
> Two suggestions appear to be indicated. The first is that under the ordinary conditions there is an upper limit to molecular magnitude, and that for most substances, more especially colloids, the molecular weight cannot exceed a value of about 20,000. The second is that our ordinary kineto-molecular conceptions no longer apply when for a given temperature the molecular magnitude exceeds a certain critical value. The latter view seems most in keeping with our present knowledge and perhaps serves to throw some light on the behaviour of colloids (30)."

The rapid acceptance of the association theory was accompanied by an equally rapid dropping of the high molecular weight or polymer concept. Olby (31) has stated that three developments made the theory attractive as an explaination for the behavior of polymers. First, he sates, was Alfred Werner's introduction of the concept of two kinds of combining forces--Hauptvalenzen or primary valence forces, and Nebenvalenzen or secondary forces (32). When applied to cellulose, proteins, or rubber, the molecules were envisioned to be united by primary forces (covalent bonds) but still possess degrees of "residual affinity" whereby several molecules can further unite to form aggregates (colloids).

Second, he says, was the expansion of colloid science into biology. Since the fledgling science bridged the gap between the physical sciences and biology, it received reputable recognition.

The third development was the seeming experimental support of the association theory by X-ray crystallography. This support was based on the then accepted idea that the molecular cannot be larger than the unit cell of the crystal. Although it is obvious to us that this is untrue, the idea was then "obviously true". In the early days of X-ray crystallography, the majority of research was performed by mineralogists. And it was in their studies of low molecular weight inorganics that the ideas about the relationship of molecular and crystal cell size was developed.

Yet, even as the association theory was at the peak of its acceptance, the pieces to a generally accepted, high molecular weight model were being formed. (Unable to resist the temptation) the high molecular weight concept bounced back with the work of Carl Harries in 1904.

The Chemistry of Rubber

The first known mention of rubber is found in the 1511 writings of Pietro Martyre d'Anghiera, but until the late Eighteenth Century, it remained pretty much a curiosity item. Its name, rubber, came from the discoverer of oxygen, Joseph Priestly, who reported in 1770 using it to "rub out" black pencil marks. But its application in large scale commerce was not practical until much later.

It was well known at the turn of the century that rubber has the empirical composition, C_5H_8. Michael Faraday elucidated its composition in 1826 by careful elementary analysis. His work, an effort of extreme complexity, has been diminished by the years, but it regains its stature when you recall that over thirty years passed before the next major step was performed. In those thirty years rubber was blended, dissolved, and even vulcanized (by Charles Goodyear in 1839), but it was in 1860 that its major chemical component was discovered. This important finding was made by Greville Williams. He named the product of the destructive distillation of rubber, isoprene.

The earliest mention of the polymerization of isoprene was also made by Williams in 1860 (33). He noted the formation of a "white, spongy mass" when isoprene was left in a bottle with oxygen. Afterwards G. Bouchardat observed that isoprene could be converted to a sticky mass by the action of either carbon dioxide or cold aqueous hydrochloric acid (34). Tilden in 1882 (35) and independently Wallach in 1887 (36) were the first to prepare elastomers of isoprene, but little else was known of their structures. Gladstone and Hibbert (37) cryoscopically obtained molecular weight values of 6000 to "at least 12,000" in 1889, but these values were too large to be accepted by investigators of the day.

Harries showed that the degradation of rubber by ozone yielded chiefly levulinic acid and aldehyde (38). This fact, he concluded, indicated that rubber was made up of the repeating unit:

$$-CH_2-\underset{\displaystyle CH_3}{\underset{|}{C}}=CH-CH_2-$$

In later work, he reported the ozonide of rubber to have the empirical formula $C_{10}H_{16}O_6$, and accordingly assigned the structure shown below.

$$\begin{array}{ccc} & O-C(CH_3)-CH_2-CH_2-CH-O & \\ O & & O \\ & O-CH-CH_2-CH_2-C(CH_3)-O & \end{array}$$

From these results he assigned a cyclooctadiene formula to rubber, and concluded, in the tone of the times, that the rubber molecules combined through the action of "partial valence" into much larger aggregates (39).

As Harries systematically studied rubber other workers also degraded and reformed this material. By 1910, S. S. Pickles had proposed that rubber was composed of covalently bound chains of isoprene, and that variations in the chains accounted for differences in the properties of rubbers (40). Pickles was the first to assign a chain structure of rubber on the basis of the properties of the chemically modified material. He noted that saturation of the double bond with bromine did not destroy the "colloidal nature" of the material. In a remarkably accurate proposition of structure, he made but one error. He assumed that the chain ends combined to form a ring of eight isoprene units. As we shall see later, he was not alone in this ascertation.

Harries rejected Pickles formula in 1911 (41), but in his subsequent work he expanded the size of his ring formula to include five and, eventually, seven isoprene units. Although incorrect in many assumptions, this work provided the background necessary for more probing thought about the structure of materials, and for the rebirth of the high molecular weight concept.

Organic Chemistry, X-ray Crystallography, and Modern Polymer Theory

Hermann Staudinger, on developing a new and simple preparation of the monomer, studied the polymerization of isoprene as early as 1910 (42). Stimulated by the differences in physical properties between his synthetic rubber and natural rubber, he turned his full attention to the study of polymers. In the fashion of the established German school of organic chemistry, Staudinger studied a wide range of materials we now know to be polymeric. His shift in research interest was quite courageous as it caused a stir in Germany. He was, as a result of his work which included the discovery of ketenes, an established, reputable synthetic organic chemist.

Up until Staudinger entered the field most polymer preparations were isolated events. As examples, in 1872 Baumann described the preparation of an insoluble mass when vinyl chloride was exposed to sunlight (43) just as Simon had formed a jelly of styrene in 1839 (44). Staudinger, however, systematically prepared the materials and studied their preparation as well as properties. By 1920, he was convinced the association theory was incorrect.

In a classic paper titled "Uber Polymerisation" (45) he summarized his findings, and proposed formulas for polystyrene and polyoxymethylene (paraformaldehyde) that were linear, long chains.

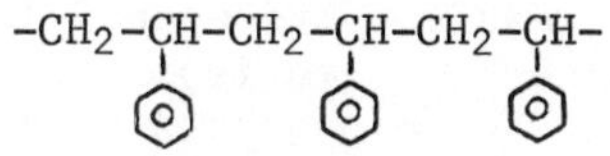

$$-CH_2-O-CH_2-O-CH_2-O$$

He even advocated a chain structure for rubber, and claimed that its colloidal properties were due entirely to high molecular weight. It is interesting to note that these structures proposed in 1920 are still applicable today.

At about the same time that Staudinger was publishing his findings regarding high molecular weight substances, R. O. Herzog and W. Jancke demonstrated that at least a part of a cellulosic fiber was crystalline (46). This was an important finding because of its subsequent interpretation by Michael Polanyi, an associate of Herzog and Jancke. Employing the new technique, X-ray diffraction, they obtained a powder pattern which was neither clear spots, nor powder rings, but something between the two--smeared points placed symetrically in groups of four. Unable to decipher the diagram, Herzog assigned the task to Polanyi.

Polanyi's interpretation (47) was the second important step in the resolution of modern polymer theory for it marked the beginning of the use of X-ray diffraction in the investigation of polymer structure. Polanyi's conclusion was that the X-ray diffraction spots were in agreement with either long glucosidic chains or rings consisting of two glucose anhydride units. Polanyi unfortunately made it clear that X-ray data alone could not distinguish between these two possibilities. It has been charged that his guarded, cautious language contributed to the idea that the small basic unit of the lattice of crystalline cellulose was proof of its low molecular weight (48). None-the-less developments after 1920 proceeded along two lines, chemical investigations lead by Staudinger and X-ray diffraction studies.

In the next few years Staudinger repeatedly demonstrated that polymerization lead to long chains of primary or covalently bonded monomers. He also showed these synthetic materials often resembled natural materials in many significant chemical and physical properties.

Staudinger, like Pickles in 1910, chemically modified rubber and noted its failure to lose colloidal properties as evidence of chain structure (49). His experimental proof was impressive for he had catalytically hydrogenated natural rubber and then thoroughly studied the properties of the saturated product. He reasoned that the disappearance of the double bonds of natural rubber should result in a loss of "residual valence", and failure to do so was conclusive. His opponents expressed doubts about the validity of the experiment, asserting that he had not performed a true hydrogenation. They pointed out earlier hydrogenzation experiments in which distillable products had been isolated. Although it was discovered later that these products were caused by reaction conditions which cracked the chain structure, doubts about the validity of Staudinger's work were cast. Undaunted, he continued. Between 1922 and 1930 Staudinger published more than nineteen papers on the chemistry of rubber alone.

During this time, the debate over the existence of long covalent chains became heated among physical chemists. Polanyi's view that Herzog's and Jancke's powder diagrams could be explained by long chains was not well received. Herzog, for one, disagreed with the concept, prefering to equate a small unit cell with low molecular weight. The division of ideas between Herzog and his assistants, Polanyi, Herman Mark, Rudolf Brill, and Karl Weissenberg, grew. Herzog, by 1925, had clearly defined these limits settling on a degree of polymerization of "only two, four, or a (slightly) higher number (50), but to Herzog's credit, the research continued.

In 1923, Brill obtained some excellent diffraction patterns on silk fibroin and concluded that there were eight amino acid residues in its unit cell (51). Further challenging Herzog, Weissenberg discussed the possibility of long chains in 1925 (52). While at the same time J. R. Katz discovered that the diffraction pattern of stretched rubber indicated that partial alignment of its molecules suggested the probability of high molecular weight (53). This point was stressed in the quantitative interpretation of Katz's findings by E. A. Hauser and Mark (54).

In 1926, Sponsler and Dore presented a complete picture of the cellulose molecule with a model of its elementary cell (55). Making use of Haworth's brilliant deduction of the 1,4 bonded glucose ring (56), they described a pyranose ring, with the side chain, $-CH_2OH$, being turned alternately to the left and to the right, and the units being joined by primary valencies to form a chain molecule very much longer than the unit cell.

The Events of 1925 to 1930

Hermann Staudinger received the Nobel Prize in Chemistry in 1953 for his work on macromolecules. The award not only acknowledged the contributions of the man who first proposed use of the name "macromolecules" (57), but recognized the field of macromolecular chemistry. In his address to Staudinger for the award of the Nobel Prize, A. Fredga said:

> Professor Staudinger. Thirty years ago, you adopted the view that a chemical molecule is able to reach almost any size . . . . It is no secret that for a long time many colleagues rejected your views which some of them even regarded as abderitic. Perhaps this was understandable. In the world of high polymers, almost everything was new and untested. Long standing, established concepts had to be revised or new ones created. The development of macromolecular science does not present a picture of peaceful idylls. (58)

Fredga was doubtlessly referring to the conflict between the advocates of the association theory and those who supported the long chain concept. The conflict of ideas came to a head in the period 1925 to 1930. In this period the respective protagonists presented their ideas in two important conferences and several decisive papers. By the end of this period resistance to the macromolecular viewpoint was reduced to a few holdouts, but resolution of the facts, as Fredga indicated, did not come easily.

A glimpse of the stormy events of 1925 to 1930 was seen on the occasion of Staudinger's farewell address to the Zurich Chemical Society. Staudinger lectured at the meeting on the existence of thread-like macromolecules consisting of a long series of "Kekule'" or covalent bonds. Since his model was in direct conflict with many X-ray crystallographer's concept of a small unit cell (and thus low molecular weight), his lecture stirred a controversy. According to eyewitnesses, many notable scientists, including Karrer, Niggli, Wiegner, Scherrer, and Ott, tried in vain to convince him that his ideas conflicted with exact scientific data. The meeting, it is reported, ended when Staudinger shouted, "Hier stehe ich, ich kann nicht anders" (59).

The real effect of this encounter was to make the debate between the association and polymer factions well known. Aware of growing interest, Richard Willstaetter arranged a symposium on the topic at the September, 1926, meeting of the "Gesellschaft Deutscher Naturforscher und Arzte" held in Dusseldorf. The meeting was a classic showdown between Max Bergmann and Hans Pringsheim, and Staudinger and Mark.

At the meeting, Bergmann and Pringsheim presented impressive and lucid papers declaring that the classic structure theory of Kekule' was inappropriate to explain the complex carbohydrates and proteins. Bergmann cited "psuedo-high molecular weight" inorganic complexes, and Pringsheum discussed inulin. The examples lead the respective investigators to conclude that the properties of these compounds were due to a combination of primary and aggregating valencies, and that psuedo-high molecular weight is a result of the latter forces (60).

In the third presentation, Mark, a leading expert in the area of structural analysis by X-ray crystallagraphy, expressed the opposite view. Comparing hexamethylenetetramine and cellulose, he proposed that cellulose consists of small units held together by forces "comparable by type and magnitude to the inner molecular forces". Mark concluded, "The whole crystallite appears as a large molecule" (61).

Staudinger, the final speaker, presented a broad array of data on polymerization, hydrogenation, comparisons of viscosity, melting points, and solubility of polymers. He pointed out that in the conversion of polystyrene to hexahydropolystyrene, and polyindene into hexahydropolyindene, the products retained their high molecular weight properties. Again, he maintained this proved "the monomers are united by main valencies" (62).

Reaction to the presentations was varied. Chairman Willstaetter spoke out for the high molecular weight advocates. declaring, "Such enormous organic molecules are not to my personal liking, but it appears that we all shall have to become aquainted to them". On the other hand, another attendee reportedly said, "We are shocked like zoologists would be if they were told that somewhere in Africa an elephant was found who was 1500 feet long and 300 feet high" (63). In general, the data presented supporting high molecular weight was still not sufficently convincing, nor was the decisive value which X-ray spectrography could have for the subject understood (64).

Activity in the field was, however, expanding. For example, in 1927 Drew and Haworth (65) obtaind a crystalline polymeric powder by the action of hydrogen chloride on the lactone of 2,3,4-trimethyl-l-arabonic acid. Citing the increase in melting point and molecular weight, and loss of specific optical rotation, they ascribed a cyclic, high polymer structure to this polyester.

At the same time, Mie and Hengstenberg, working in close collaboration with Staudinger, confirmed the chain structure of polyoxymethylene by X-ray analysis (66,67). They demonstrated that the X-ray diffraction pattern of polyoxymethylene has a characteristic interference pattern which varies with the number of CH_2O units in the chain. Although their method failed when applied to higher polymers, they were able to confirm a chain structure for materials with ten to twenty CH_2O groups.

A year later Hengstenberg and Mark moved to the Ludwigshafen laboratory of I. G. Farben, and combined efforts in studying fiber structures by X-ray analysis. Right off, they observed low angle diffraction spacings in cellulose suggestive of very large unit cells. They proposed cellulose molecules of at least 600Å, which corresponds to a chain of 120 glucose residues (68).

At the same time K. H. Meyer and Mark (69) proposed an important structure for cellulose which is best described as a compromise between the aggregates of the association theory and Standinger's macromolecules. In an extensive paper, they carefully developed the idea of cellulose chains consisting of so called "primary valence chains". They further proposed that the primary valence chains were aggregated by molecular forces such as hydrogen bonding and van der Waal's forces. Their model, which became a standard, combined the important features of both concepts by proposing micelles of long, not short, molecules. The physical properties of cellulose were attributed to these forces, for example, tensile strength to the primary valence bonds and insolubility to the secondary forces.

The principles were refined by Meyer in a second paper (70). In it he proposed that the micelles occurred at regular intervals. He also included an explanation of the elasticity of rubber based on the assumption that the molecular chains tended to roll together in knots in unstretched rubber, but line up when stretched. This explanation was especially elucidating since it agreed well with Katz's discovery (53) that amorphous rubber crystallizes when stretched.

Between 1925 and 1930 a large number of additional polymers were prepared and characterized as high molecular weight. A list of some of these polymers includes vinyl acetate (71, 72), methyl orthosilicate (73, 74, 75), ethylene oxide (76), acrylic acid (77), ethylene (78, 79), vinyl chloride (80, 81), adipic anhydride (82), and decamethylene dibromide (83). In many instances the authors cited the polymer models of both Staudinger, and Meyer and Mark to explain their experiments. This was not unreasonable since Staudinger's macromolecule, and Meyer's and Mark's micelles differed from each other very little. At the same time by embracing the concept of long chains, they differed substantially with the advocates of the association theory. However, Fredga pointed out these were not years "marked by peaceful idylls".

Shortly after the publication of Meyer's paper Staudinger denounced the work of Meyer and Mark as what he called the "New Micelle Theory" (84). In a short time Staudinger and Meyer embarked on an exchange of polemical letters and papers which lasted for more than ten years. Mark, caught between recalled:

> Even the champions of the long chain aspect did not agree with each other, as they easily could have done because instead of concentrating on the essential principle, they disagreed in specific details and, at certain occasions, they argued with each other more vigorously than with the defenders of the association theory. (48)

Yet as the many sided debate went on, Wallace Carothers started a series of investigations in 1928 which would eventually establish the macromolecular concept. His objective from the beginning was to prepare polymers of known structure through the use of established reactions of organic chemistry (85). In the brilliant years before his untimely death in 1937, he studied the preparation and properties of polyesters, polyanhydrides, polyamides, and polychloroprene (28). As a result of his studies, he restated and extended the concepts of Staudinger, and Meyer and Mark, with such careful reasoning and massive documentation that by his death the chain concept was accepted without further criticism (86).

Tying the Ends

In retrospect it is probable that acceptance of macromolecular theory was slowed by of all things the thoroughness of the procedures of the time. An example is the problem of accounting for the structure of the ends of the polymer molecules.

The standard procedure of organic chemistry at that time was to prepare (or isolate) pure substances, and characterize these substances by elementary analysis and molecular weight determinations. This approach was very successful when applied to most lower molecular weight organic substances, but when used dogmatically in studying polymers, it forced the investigators into unfounded conclusions.

As we have already discussed, Harries and Pickles assigned cyclic structures to chains of isoprene in natural rubber. They had to. It would have been inexcusable for them to ignore their inability to detect end groups, or otherwise account for the end groups in the empirical formula.

Similarly Stable and Posnjak (87) proposed cyclic formulas of four, five, or possibly more structural units for polystyrene. Lebedev (88) first assigned the cyclooctadiene structure to polybutadiene shown below before expanding the concept of the ring to include several isoprene units.

$$\begin{array}{l} CH_2\text{-}CH\text{=}CH\text{-}CH_2 \\ | \qquad\qquad\quad\ | \\ CH_2\text{-}CH\text{=}CH\text{-}CH_2 \end{array}$$

Even Staudinger (89) assigned cyclic structures to poly-indene and polystyrene. He visualized that these polymers would be stretched out into double threads with closed ends in which the two halves of the ring would lie parallel. He called this his "bifilar" hypothesis.

Resolution of the chain end problem eventually came from its source--organic chemistry. Employing stepwise polymer preparation and careful titration of the end groups Lycan and Adams found high molecular weight fractions in polyesters derived from ω-hydroxydecanoic acid (90). Similarly Carothers and Dorough compared the end group determination molecular weight of poly(ethylene succinate) with ebullioscopic determinations (91). They found the values to be of comparable magnitude, and concluded that the facts were incompatible with a cyclic structure.

The need to evoke cyclic structures eventually passed, but failure to recognize the very high molecular weights of certain materials such as the addition polymers further slowed the process. In these materials the molecular weight so diluted the (then) unidentified end groups that detection was impossible. Understanding of this structural feature was achieved in 1937 by P. J. Flory. Flory proposed the chain transfer step, or a reaction mechanism in which the growing free radical might be saturated with an atom of another molecule (92).

```
                H H
Polymer -C-C•  +  H-R' ———>
                H R

                H H
Polymer -C-C-H  +   R'•
                H R
```

This second molecule might be a monomer, polymer, or solvent. Because of chain transfer the end of one polymer chain might be a hydrogen atom, and the beginning of the next the radical formed by removing the hydrogen atom from the solvent molecule. In the same paper, he proposed the two most probable chain termination reactions, mutual combination and disproportionation.

Elucidation of the nature of the end groups was important in the development of polymer science, but a more complex step was expression of the magnitude of molecular size.

Staudinger was probably the first to recognize this, for in 1928 he proposed that synthetic macromolecules were polydisperse and their molecular weights would have to be expressed as average values (93). He also recognized the dependence of physical properties on molecular weight, and pursued this dependence as a measure of molecular weight determination. The result was his application of solution viscosity.

Since that time a great deal of practical use has been made of solution viscosity measurements, and a large literature has grown up around interpretation of these measurements (a review of these developments is given in the following chapter). Unfortunately it has been concluded after much discussion that an understanding of average molecular weight cannot be developed in an absolute sense from this method.

The understanding of average molecular weight as a precisely defined structural feature began with the work of Lansing and Kraemer in 1935 (94). They drew the first sound distinction between different kinds of average molecular weights to be expected from different methods of measurement. For example they defined the number-average molecular weight ($\overline{M}_n$) as the weight of the whole divided by the number of molecules in it, thus:

$$\overline{M}_n = \frac{\sum_{i=1}^{\infty} NiMi}{\sum_{i=1}^{\infty} Ni}$$

In distinction to the number-average, the weight average ($\overline{M}_w$) defined by Lansing and Kraemer gives extra weight to large molecules. This average is expressed as:

$$\overline{M}_w = \frac{\sum_{i=1}^{\infty} NiMi^2}{\sum_{i=1}^{\infty} NiMi}$$

The precise definition of average molecular weight was a major unifying factor, for only with the use of these definitions were many experimental ambiguities cleared up. As a simple example, cellulose could be defined by either $\overline{M}_n$, which is sensitive to lower molecular weight parts (or impurities), or by $\overline{M}w$, which reflects the presence of the very high molecular weight components. Lansing's and Kraemer's work, combined with the sophistication of analytical techniques, provided an important fundamental development.

In reviewing, the development of a useful polymer theory was a prerequisite for the expansion in research, and explosion in use, of these unique materials. The development occured in several distinct phases. It was marked by several landmark events and papers, and it can claim its conception as the product of remarkably few workers. It is doubtful that the army of scientists, technicians, and engineers involved in polymer research, much less the lay recipients of the wealth and benefits of the resultant technology, realize the impact that so few have had on their lives. We are reminded in a time of diminishing influence of the individual, that singular contributions have been, and should continue to be, an important part of science.

Acknowledgements

A ordering of events such as those in a review of polymer science is necessarily influenced by the opinions of those who have gone before. In this spirit the author desires to recognize the reviews of Flory (10), McGrew (86), and Olby (31). However, the best indicator of the thinking of those times can be obtained by the literature of the day. A paper by Carothers titled, simply, "Polymerization" (95) is representative, and was used extensively in preparation of this paper.

Literature Cited

1. Hafner, K. Angew. Chem. Int. Ed. Engl., 1979, 18, 641.
2. Lourenco, A.-V. Ann. chim. phys., 1860, 51, 365.
3. Lourenco, A.-V. Ann. chim. phys., 1863, 67, 293.
4. Kraut, K. Ann., 1869, 150, 1.
5. Gerhardt, C. Ann., 1853, 87, 159.
6. Schiff, H. Ber., 1882, 15, 2588.
7. Klepl, A. J. prakt. Chem., 1882, 25, 525.
8. Klepl, A. J. prakt. Chem., 1883, 28, 193.
9. Piutti, A. Ber., 1883, 16, 1319.
10. Flory, P. J. "Principles of Polymer Chemistry"; Cornell University Press: Ithaca, NY, 1953; p. 6.
11. Hlasiwetz, H. and Habermann, J. Ann. Chem. Pharm., 1871, 159, 304.
12. Musculus, F. and Meyer, A. Bull. soc. chim. France, 1881, 35, 370.
13. Raoult, F. Compt. rend., 1882, 95, 1030; ibid, 1885, 101, 1056.
14. van't Hoff, J. H. Z. physik. Chem., 1887, 1, 481.
15. Brown, H. T. and Morris, G. H. J. Chem. Soc., 1888, 53, 610; ibid, 1889, 55, 465.
16. Lintner, C. J. and Dull, G. Ber., 1893, 26, 2533.
17. Rodewald, H. and Kattein, A. Z. physik. Chem., 1900, 33, 579.
18. Hoesch, K. "Emil Fischer; Berlin, 1921; p. 389; lecture on proteins given on January 6, 1906 (cited in ref. 48).
19. Fischer, E. Ber., 1913, 46, 3253.
20. Odling, W. "Thomas Graham" in Farber, E. Ed. "Great Chemists"; Interscience Publ.: New York, 1961; pp. 551-571.
21. Thiele, J. Ann., 1899, 306, 92.
22. Rohm, Dissertation, Tubingen, 1901 (cited in ref. 77).
23. Schroeter, G. Ber., 1916, 49, 2697; ibid, 1919, 52, 2224.
24. Pringsheim, H. Naturwissenschaften, 1925, 13, 1084.
25. Hess, K. Ann., 1924, 435, 1.
26. Bergmann, M. Naturwissenschaften, 1925, 13, 1045; Ber., 1926, 59, 2973.
27. Abderhalden, E., Naturwissenschaften, 1924, 12, 716.
28. Mark, H. and Whitby, G. S., ds. "Collected Papers of Wallace Hume Carothers on High Polymeric Substances"; Interscience Publ.: New York, 1940; p. 123.
29. Svedberg, T. "Colloid Chemistry: Wisconsin Lectures"; 2nd. Edition, New York, 1928 (cited in ref. 31).
30. Crompton, H. Proc. Chem. Soc., 1912, 28, 193 (cited in ref. 31).
31. Olby, R. "The Path to the Double Helix"; University of Washington Press: Seattle, WA, 1974; pp. 1-37.
32. Werner, A. Ann., 1902, 322, 261.
33. Williams, G. Proc. Roy. Soc., 1860, 10, 517.

34. Bouchardat, G. Compt. rend., 1879, 89, 1117.
35. Tilden, W. A. Chem. News, 1882, 46, 120.
36. Wallach, O. (cited in ref. 40).
37. Gladstone, J. H. and Hibbert, W. Phil. Mag., 1889, 28, 38.
38. Harries, C. Ber., 1904, 37, 2709.
39. Harries, C. Ber., 1905, 38, 1195.
40. Pickles, S. S. J. Chem. Soc., 1910, 97, 1085.
41. Harries, C. Ann., 1911, 383, 184.
42. Staudinger, H. and Kleuer, H. W. Ber., 1911, 44, 2212.
43. Baumann, E. Ann., 1872, 163, 308.
44. Simon, E. Ber., 1839, 31, 265.
45. Staudinger, H. Ber., 1920, 53, 1073.
46. Herzog, R. O. and Jancke, W. Ber., 1920, 53, 2162.
47. Polanyi, M. Naturwissenschaften, 1921, 9, 288; ibid, 337.
48. Mark, H. F. in Milligan, W. O., Ed. "Proceedings of the Robert A. Welch Foundation Conferences on Chemical Research. X. Polymers"; Robert A. Welch Foundation: Houston, 1967; pp. 19-58.
49. Staudinger, H. and Fritsch, K. Helv. Chim. Acta, 1922, 5 785; ibid, 1922, 6, 705.
50. Herzog, R. A. Helv. Chim Acta, 1928, 11, 529.
51. Brill, R. Ann., 1923, 434, 204.
52. Weissenberg, K. Ber., 1926, 59, 1526.
53. Katz, J. R. Naturwissenschaften, 1925, 13, 1411; Ergeb. exakt. Naturw., 1925, 4, 169.
54. Hauser, E. A. and Mark, H. Kolloidchem. Beih., 1926, 22, 63.
55. Sponsler, O. L. and Dore, W. H. Coll. Sym. Mon., 1926, 4, 174.
56. Haworth, W. N. et al J. Chem. Soc., 1916, 109, 1314; ibid, 1920, 117, 199; ibid, 1923, 123, 294.
57. Staudinger, H. Ber., 1924, 57, 1203.
58. Fredga, A. "Les Prix Nobel en 1953"; Stockholm, 1954; p. 27 (cited in Staudinger, H. "From Organic Chemistry to Macromolecules; Wiley-Interscience: New York, 1970; p. 7.).
59. Frey-Wyssling, A. Mikroskopie, 1964, 19, 2 (cited in ref. 31, p. 7).
60. Bergmann, M. Ber., 1926, 59, 2973.
61. Mark, H. Ber., 1926, 59, 2982.
62. Staudinger, H. Ber., 1926, 59, 3019.
63. Mark, H. F. "History of Polymer Structure" in Hummel, D. O., Ed. "Proceedings of the 5th European Symposium on Polymer Spectroscopy"; Verlag Chemie GmbH: Weinheim, 1979.
64. Katz, J. R. in "The Phenomena of Polymerisation and Condensation"; Gurney and Jackson: London, 1936; p. 77-93.
65. Drew, H. D. K. and Haworth, W. N. J. Chem. Soc., 1927, 775.

66. Staudinger, H., Johner, H. and Signer, R., Mie, G., and Hengstenberg, J. Z physik. Chem., 1927, 126, 425.
67. Hengstenberg, J. Ann. Physik., 1927, 84, 245.
68. Hengstenberg, J. Z. Krystal. Miner., 1928, 69, 271.
69. Meyer, K. H. and Mark, H. Ber., 1928, 61, 593.
70. Meyer, K. H. Z. Angew. Chem., 1928, 41, 935.
71. Staudinger, H., Frey, K., and Starck, W. Ber., 1927, 60, 1782.
72. Whitby, G. S., McNally, J. C., Gallay, W. Trans. Roy. Soc. Can., 1928, 22, 27.
73. Konrad, E., Bachle, O., and Signer, R. Ann., 1929, 474, 276.
74. Kipping, F. S., Murray, A. G., and Maltby, J. G. J. Chem. Soc., 1929, 1180.
75. King, G. J. Oil Colour Chem. Assoc., 1930, 13, 28.
76. Staudinger, H. and Schweitzer, O. Ber., 1929, 62, 2395.
77. Staudinger, H. and Urech, E. Helv. Chim. Acta, 1929, 12, 1107.
78. Stanley, H. M. J. Chem. Soc. Chem. Ind., 1930, 49, 349.
79. Lebedev, S. V. and Koblyanskii, G. G. Ber., 1930, 63, 1432.
80. Flumiani, G. Z Elektrochem., 1926, 32, 221.
81. Standinger, H., Brunner M., and Feist, W. Helv. Chim. Acta, 1930, 13, 805.
82. Hill, J. W. J. Am. Chem. Soc., 1930, 52, 4110.
83. Carothers, W. H., Hill, J. W., Kirby, J. E., and Jacobson, R. A. J. Am. Chem. Soc., 1930, 52, 5279.
84. Staudinger, H. Ber., 1928, 61, 2427.
85. Adams, R. "Wallace Hume Carothers" in Farber, E., Ed. "Great Chemists"; Interscience Publ.: New York, 1961; pp. 1601-1611.
86. McGrew, F. C. J. Chem. Educ., 1958, 35, 178.
87. Stobbe, H. and Posnjak, G. Ann., 1910, 371, 259.
88. Lebedev, S. V. and Skovronskaya, N. A. J. Russ. Phys. Chem. Soc., 1911, 43, 1124.
89. Staudinger, H. and Breusch, F. Ber., 1929, 62, 442.
90. Lycan, W. H. and Adams, R. J. Am. Chem. Soc., 1929, 51, 625.
91. Carothers, W. H. and Dorough, G. L. J. Am. Chem. Soc., 1930, 52, 711.
92. Flory, P. J. J. Am. Chem. Soc., 1937, 59, 241.
93. Staudinger, H. Ber., 1928, 61, 2428.
94. Lansing, W. D. and Kraemer, E. O. J. Am. Chem. Soc., 1935, 57, 1369.
95. Carothers, W. H. Chem. Rev., 1931, 8, 353.

RECEIVED May 11, 1981.

4

The Science of Macromolecules: Evolution and Early Elaborations

ROBERT SIMHA

Case Western Reserve University, Department of Macromolecular Science, Cleveland, OH 44106

The occurence of the Symposium and its publication are most welcome occasions on at least two grounds. One, of course, to honor the man and his manifold contributions. Secondly, they induce us to think of our science in a retrospective and perspective manner, more so than we are wont to ordinarily do in the course of our particular research activities.

What we want to do then is to discuss some crucial developments of the past, keeping in mind however that neither the recollections of an eyewitness, an essay of the historian of science, or finally a complete survey are to be expected. Instead, we merely ask: how did the two corner stones, namely the recognition of the macro or chainmolecule and the notion of chain flexibility evolve? What were some of the first applications, based on these concepts? An even sketchy consideration of these matters allows us to gage the flavor of the scientific methodology and at the same time, the impact of Hermann Mark's thoughts and work during that period.

The student of the early literature (1) informs us that notions of large particle weights, exceeding the molecular weights familiar to the classical organic chemist by orders of magnitude, were extant already in the nineteenth century. But we have known in other instances that there can be a sometimes extensive stretch in time between the hypothesis and its elevation to the rank of a well founded theory. Recall the classical examples of the atomic theory, or in more recent times the quantum, designated by Max Planck originally as a hypothesis rather than a theory. And so it has been in the present instance.

We can then begin with the mid twenties of our century, when two opposing views had crystallized. Both accepted the existence of large particle masses, although concrete numerical values were hardly accessible at that time. The point of dispute was: aggregate vs. giant molecule. The issue is illustrated by the case of natural rubber (see Table I). Isoprene had been recognized long before as the basic constituent. Does it form some low molecular weight structure, possibly a ring? Do a

0097-6156/81/0175-0045$5.00/0

TABLE I

Natural Rubber: Two Structural Views

1. Isoprene → "Basic Unit" (?)
   "Basic Unit" → Colloidal Particle
   Secondary Forces (?) → Solvent Effects

2. Isoprene → Long Chain
   Chain (Macro) Molecule
   Chemical Bonding → No Solvent Effects

number of these rings combine to form a colloidal particle by the intervention of secondary forces, in this instance possibly arising from the presence of double bonds? If so, the medium ought to play a significant role in determining the particle mass. Alternatively, could isoprene molecules contrive to form long chain structures, and could these structures be responsible for the colloidal characteristics observed in solution? The existence of such a chain or macromolecule implies chemical bonding of small units and hence no solvent effects on the particle mass, provided of course we operate in sufficiently dilute solution.

What were then the approaches used to settle the issue? In view of the outcome which was decided more than forty years ago, we need to deal here only with the methodologies supporting the existence of the macromolecule. One was from the organic chemical direction. Hermann Staudinger showed that the elimination of double bonds by hydrogenation did not eliminate the characteristic colloidal features (2). During the same period, he also investigated a synthetic polymer, namely poly(oxymethylene) (3). He was able to produce by controlled chemical degradation a homologous series of pure oligomeric poly(oxymethylene diacetates) with $n \leq 17$ and demonstrate the systematic variation of their physical properties. He concluded that the repeat units of the original particles were linked to each other so as to form long chains. By analogy he reasoned, other polymers would be similarly constituted.

The other approach originated from the physicochemical side, specifically X-ray spectroscopy of semicrystalline systems, in particular cellulose. Hermann Mark concluded that the micro-building blocks contained but a small number of structural residues and that these were held together at distances corresponding to valence bonds and with forces of the intensity corresponding to valence forces (4).

From these considerations there evolved the concept of "primary valence chains" in cellulose, held together in bundles, or micelles (crystallites) by secondary forces, as propounded by Meyer and Mark (5). This view was then extended to encompass other high polymers as well. It should be noted however, that Freudenberg had already proposed a chain structure for cellulose, based on degradation experiments (6). If the micelles were to survive as entities in solution, then a particle mass determination by whatever method would yield a lower limit for the mean chainlength of the constituent chains. The fact that one was in any case dealing with a distribution of masses, was recognized by all concerned. However the quantitative consequences of this feature for the interpretation and comparison of data derived by means of different experimental methods remained to be appreciated, until reliable solution methods for large (molecular) weights became available.

We recognize at this point the common ground between the investigators at the University of Freiburg on the one hand and the Ludwigshafen laboratory of the I. G. Farben on the other. That is, both supported, with the aid of different methodologies, the second view illustrated in Table I; this at a time when the opposing view was still defended by a respectable group of investigators, organic as well as physical chemists. This is not the place to inject personal matters into the discussion and consider reasons why the expected cooperation did not materialize. The reader interested in the history of the ensuing controversies should consult the original literature of the late twenties and thirties and Ref. 1. Let it only be said that Mark attempted to mediate and to emphasize scientific aspects and the basic common ground. One might speculate that cooperation would have accelerated the start of the second stage, i.e., applications (see below).

To revert to the main trend of our discussion, the issue to be considered at that stage, was characterization in solution. The unequivocal theoretical basis for colligative properties had been provided by the classical thermodynamics of solutions. Svedberg's sedimentation equilibrium in the ultracentrifuge had only recently begun to make its impact, primarily with protein solutions. The theory of Rayleigh scattering in solution was all there by 1930, but its realization as a tool for determinations of large masses had to await another decade. These large masses rendered the use of colligative properties, including osmotic pressure, a difficult experimental task.

An experimentally comparatively easy procedure however was and is the determination of the viscosity increment in solution. Indeed these increments had been shown earlier to be much higher in suspensions of starch or rubber than for, say, a sugar solution. In 1929 Staudinger (7) reported a series of viscosity studies on natural and guttapercha solutions. No (significant) variations in the relative viscosities with changing solvent medium were observable. This in contrast to the reduction brought about by high temperature treatments. Staudinger concluded that degradation of individual macromolecules was involved. Moreover, the results for the degradation products suggested a proportionality between viscosity increments and osmatic molecular weights in a range below $M \simeq 10^4$. From these and further viscosity studies there evolved Staudinger's famed viscosity relation (8):

$$(\eta-\eta_o)/(\eta_o c) = \eta_{sp}/c = K_s \times M \tag{1}$$

first enunciated for polystyrene solutions. The subscript o refers to the solvent and the parameter K_s varies with the solvent and solute, but, the important point, is to be independent of molecular determinations of M in the low, so-called "hemi-colloidal" range. The significance of this relation for further developments can not be overemphasized. We shall revert to this matter but have first to recall the status of viscosity theory (9). Einstein's methodology was to be the guiding spirit of all subsequent work on solution viscosity of colloidal suspensions and polymer solutions up to the present time. He considered a suspension of N compact spheres each of volume v_p in a volume V and obtained the result:

$$\lim_{c\to 0}(\eta-\eta_o)/\eta_o = [\eta]c = f \times Nv_p c = f(c/M)N_A v_p \tag{2}$$

where c stands for the concentration in weight per unit volume, and f is a geometric factor depending on particle shape and hence on molecular or particle mass M. For the Einstein spheres f is a pure numeric, viz. 2.5, when the intrinsic [η] is expressed as cm^3/g. It is worth recalling the immediate attention eq. (2) received by colloid chemists, which ultimately led to the detection of a numerical error in the original derivation.

The above result was invoked by Mark and Fikentscher (10). To make the necessary extension of eq. (2) to higher concentrations, they write:

$$(\eta-\eta_o)/(\eta_o c) = A_1 + A_2 c \tag{3}$$

where the Einstein term A_1 has been augmented by an interaction term. The A_i depend on particle shape in a manner which could not be explicitly formulated at the time. Equation (3) is patterned after the virial expansion of the equation of state in a gas or of the osmotic pressure equation of a solution. This intuitive empirical relation was one of the motivations in R. Simha's doctoral dissertation to derive the first extension of Einstein's theory by the introduction of hydrodynamic interactions (11). Mark and Fikentscher derived from eq. (3) a relation between molecular weight and concentration of a series of equiviscous suspensions.

Staudinger also made use of Einstein's result. To account for higher concentrations, he proposed an exponential relation which however is not obeyed by polymer solutions. Reverting to dilute solutions, he recommended for the concentration below which eq. (1) would become valid, the so-called base molar solution (8), i.e. c of the order of fifty to hundred g/ℓ, hardly a dilute solution. The extrapolation of η_{sp}/c to zero concentration was adopted only subsequently. Staudinger then offered a derivation (8, 12) based on Einstein's basic ideas. To obtain the parameter f in eq. (2), be represented the solute as a rod, rotating in the plane of shear. Now in a homologous series of such rods of length ℓ and thickness d we have:

$$\ell \propto M \; ; \; d \propto M^0$$

The cylindrical hydrodynamic volume swept out by the particle, $\ell^2 d$, is accordingly proportional to M^2 and thus the viscosity increment proportional to M, i.e. eq. (1) is obtained. Staudinger's viscosity relation has exerted a great invluence on both experimental and theoretical research. As is well established, eq. (1) is not verified by either for the type of polymers under discussion here. The investigations resulting in this inclusion tend to confirm the basic concepts of macromolecular science mentioned at the beginning and thus may also be regarded as one of the first applications of these concepts.

Returning to Staudinger's derivation, it must be revised on two grounds. First the kinematics of motion is three rather than two-dimensional and the hydrodynamic volume spherical rather than cylindrical, i.e. $\propto \ell^3$. The detailed calculation for thin ellipsoidal particles (13) shows an approximate proportionality of the intrinsic viscosity with $M^{1.7}$, a considerable difference from eq. (1) for large M.

While the underlying model is appropriate for rigid macromolecules, it is not applicable to the types of systems, such as cellulose, cellulose derivatives, polyisoprene or vinyl polymers, for which Staudinger had intended it. The reason is chain flexibility which arises from an at least limited freedom of rotation of bonds, connecting the repeat units of the chain, a freedom prevailing for single C-C and other linkages as well. Quantitative information for small organic structures had been generated by a variety of spectroscopic and thermochemical experimentation, and the measurement of dipole moments in polar molecules (14). The consequences of this relative internal freedom of thermal motion for the conformations of a chain containing a large number of bonds are profound. It was realized by several of Staudinger's contemporaries (14) although not by him, that a statistical distribution of possible bond configurations necessarily results in a statistical distribution of overall chain conformations. The average represents a loose coil rather than an extended structure with dimensions increasing as $M^{(1+p)/2}$ and p at most a small number (15, 16, 17). This element of chain flexibility was soon recognized to pervade the physical properties in solution as well as in the amorphous bulk state. Reverting to the viscosity problem, eq. (1) must be replaced by the two-parameter expression

$$[\eta] = KM^a \tag{1a}$$

where K is a constant for a given solute-solvent pair and the exponent does not assume a universal value for all systems but is in general less than unity. An equation of this form with a = 2/3 was first derived by W. Haller (18). It has since then entered the literature as the Houwink-Mark relation. It should be noted in this connection that even in Staudinger's laboratory deviations from eq. (1) were observed for poly(vinyl acetate) and polyacrylates (19). These were qualitatively consistent with eq. (1a), that is, osmotic molecular weights were found to be considerably larger than expected from eq. (1). We have devoted the considerable amount of attention to the viscosity problem which is its due in the development of macromolecular science. Staudinger's work has exerted a great influence, in a different manner than Einstein's of course, by initiating and catalyzing extensive experimental and theoretical research. This has aimed at obtaining an acceptable picture of macromolecular dimensions in (infinitely) dilute solution, at a calibration of eq. (1a), and finally at providing a theoretical basis for this equation, based on Einstein's orginal ideas. At the same time, solution viscosity represents an early application of the notion of the flexible macromolecule to a material property.

The importance of thermodynamic methods for the characterization of the macromolecule was obvious, as was the particular role among colligative properties played by osmometry. Beyond this and accepting the second view illustrated in Table I, one was led to inquire into the consequences of this view for the thermodynamics of macromolecular solutions. By the mid nineteen thirties a series of experimental studies on solutions of oligomers and polymers were being undertaken, of which those carried out in Kurt H. Meyer's laboratory in Geneva should be particularly recalled because of their detailed and systematic character (20). The crucial recognition was the enhanced entropy of mixing arising from the disparity in size between solute and solvent, clearly even if only qualitatively discussed by Meyer (Ref. 20, p. 586). Quantitative results for idealized models were derived first by the English school of statistical mechanics (21, 22).

Finally, we turn from solutions to the bulk state of amorphous polymers, specifically the thermoelastic properties of the rubbery state. The contrasting behavior of rubber, as compared with other solids, such as the temperature decrease upon adiabatic extension, the contraction upon heating under load, and the positive temperature coefficient of stress under constant elongation, had been observed in the nineteenth century by Gough and Joule. The latter was able to interpret these experiments in terms of the second law of thermodynamics, which revealed the connection between the different phenomena observed. One could conclude the primary effect to be a reduction of entropy rather than an increase in energy. This is as far as phenomenological theory could go. It became now possible to aim at a molecular theory, based on a statistical mechanics of bulk polymers, and thus to refer to a rubbery state of matter, in the same sense as, for example, to the liquid state. Initial attempts to relate the deformation of rubber to the thermal motion of rigid rods (23) and to a helical conformation (24) were superseded by considerations of a realistic chain structure first advanced by Meyer, v. Susich, and Valkó (25). The essential idea is an entropy reduction, resulting from a reduction of the number of conformations of a coiling chain upon deformation. A quantitative evaluation of this picture by Guth and Mark followed soon thereafter (16). A model of a cross-linked rubbery network was first analyzed by W. Kuhn (26).

The applications of the basic tenets in macromolecular science discussed here have been concerned with certain physical properties. We desist from a discussion of the kinetics and statistics of formation and decomposition of high polymers by the various chemical routes, which evolved at about the same time.

The lines of research sketched introduced a heroic age of macromolecular science during the following two decades or so. In this period the primary emphasis in physical research remained on the plane of general phenomena rather than specific polymers per se. That Hermann Mark with his collaborators of backgrounds ranging from organic chemistry to theoretical physics, exerted a strong and characteristic influence is evident from even a cursory examination of the period's literature. Mark's interests nevertheless were not confined to the topics of this article. In later years, his more urgent concerns turned to other directions in the polymer arena. However throughout, he has continued as an active witness of developments the origins of which can be traced back to those early years. One needs to recall only typical areas such as the chain dynamics of the single chain, dynamics and rheology of moderately and highly concentrated solutions and the melt, the statistical thermodynamics of bulk polymers, the thermoelasticity of the rubbery melt and gel, the vast structural efforts on synthetic and biopolymers, and the extensive kinetic studies. In the course of his long scientific career Hermann Mark has witnessed all this and he will, of course continue to do so.

Literature Cited

1. Useful references and documentary material are found in Priesner, Claus, "H. Staudinger, H. Mark and K. H. Meyer: Thesen zur Größe und Struktur d. Makromoleküle"; Verlag Chemie: Deerfield Beach, Fla., 1980.
2. Staudinger, H. Ber. 1925, 57, 1203.
3. Staudinger, H. Helv. Chim. Acta 1925, 8, 65.
4. Mark, H. Ber. 1926, 59, 2982.

5. Meyer, K. H.; Mark, H. Ber. 1928, 61, 593.

6. Freudenberg, K.; Braun, E. Liebigs Ann. 1928, 460, 288.

7. Staudinger, H.; Bondy, H. F. Liebigs Ann. 1929, 468, 1.

8. Staudinger, H.; Heuer, W. Ber. 1930, 63, 222.

9. Einstein, A. Ann. Physik 1906, 19, 289; 1911, 34, 591.

10. Mark, H.; Fikentscher, H. Koll.-Z. 1929, 49, 135.

11. Guth, E.; Simha, R. Koll.-Z. 1936, 79, 266.

12. Detailed discussions and references are found in Staudinger, H., "Die hochmolekularen organischen Verbindungen"; J. Springer Verlag: Berlin, 1932.

13. Simha, R. J. Phys. Chem. 1940, 44, 25; J. Chem. Phys. 1945, 13, 188.

14. A detailed survey of the literature of the period is given in Mark, H. "Physical Chemistry of High Polymeric Systems"; Interscience Publ. Inc.: New York, 1940.

15. Eyring, H. Phys. Rev. 1932, 39, 746.

16. Guth, E.; Mark, H. Monatsh 1934, 65, 93.

17. Kuhn, W. Kolloid-Z. 1934, 68, 2.

18. Haller, W. Koll.-Z. 1931, 56, 257.

19. Staudinger H.; Warth, H. J. Prakt. Chemie 1940, 155, 261

20. See Meyer, K. H. "Natural and Synthetic High Polymers"; Interscience Publ. Inc.: New York, 1942.

21. Guggenheim, E. A. Trans. Faraday Soc. 1937, 33, 151.

22. Fowler, R. H.; Rushbrooke, G. S. Trans. Faraday Soc. 1937, 33, 1272.

23. Wöhlisch E.; de Rochemont R. Z. Biol. 1927 85, 406.

24. Mark, H.; Valkó, E. Kautschuk 1930, No. 10, 210.

25. Meyer, K. H.; v. Susich, G.; Valkó, E. Kolloid-Z. 1932, 59, 208.

26. Kuhn, W. Kolloid-Z. 1936, 76, 258.

RECEIVED August 19, 1981.

5

Polymer Science Through the Fifties

C. S. MARVEL

University of Arizona, Department of Chemistry, Tucson, AZ 85721

I shall confine my remarks to the development of polymer science in America. There was considerable work before that in Europe, but that was before chemistry of polymers became sophisticated and real progress was made in the field. In about 1910 industrial work on polymers began to be carried out, although the understanding of polymers was very meager. The industrial work was done mainly on natural products such as horn and hoof and they were converted into useable materials by various means without much understanding of the real chemistry involved.

Horns and hooves were the raw materials for the early polymer preparations. These materials were ground up and treated in various ways so that they could be fabricated into such items as combs to use for ladies' hair, and other specialty things of that sort. The next development was the use of cellulose from cotton or from wood as the raw material which was studied for making films and fibers. Work on the cellulose structure had provided information that it was a hydroxylated product, and by converting the hydroxyls to esters, the natural cellulose could be turned into a soluble material, which was spun into fibers and cast into films to make the first cellulose rayon-type material and cellulose films.

One of the earlier methods was to treat cellulose with sodium hydroxide and carbon disulfide to obtain xanthate esters which could be dispersed in water and cast into sheets or spun into fibers. Subsequent treatment with acid decomposed the xanthates and gave regenerated cellulose, either in fiber or film form. The fibers were called viscose rayon and the films were named cellophane. Cellophane is still used as a wrapping film and some of it is still manufactured by the xanthate process. By treatment with nitric acid, cellulose was converted to a trinitric acid ester, which could be cast into units which were satisfactory for making gun cotton for a smokeless powder for either artillery shells or shotgun ammunition. It was quite insoluble but it could be converted to a jelly-like mass, which could be shaped into a desired form for ammunition use. Under milder conditions, a lower nitrate

0097-6156/81/0175-0053$5.00/0

of cellulose could be made which was called pyroxylin, and this was soluble in ether and acetone. This material could also be spun into a film or a fiber and the film was actually much used in photographic work. But it was quite hazardous because of the high flammability.

When blended with camphor, pyroxyline gave celluloid, which was used for making buttons, collars, cuffs, and so forth. Again, the flame hazard was great. Solutions of pyroxylin were used to make films to cover cuts, bruises, burns, etc. but the fire hazard was great. Later cellulose was treated with acetic anhydride and sulfuric acid to give a lower acetate which was soluble in acetone and which could be cast into films or spun into fibers. The films were much used for photographic films and were much safer than the nitrate films. Acetate rayon was made this way. It is still a product of the textile industry. The acetate quickly took the place of the nitrate in films and fibers. Some mixed esters of acetic or butyric acid were prepared to give a molding plastic which had considerable use at one time.

True synthetic polymers came into use when Bakeland came from Belgium and applied his knowledge of the formation of a moldable plastic from phenol and formaldehyde to give the product named Bakelite. This was about 1914. This product, under heat and pressure, set up to a thermo-setting resin and had useful properties especially as an insulating material for electrical items. It is still widely manufactured and used for such purposes and in the making of varnishes and finishes of one sort or another. The chemical reactions involved in the formation of Bakelite were not known at that time, but the conditions for manufacturing some varieties were well worked out so that they could be prepared on a large scale.

Soon after the Bakelite era came other synthetic plastics. The glyptals made from phthalic anhydride and glycerol were developed as compositions for use in paints and varnishes. If the reaction was carried out too long the product became intractible. But under milder conditions, other products could be obtained which could be used in making soluble products and then they could be set further after forming. It was learned that by modifying the reaction mixture with some monobasic acid to balance the hydroxyls and carboxyls in the reaction mixture, more soluble products could be obtained. Kienle of General Electric, was one of the early developers of these products. Later many other alkyd resins from other polyhydroxyl compounds and poly acids were produced for technical use.

Another of the early thermo-setting resins was the urea formaldehyde resin developed by American Cyanamid. Urea and formaldehyde gave methylol and dimethylol urea which on heating, further condensed to give the colorless resin. The technology of this process was well worked out, but the chemistry was still not completely understood. The colorless resin has been used to make many varieties of decorative materials and some essentially unbreakable dishes, among other uses.

The next products developed were the thermoplastics made by vinyl polymerization. Vinyl chloride, and vinyl acetate copolymers known as Vinylite, was one of the first produced by Carbide and Carbon. It had been worked on in Europe and had also been worked on by Ostromysslenski, a chemist with the U. S. Rubber Company. The Carbide chemists realized at an early date that to get a uniform composition it was necessary to adjust the feed ratios of the two monomers to get a proper combination in the reactor to give a constant composition to the product. The next major development of a vinyl product was in the development of Lucite or Plexiglas polymers, which are clear, strong plastics. The clarity of these plastics made them very useful for canopies for fighter pilots to protect them from bullets in battle. Their strength and clarity are two very important properties which make them still most useful. Their resistance to sunlight is very good.

All of these developments came along before chemists understood what polymers really were. Originally, they were believed to be materials held together in mass by some active forces other than normal valence forces. In about 1920 Hermann Staudinger realized that polymers were macromolecules held together by normal valence forces. He understood that polymers were really a mixture of large molecules of different molecular weight and they had properties imparted to them by this very set of circumstances. The high viscosities of these molecules were connected with the molecular weight and there was a fixed relationship of viscosities with molecular weight which could be measured to tell what progress was being made in synthesis.

It remained for Carothers, who joined the DuPont organization in 1928, to show that Staudinger's view that polymers were truly macromolecules was correct and that they could be formed by normal chemical reactions, and thus show the characteristics of polymer molecules. Staudinger's views of the natures of polymer structures were sound, but he did not really demonstrate them experimentally. It remained for Carothers to do that with his condensation polymers. Carothers fashioned polymeric materials such as polyesters and polyamides and other polymers by self-reacting molecules like hydroxy acids, amino acids or di-hydroxyl compounds and diamino compounds with dibasic acids. He clarified the ideas of recurring units in polymers, of bi-functional molecules, end groups, reactivity ratios, viscosity molecular weight relations, etc., so that further progress in the field could be more readily made. In a short period of time, about 10 years, he discovered that polyamides such as poly-hexamethylene adipamide gave a strong plastic that could be spun from a melt to give an oriented fiber, which was called Nylon.

It's interesting to recall that in the extreme excitement of this discovery, the group working at the Experimental Station at DuPont did not realize that poly-hexamethylene adipamide was a valuable polymer for fiber formation, and no patent was filed. The work did go on on polyesters which were more simple to work with, and it was while working with those that Julian Hill discovered they could be drawn into fibers. He noted that by drawing a stirring rod out of a molten polyester, he obtained a thread which, on stretching, looked silky. Examining the thread he found the polymer molecules had been oriented longitudinally and made them much more stable fibers in this way. The polyesters were too low melting to make textile fibers. However, when he applied this cold drawing to polyamides they gave good silky threads which were useful as textile fibers and he was able to make materials from polyamides which had properties that were satisfactory for textile use. It was the cold drawing patents that gave DuPont its monopoly for the start of synthetic fibers.

Other textile fibers such as polyethylene terephthalate came into production first in England and then in America. The English discovered the fiber and the Americans discovered the film. When it was drawn bi-axially, polyethylene terephthalate gave a very good film which has had wide use.

Carothers's work was mainly in the field of condensation polymers. He did make one significant contribution to addition polymers. Calcott at Jackson Laboratory had noted Father Nieuwland's work on the dimerization and trimerization of acetylene and felt these monomers should be useful in making a synthetic rubber. He acquired the patent rights from Father Nieuwland for this discovery and then began working on these products to see if he should convert them into rubber. After some years of work, he went to Carothers for help and asked Carothers if he would like to have these two monomers to study. Carothers thought they would be interesting. One of his early acts was to assign to Arnold Collins the task of purifying some of these monomers, and carrying out the distillation of the crude polymerization product from acetylene, Collins noticed a small intermediate fraction of liquid which boiled between monovinylacetylene and divinylacetylene. It was late in the day and he set this small sample aside to take care of later. When he returned on the next working day, he found that the intermediate fraction had solidified and had rubbery properties. In fact, it would bounce on the table. He examined it and found that it really was the hydrochloric acid addition product of monovinylacetylene, which had been formed along with the dimer and trimer of acetylene. When he subsequently found out how to produce it, he patented it and the first useful synthetic rubber, Neoprene, came from its polymerization. Neoprene is still the oil-resistant strong rubber of choice in industry.

While Carothers and his group were studying condensation polymers, other industrial companies were studying vinyl polymerization and getting basic understanding of that process. The discovery that vinyl polymerization developed in three distinct stages -- initiation, propagation, and termination -- did much to give an understanding of the process. Initiation could be by radical formation or by anionic process, depending on the type of monomer. Propagation was a rapid step where the activated monomer attached to the growing polymer chain. Termination could be by chain transfer, disproportionation, or coupling. The various methods of termination depended to some extent on the monomers used in the study of the polymerization.

The early work was largely on vinyl chloride-vinyl acetate copolymers. Ostromysslenski at U. S. Rubber Company did a lot of work on the vinylites; the Rohm and Haas group and DuPont were most active in the methyl methacrylate polymers (Lucite and Plexiglas). The discovery of plasticization of polyvinyl chloride by Waldo Semon of Goodrich using tricresyl phosphate and dioctyl phthalate was a very important step in the development of useful products from polyvinyl chloride so that this polymer became one of the leading vinyl products that is produced. The work of Mayo and Walling at U. S. Rubber on reactivity ratios of monomers aided greatly in making copolymers of definite composition and uniformity. Price and Alfrey also contributed much in this field.

Before the mechanism of vinyl polymerization was understood, the question of the structure of vinyl polymers was of considerable interest. Staudinger had written these polymers as having a head-to-tail arrangement of recurring units, but he had not really furnished evidence of the structure. As Carothers once said, Staudinger had assigned the structure by pronouncement. He was as usual correct, and chemical evidence was developed to establish such structures. For example, when monovinyl methyl ketone polymerized, it could produce by head-to-head, tail-to-tail reaction a 1,4-diketone. By head-to-tail polymerization it would give a 1,5-diketone. These two types have different reactions. The study of the polymer proper showed that the polymer was a 1,5-diketone. In the case of polyvinyl chloride, a head-to-head, tail-to-tail polymerization would lead to a 1,2-dihalide compound, and a head-to-tail polymerization would lead to a 1,3-dihalide. It was the latter that was produced as shown by the fact that only about 86% of the chlorine was removed by treating the polymer with copper. Later it was shown that this treatment produced cyclopropane units in the chain so that the question of head-to-tail structure was definitely established. A few polymers do have structures containing both head-to-tail and head-to-head, tail-to-tail structures.

When vinyl flouride is polymerized, a certain percentage of the flourine atoms are found on adjacent carbon atoms. When polyvinyl alcohol made by hydrolyzing polyvinyl acetate was treated with periodic acid by Flory, he found that the molecular weight was greatly reduced, indicating that there were 1,2-glycol units in the chain. But most of the polymers were regular head-to-tail polymers.

Vinyl polymerization was advanced a great deal during the government rubber program. The research groups of U. S. Rubber, Firestone, Goodyear, and Goodrich, together with university groups from Massachusetts Institute of Technology, Cornell, Case, Chicago, Illinois, and Minnesota, and the industrial groups of Esso, Bell Telephone, and DuPont, collaborated in a very effective program to produce a useful rubber. When the war cut off the supply from the Far East, within a year these research groups had made a product that was satisfactory for passenger cars. The government set up a rubber czar in Washington to see that all priorities for supplies needed were provided to this research group. Industrial companies built plants for making rubber before the exact process for its manufacture was settled. But, they were very effective and produced rubber at about 150% of the rated capacity. When the urgency was over, the government sold the plants back to industry and it is said that the whole operation was profitable.

With minor modifications that rubber is still used for passenger cars. It is not suitable for large trucks and bomber tires because of the excess of heat build-up in operation. Before the end of the rubber program, two of the companies, Firestone and Goodrich, had developed processes that produced rubber essentially like natural rubber. Firestone used a lithium catalyst for the polymerization, and Goodrich used a modified Ziegler catalyst. These materials were manufactured for a while until the oil prices became too prohibitive and the natural rubber was again used for heavy-duty tires.

Industry developed polymers from most of the cheaper raw materials during this period, and production of these products rose to the billions of pounds. Specialty polymers were also produced such as epoxy resins, flurocarbon polymers, and thiokol rubbers, which have made a comeback for use in developing fuel for spaceships. Polystyrene developed remarkably after pure styrene became available through the rubber program. ABS resins (butadiene-styrene-acrylonitrile terpolymers), came into use for making a very strong film for luggage and automobile seats for rugged wear. It is our only terpolymer which is made in large quantities in industry.

In conclusion, I want to pay my respects to Professor Herman Mark, whom we are honoring today. He has been a very important factor in the development of polymer chemistry in America. He came to Brooklyn and one of his first acts was to arrange a seminar on Saturday mornings on polymer chemistry, which had the attendance of the local polymer chemists in the area, both from industry and the universities. This was an important stimulus to all, to be brought up to date on what was new and to get together and talk over the new products. He started the first Polymer Institute of America in Brooklyn and trained many of our active polymer chemists. He founded the Journal of Polymer Science, with its Chemistry Edition, Physics Edition, Polymer Letters; the Journal of Applied Polymer Science; and the regular reports of important polymer symposiums being held throughout the world.

He was largely responsible for the founding of the Polymer Division of the American Chemical Society. He was active as a consultant in industry and university groups, thus stimulating the chemists and keeping them up to date. His contacts were international and in this way he helped to exchange information in the field around the world. His activities have been a great stimulant to the progress of polymer chemistry in America and throughout the world.

At the end of 1950, America had really arrived at the Polymer Age and progress has continued, as will be seen in the talks that will be given later in this Symposium.

RECEIVED February 5, 1981.

6

Herman F. Mark: The Geheimrat

G. ALLAN STAHL

Phillips Petroleum Company, Research and Development, Bartlesville, OK 74004

> "Once in Ludwigschafen the laboratory director, O. Seidl, told me, 'Mark, you should go into business. If you are as successful in making money as you were in spending it on research we can already congratulate ourselves.'"
>
> H.F. Mark

The years Mark worked in Berlin in retrospect set the stage for his rocket like rise to eminence in polymer science. Postwar inflation placed many of Berlin's bolder attractions out of the reach of the young scientist and his new bride, Mimi. In place of the expensive frolics, the Mark's entertained and exchanged ideas with an international set of coworkers in his Lichterfelde-West apartment.

Besides his fellow Austrians, Mark's close friends included Hungarians and Poles, as well as Germans. This broad mesh of backgrounds and styles of thinking was catalyzed by the unexcelled and readily available music, art, and architecture, and stimulated by such scientists as Schroedinger, Haber, Hahn, V. Laue, Einstein, Planck, Nerst, and Schlenk. The result was it expanded his perspective, taught him how to communicate and lead a diverse group of scientists despite problems of differing nationality and proclivity, and enhanced his natural sagacity as only a broad range of experiences can.

A broadening in Mark's intellect is shown clearly in his publications of this period. The topics in 1926 and 1927 alone ranged from atomic structure and quantum theory (1), and Compton radiation (2) to the scattering of x-rays by an ideal gas (3) and mineral structure (4). The shear diversity of his contacts and interests made him qualified for expanded responsibilities.

0097-6156/81/0175-0061$6.75/0

Professor Fritz Haber, Director of the Kaiser Wilhelm Institute, no doubt had this in mind when he called Mark to his villa during the summer of 1926. In a visit reminscent of Haber and Schlenk's meeting which brought Mark to the Institute a few years before, Haber outlined Mark's achievements and described a new opportunity for advancement, this time in the chemical industry. The position was as an Assistant Director of Research under Kurt H. Meyer with the giant I.G. Farbenindustrie.

Research at I.G. Farben at that time was conducted in five large research laboratories.Inorganic chemistry was the primary research area at Bitterfeld under L. Pistor, while pigments and pharmaceuticals were studied at Leverkusen under C. Duisberg and at Hoechst under E. Schmidt. Research of ammoniacal fertilizers was conducted at Oppau under A. Mittasch. The fifth laboratory was the central laboratory in Ludwigschafen, headed by Meyer.

Meyer, who was also an executive of the company, sought an energetic, diverse individual to direct the laboratory in a concentrated study of fibers and films. In the period ensuing the First World War, the I. G. factories had switched from guncotton manufacture to cellulose acetate and viscose, but too little was known of the basic characteristics of these materials to make substantial improvements on their properties. Further, it was believed that synthetic fibers were possible, but their preparation was not practical at that time. The farsighted Meyer recognized the business potential that an understanding of the basic nature of fiber-forming substances could tap.

When he interviewed with Meyer in 1926, Mark outlined a typically thorough program. He proposed a team of organic and physical chemists, and physicists who would evaluate the influence of structure on properties such as rigidity, elasticity, melting point, and water absorption. Work, he proposed, would shift toward new material development and into the manufacturing facilities to evaluate the effects of processing on structure as their knowledge base expanded. "All of this," Mark told Meyer, "can be accomplished if I have the necessary apparatus, equipment, and a few able physicists, physical chemists, and, of course, a good organic chemist."

Meyer approved and hired Mark. In doing so he began simultaneously a very close, life-long friendship and a professional relationship which would help bring I. G. Farben and polymer science to new heights. Mark moved from Berlin to Mannheim on January 1, 1927.

In assuming his duties, Mark began an intensive period of plant visits, laboratory design, and personnel hiring. Six months afterward, he had hired a number of excellently trained scientists such as physicists J. Hengstenberg from Freiburg, and R. Wierl and K. Wolf from Munich. In addition, he brought physical chemists H. Dohse, G. V. Susich, E. Valko, H. Kalberer,

M. Dunkel and H. Fikentscher, and, of course, "a good organic chemist" (or two) C. Wulff and H. Hopff. The group eventually numbered about fifty with fifteen scientists, and were organized in three sections--synthesis, characterization, and applications--with Mark as the coordinator. In addition by mid-1927, the I. G. could boast of the best X-ray diffraction equipment on the continent.

Mark's inclination based on his research at the Fiber Research Institute was that "colloidal" substances such as cellulose and rubber were of high molecular weight. Work began with this cornerstone concept. In the beginning research was mostly concerned with testing viscose and cellulose acetate fibers. Before long a great deal of research was conducted on materials subsequently called addition polymers.

The synthesis section systematically prepared new monomers, polymers, and an ever increasing number of copolymers. At the same time, the characterization and applications sections tested the polymers in order to ascertain which were worthy of larger scale experiments, scale-up, and patent protection. They also performed the work required to satisfy production details. These efforts, directed by Mark's personal hands-on style of management, were the first serious attempts at commercialization of polystyrene, poly(vinyl chloride), poly(methyl methacrylate), and synthetic rubber.

Work with these materials would have been less productive had Mark have not simultaneously expanded the original studies to include a large number of related and seemingly unrelated natural substances. An incomplete, chronological review of Mark's publications of 1928 brings to light the methodical thoroughness of their research.

Early in the year, Mark published two fundamental papers dealing with the foundations of crystal structure analysis (5) and Naegeli's micellular concept (6). Shortly afterward Hengstenberg and Mark (7), and Meyer and Mark (8) expanded the existing views and proposed crystallites of cellulose and rubber, and silk respectively. The proposed crystallites were much larger than the commonly accepted (colloid-) association theory would permit. This assertiveness was simultaneously supportive of Staudinger's high molecular weight concept or macromolecules, and instructive regarding the properties of related natural and synthetic materials. In this same year, Mark published papers on chitin (9), rubber (10), and started an exhaustive study of the structure of cellulose.

It is interesting to note that in their first paper on cellulose (11) Meyer and Mark proposed a structural unit cell model which is classic and accepted, for the largest part, even today. They proposed a cellulose crystallite in which all atoms were bound to each other in very long chains by primary valence forces, and the chains in turn aggregated in larger combinations by secondary forces. The simplicity of this concept is misleading. It represented a compromise between the association theory of molecular interaction and macromolecules, while actually embracing the latter. In the following years, Mark published six other papers on cellulose, starch, and sugar. Meyer and Mark's contributions to the development of a polymer theory and relationship with Professor Hermann Staudinger are the subject of a subsequent section.

Mark's contributions while at the I. G. were not limited to the emerging field of polymer science. In those five years, he also took part in studies of X-ray optics and continued his study of the X-ray structure of metals and metal salts. Other seemingly unrelated paper were published on the width of X-ray emission lines (12), Schlenk isomerism (13), the structure of aromatic compounds (14), and a special "hobby" the optical Stark Effect (15, 16). Regarding this latter work, Mark relates that his supervisors tolerated the research commenting that "as long as they are doing something decent and important" it was okay "as sport doesn't cost much money".

In 1930, R. Wierl and Mark studied N. Davidson and J. Germer's experiments on electron diffraction. Employing their wide experience in instrumentation, they promptly constructed an improved electron scattering apparatus. With this instrument, they determined the interatomic distances in a number of molecules and published a series of papers on the technique and their findings (17, 18, 19). Mark's contributions to the field of crystal structure are discussed in a later chapter of this volume and will not be covered in more detail here (see Pauling, L. "Herman Mark and the Structure of Crystals", this volume.).

Throughout the Ludwigschafen years Mark and his associates maintained close contact with the Universities at Heidelberg and Freiberg and technische hochschulen at Darmstadt and Karlsruhe. In Heidelberg they consulted with Karl Freudenberg, who was well known for his fundamental contributions toward the acceptance of the chain structure of cellulose. As we shall discuss later, they colloborated with Hermann Staudinger at Freiberg. Staudinger, in addition to his organic chemical preparations, ketene studies, and ground breaking work in macromolecules, also produced a large number of highly qualified assistants, several of whom later worked for I. G. Farben. Close contact was maintained with the schools at Darmstadt where E. Berl was actively involved in the evaluation of the technical properties of cellulose and its derivatives, and Karlsruhe. Mark was an associate professor at Karlsruhe, and accordingly observed G. Bredig and A. Reis' studies of the physical chemistry of colloids and crystals.

Figure 1. I. G. Farben scientists with Professor J. R. Katz (seated) in Ludwigschafen in 1929; Dr. H. F. Mark is standing at the far right.

The indefatigable Mark could in these years be found at work at nearly any hour. From his morning shave frequently taken in his office to a late evening conference in a local wine cellar, Mark lectured, wrote, coaxed, pushed, defended, and shoved his researchers to seldom matched levels of production. One coworker at Ludwigschafen recently commented, his "astounding amount of work, especially his contributions in books and papers was unthinkable without economy of work and many hours of overtime" (20). The same can be said for the whole laboratory.

Through their efforts, I. G. Farben was the world leading corporation in commercializing new polymers and copolymers. A list of their polymers which were commercialized in the era includes polystyrene, poly(vinyl chloride), poly(methyl methacrylate), poly(vinyl alcohol), and the first synthetic rubbers, Buna-N and Buna-S. Among many other accomplishments, the laboratory was a leader in knowledge of polymer structure as well as copolymerization, first (after a Mark visit to the Frankfurt Arboretum) in developing peroxide polymerization initiators, at the forefront in the development of new equipment ranging from X-ray and electron diffraction apparatus to fiber spinning nozzels, and one of the finest X-ray analysis laboratories anywhere.

Mark's personal record between 1927 and 1932, his years with I. G. Farben, indicates over eighty publications, seventeen patents (issued to I. G. Farben), and three books. Two coauthored by Meyer and Mark, "Aufbau der Hochpolymeren Substanzen" (21) and "Der Aufbau der Hochpolymeren Organischen Naturstoff" (22), were printed several times, and were considered texts of the day. "Die Experimentellen und Theoretischem Grundlagen der Elektronenbeugung" (23) was well accepted. Mark also published a treatise on the "Physik und Chemie der Cellulose" (24) during this period.

The charmed times ended during the summer of 1932. The political situation in Germany swung and favored an imminent assumption of power by the National Socialists Party. Mark's position, although personally apolitical, was precarious because of his Jewish heritage. A director of the Ludwigschafen and Oppau works, F. Gaus, advised him to look for an academic position. Gaus further agreed to support Mark in an academic position for three years after leaving the I.G. to ease discomfort during the period of establishment. Mark accepted this generous, altruistic offer, and after considering several attractive offers, all unfortunately in Germany, he accepted a position at the University of Vienna. In October, 1932, Mark, his wife, and two sons moved to Vienna. About the same time his close friend, collaborator, and former supervisor, K. H. Meyer, moved to Geneva.

"Because of small differences in neighboring opinions, a larger point of view often does not get enough attention."

H. F. Mark

"Mark will NEVER have a priority fight. He is only unhappy if a person allows something to die on the vine."

E. Proskauer

When Herman Mark first evaluated the crystal structure of rubber (with E. A. Hauser) and cellulose (with J. R. Katz) in 1924 and 1925, it was generally accepted that these materials were low molecular weight or monomeric. The unusual properties of these substances, now known to be related to high molecular weight, were then attributed to aggolomeration or "association" of the low molecular weight precursors. A common explanation for the associations were secondary forces such as Johannes Thiele's partial valences.

This concept, the association theory, was not lacking in either supporters or basis. It grew from the colloid theory of Thomas Graham, was reinforced by Thiele, and took its ultimate expression from three important scientific developments of the early twentieth century. They were, according to Olby (25), the introduction of A. Werner's concept of two kinds of combining forces--primary and secondary valences, the extension of colloid science into biology (thus gaining it reputable recognition), and the seeming support of the low molecular weight concept by X-ray crystallography.

This latter support was based on the then accepted idea that the molecule could not be larger than the unit cell of the crystal. In addition, supporters noted that the colloidal properties of many substances such as soaps are based on the creation of large aggregates of small single molecules.

There was, therefore, good reason in the 1920's to assume an analogous foundation for rubber, cellulose, and other related materials. Not everyone agreed. Workers in preparing "rubber like" substances from isoprene had, since the turn of the century, been explaining the resultant properties by use of ever expanding combined isoprene rings. C. Harries and S. S. Pickles were independently predicting up to seven and eight isoprene units in rubber, but the point of view was less commonly accepted than the association theory.

The death knell for the association theory began in 1920. That year H. Staudinger published an important paper called "Uber Polymerisation" (26). The paper was a summary of his findings over several years regarding high molecular weight. The concepts put forth in that paper, the proposed linear chain formulas for polystyrene and polyoxymethylene for example, are still valid today. As a result of his many contributions to the development of polymer or (as he called it) macromolecular theory, Staudinger was awarded the Nobel Prize in Chemistry in 1953. Acceptance of Staudinger's ideas did not come quickly, and Mark and his associate, K. H. Meyer, played important roles. Their contributions and relationship with Staudinger is an interesting bit of scientific history (27).

In 1926 Staudinger was totally involved in defending his ideas regarding the macromolecule. On the other hand, Mark, at the Fiber Research Institute in Dahlem, was riding the crest of acclaim as an outstanding crystallographer and expert in molecular structure. The two were requested to present papers by R. Willstaetter at a special symposium of the "Gesellschaft Deutscher Naturforscher und Arzte".

The symposium was held on September 23, 1926, in Dusseldorf. It was a classic showdown between association theory constituents, M. Bergmann and H. Pringsheim supported by E. Waldschmidt-Leitz, and Staudinger. Mark was invited by Chairman Willstaetter at the prompting of his close friend F. Haber to address the problem of molecular weight and elementary crystal cell size. Quantification, as Willstaetter, Haber, and Mark recognized, indicated no more than four residues in the unit cell of cellulose and silk. If in fact the molecule could be no larger than its elementary cell, this was weighty evidence against the existence of very large molecules. (Texts of the Bergman, Mark, Waldschmidt-Leitz, Pringsheim, and Staudinger presentations were published. The interested reader should consult: Ber., 1929, 59, 2973, 2982, 3000, 3008, 3019..)

Bergmann opened the meeting and brought forth the argument that the classic theory of Kekule' chains was inappropriate to explain the macroproperties of inulin and in general proteins. As an example he referred to "psuedo-high molecular weight" inorganic complexes. In a similar fashion, Pringsheim discussed the nature of inulin and other polysaccharides. Bergmann and Pringsheim cited the work of P. Karrer, K. Hess, and R. Pummerer and referred to primary and secondary bonding as proposed by Werner.

Staudinger countered with a voluminous review of his work on rubber and synthetic polymers. He supported his contentions of high molecular weight with solution viscosity data and results of hydrogenation experiments on rubber, polystyrene and polyindene. The hydrogenation experiments were designed to demonstrate retention of properties despite loss of "associable" unsaturation. His interpretations were not widely accepted because of the elementary cell - molecular weight ambiguity. It was anticipation of this stalemate which prompted Mark's invitation.

Mark's paper, titled "Die roentgenographishe Ermittlung der Struktur organischer besonders hochmolekularer Substanzen", systematically compared the analysis of hexamethylenetetramine

Figure 2. Professor Hermann Staudinger at Freiberg im Breisgan on July 14, 1953.

with cellulose. In a lengthy presentation, he concluded that cellulose is held together by forces "comparable in type and magnitude to the inner molecular forces" and "the whole crystallite appears as a large molecule".

Reaction to the symposium was varied. A number of persons (including Chairman Willstaetter) were swayed in favor of large molecules, while others were incredulous. One result of the meeting was extremely important. After Mark's presentation the problem of relating the small elementary cell to high molecular weight was no longer an obstacle.

Shortly after the Dusseldorf meeting, Mark and Staudinger started exchanging letters. On January 5, 1927, only a few days after starting his new position at the I. G. Farben Ludwigschafen laboratory, Mark wrote and asked Staudinger for samples of polyoxymethylene decomposition products for new X-ray investigations. Staudinger declined on the grounds that he was already involved in similar research with Hengstenberg (who joined Mark in Ludwigschafen the following year) and Mie. The Mark-Staudinger relationship was excellent and their contact by letters and consultation steady.

In 1928, K. H. Meyer and Mark published an extensive paper on the crystal structure of cellulose (11). They proposed a structure which is nearly the same as that accepted today. According to their concept "part of the cellulose is amorphorus but can be chemically degraded; therefore, we do not doubt that the amorphorus part consists of long chains. We even consider it probable that at the surface of the crystallite, the chains get gradually disordered and form an amorphorus bark. We cannot say that the length of the chain is actually determined by X-ray diagrams. They can be much longer and gradually degenerate into the bark." They attributed the strength of cellulose fibers to covalent bonding of the molecules, and "insolubility and tenacity to the whole structure, only when the main-valency chains themselves are of sufficent length". Their model is a compromise between the association theory and Staudinger's macromolecules.

Staudinger disagreed with Meyer and Mark only on two points. He believed their estimate of the main chains was too short and denied the existence of micelles. These differences were slight, and as Priesner reports (27), there was interest on both sides for an extensive exchange of ideas. While differing on these points, both concepts were substantially different from those of the association theory advocates.

Then, a few months afterward Meyer repeated and generalized the cellulose structure to include crystalline knots at regular intervals (28). This model was useful in explaining not only the properties of cellulose, but those of rubber as well. Staudinger called the concept unusable, incorrect, and dubbed it "the New Micelle Theory".

Shortly afterward, Staudinger formally criticized Meyer and Mark's work. It appears that it was the second publication which brought out Staudinger's objections. In a letter to Mark written October 31, 1928, Staudinger wrote: ". . . finally, I also have written a few papers opposing the views of K. H. Meyer. I do not agree with him on two points. First, in my opinion, the statements of K. H. Meyer do not represent anything new, but rather coincide in general with the opinions I have advocated for years and established experimentally. Second, I do not believe that the introduction of 'primary valence chains', instead of macromolecules, solves any problem." Staudinger also stated these views publicly (29).

In his review of these events, Priesner (27) (the only published scholar of Staudinger's letters) reports that of the supporters of the high molecular weight concept only Meyer and Mark were reproached for their views. Yet the "New Micelle Theory" represented an independent perception of the ideas. Further, Priesner asserts that Staudinger's objections regarding the "primary valence chains" must be evaluated in the language of the time. He says that Staudinger's designation for what we now call polymers--macromolecule--was at best on the borderline of acceptability. At that time, its definition inferred that all the molecules were exactly alike. Since the chains of a polymer, rubber for example, are structured the same and may not exceed certain lengths, but are not exactly the same number of monomer units, the definition did not fit. "Primary valence chain" sidestepped this problem.

None-the-less the problem was of wording, for Staudinger, Meyer, and Mark understood that the polymers were polydisperse. Indicative of this, Staudinger wrote in his reproach (29) that "in the high polymers we do not have materials with molecules having the same molecular size, but mixtures of different lengths . . . One cannot speak of molecular weight but only of average molecular weight".

Mark responded to Staudinger on November 2, 1928. Among other things he wrote, "I am sorry to see from your letter that you feel annoyed by the statements of Professor Meyer. I am convinced that it was the last thing Professor Meyer intended to do. I have, in our joint research and especially in my Hamburg lecture, emphasized the importance of your beautiful work. Introduction of the term, primary valence chain, is very purposeful since it refers to structures which are not identical . . . rather average lengths. If we add this fact to your macromolecules, then both concepts become identical.

"Because of this I prefer not to emphasize the differences . . . we mean the same thing. I believe that we should advocate this point of view together and not emphasize some slight differences in perceptions. The high polymer camp could easily make the mistake very well known in politics: because of small differences in neighboring opinions, a large point of view often does not get enough attention."

Figure 3. Professor Kurt H. Meyer in Geneva on July 19, 1951.

Meyer (30) did not believe that he could leave Staudinger's statements unanswered. He restated their agreement regarding the existence of long molecular chains; however, objected to Staudinger presenting his (Meyer's) opinions as his own. Meyer added that he had referred to Staudinger's work as often as necessary.

Staudinger countered by publishing a "Schlusswort" or final word (31), a copy of which he submitted to Meyer before publication. In this "Schlusswort" he reemphasized his position and answered Mark's request for cooperation. It is interesting that Mark wrote an extensive letter to Staudinger asking him to omit this latter part, since otherwise he too would have to respond, Staudinger's reply to Mark is not known, but the published "Schlusswort" indicates his compliance with Mark's request.

The debate might have ended here had Naturwissenschaften not have published a condensation of Meyer's Zeitschrift fuer Angewandte Chemie paper (28). Staudinger, then wrote in the former publication (32), "An article by K. H. Meyer . . . is a condensation of a previous article. The opinions in these publications about the structure of high molecular weight compounds are presented in a manner I have advocated for years in many publications and speeches. I welcome that the author shares the opinions . . . since no reference was made to my work, I have to assume that K. H. Meyer believes this is common knowledge."

In this paper Staudinger objected to the use of "primary valence chains" and micelles. However, he did not exclude the possibility of molecular associations. Although he denied the possibility of micelles, he discussed crystallites. The crystallites and the micelles of Meyer and Mark were identical.

Even though Mark's studies of high molecular weight substances expanded during this time, his participation in the polemics regarding his (and Meyer's) and Staudinger's opinions ceased. In the ensuing years Mark has repeatedly maintained that he shared the opinions of Meyer. Further, he says that Meyer encouraged him to participate. But for Mark, these were times of new studies and new commercial developments. At no other time has Mark's basic temperment shown more clearly. He desired only to move forward, not to rehash past developments.

A half century later Mark said two developments prevented his "more active" involvement. One was his and Meyer's belief that Staudinger had completely established his priority in proposing long chains. The second was the work of W. H. Carother's which convinced him in 1929 that the long chain connection of natural and synthetic polymers would soon be irrevocably resolved.

> "One day in Ludwigschafen, the great scientist Mittasch asked me, 'What are you doing?' I said I was measuring viscosity. 'Why?' he asked. I told him that we desired to determine the molecular weight of a polymer. About a half year later I saw him at a conference. He asked me, 'Say Professor, are you still working with those highly contaminated solutions?'"
>
> H. F. Mark

Herman Mark and (independently) Herman Staudinger were the first to suggest the viscosity of a polymer solution was related to the molecular weight of the polymeric solute. In doing so, they were again going heads-on against the beliefs of the association theory advocates. For according to the latter group's studies, based for the greater part on colloids, there was no linear relationship between viscosity and molecular weight of any solution.

Mark, at the I. G. Farben Ludwigschafen laboratory, became curious about the possibility of such a relationship in 1927. His interest was more than passing as fiber production at I. G. Farben was commonly practiced in the factories by spinning from high viscosity solutions. This application resulted in several scientists being assigned the study of the rheological behavior of solutions. It was only natural that Mark, from the vantage of the coordinator of several groups, notice the relationship. The practical nature of the studies prerequisited slow publication of the findings.

Towards the middle of 1929, Mark was clearly close to establishing a viscosity equation. He and H. Fikentscher published a somewhat complex relationship of viscosity, and molecular volume (33). It was based on the Einstein relationship of viscosity and solute concentration.

At about the same time, Staudinger derived his well known "law of viscosity". His work was formulated in 1929 and published in 1930 (34, 35). Also based on the Einstein relationship, Staudinger's equation was a direct relationship between the specific viscosity and the polymer molecular weight.

$$\eta_{sp} = \text{constant} \times M$$

In deriving his equation, Staudinger assumed the macromolecules were straight, rigid rods. This assumption is, with hindsight, wrong. Its accuracy because of inaccuracies is ironically good.

Use of Staudinger's assumptions results in the use of a molecular volume which is too high. At the same time, molecules in his model were expected to rotate in only one plane. In reality rotation may occur in any plane, and this limitation results in the use of too small a molecular volume. The applicability of Staudinger's equation was, therefore, based on compensation of the errors.

In both Ludwigschafen and Freiburg, work quickly shifted to determining molecular weight by viscosity measurements after experimentally establishing the relationship. Staudinger's laboratory without the restrictions of practical application published more frequently. Mark only published on the subject once again before leaving I. G. Farben in 1932 (36).

In Vienna in the fall, Mark returned to the problem employing the same approach as when at I. G. Farben. Summarized, he believed that molecules of such large dimensions were capable of assuming many different conformations through rotation about valence bonds and had to be dealt with statistically just as gases are. Convinced that he was "in over his head" mathematically, he turned to a fellow Viennese, H. Thirring, for advise. Thirring recommended Eugene Guth.

In 1933 and 1934, Guth and Mark (37, 38) and Kuhn (39) independently studied the problem and arrived at similar solutions. These investigations furnished the background necessary to more clearly define the relationship of molecular weight and high viscosity.

In the following four years Mark successively reported on the viscosity and molecular weight of cellulose (40), Staudinger's Law (41), high polymer solutions (42), and the effect of viscosity on polymerization rates (43). Confident of his findings, he proposed (at the same time as R. Houwink) the general viscosity equation now known as the Mark-Houwink Equation (44, 45).

$$\eta_{sp} = \text{constant} \times M^{a}$$

Mark and Houwink were the first to formulate the equation in the power form and to demonstrate its validity by means of empirical values. In reality, the Mark-Houwink Equation is simply the Einstein viscosity equation, which assumed spheres, transferred to particles with size dependent particle density.

The importance of the equation cannot be dismissed as simply. Since its publication, a great deal of use has been made of solution viscosity measurements, and a large literature has grown up around interpretation of the measurements. It is probably safe to say that no one working in the research or development of either natural or synthetic polymers will not at one time or another, knowingly or unknowingly, use the equation. One of the early goals of Mark and Staudinger, determination of molecular weight by viscosity measurements, cannot be accomplished in an absolute sense by this method.

> "In Austria one has to (and always has had to) search for areas of research for which the economic necessity is

> relatively small. An example is Pregl and his microchemistry. Microchemistry is a lot cheaper than macrochemistry."
>
> H. F. Mark

Mark faced a stern challenge when he became Director of the First Chemistry Institute at the University of Vienna in the fall of 1932. Modernization of the laboratory equipment and teaching were new problems to be faced alongside his research and administrative responsibilities. His toughest single task, however, was building a successful program of study in a relatively new, non-traditional field. His predecessor, Professor Rudolf Wegscheider, had led the Institute along the less radical lines of classical thermodynamics and kinetics. The area of study Mark advocated, polymers, was somehow less acceptable.

The field was new. Further, it was discredited by the intense debate raging between its advocates. If not enough, the booming commercialization of polymeric materials made the field less fit for "proper and respectable" study.

Determined, Mark quickly and diplomatically came to an agreement with his Vienna colleagues. Speaking first to E.Spaeth, the Second Chemistry Ins.Director (and successor of W. Schlenk when the latter took Mark to Berlin in 1921), then A. Franke, A. Kailan, V. Klemenc, and F. Pollak, he received their guarded approval of his plans. Then without delay, he launched a three pronged effort. His strategy was simple. As a part of modernization, he would bring quality scientists to the institute, initiate a series of "special lectures" presented by both staff and visiting scientists, and concentrate research in the less studied area of polymerization mechanisms.

Generously dispersing research funds received from I. G. Farben, the Austrian Ministry of Education, and the University, Mark soon assembled a group of skilled and motivated scientists. A list of these associates includes physicists E. Broda, E. Guth, and R. Simha; physical chemists F. Eirich, P. Gross, O. Kratky, F. Patat, E. Suess, and H. Tschamler; and organic chemists J. W. Breitenbach, H. Dostal, R. Raff, and A. von Wacek. These scientists took over responsibilities as the program grew, slowly freeing Mark. As a result of good management of talented associates, Mark's activities at Vienna were no less diverse than in Ludwigschafen and definitely internationally flavored.

His plan to concentrate research in the area of polymerization mechanisms was in the beginning more requirement than choice. On arriving in Vienna, Mark quickly realized that less experienced hands could not prepare monomers and polymers at the speed and purity necessary to compete with the Ludwigschafen and Freiburg laboratories. Thus early on, research was directed toward observing polymerization, copolymerization, and reaction

Figure 4. Professor Herman Mark and models of flexible (Mark) and rigid (Staudinger) macromolecules. The photograph was taken in Brooklyn on October 21, 1979.

Figure 5. Several members of the First Chemistry Institute in Vienna in 1934. (left to right) C. Rogowin, H. F. Mark, Mark's sister Elizabeth, I. Saito, and E. Guth.

rate characteristics. During the first years work centered about recording data, delaying interpretation of the results. Interpretation began in the mid-1930's by which time the Institute and its personnel were ready. They were by the late 1930's considered among the best in the world.

In Vienna, Mark published a number of fundamental papers. Their topics include polymerization mechanism (46, 47, 48), thermal polymerization (49, 50), polymerization kinetics (51), the effect of oxygen on polymerization (52), and measurement of molecular weight distribution (53). Guth and Mark expanded their modeling of extended and balled thread molecules to include rubber. The result of their studies was a series of very important papers in which the thermal effect on expansion and relaxation of rubber is explained (54, 55, 56).

Incredibly, a review of Mark's publications also shows that he authored six papers on surface characteristics and dying, five on the X-ray structure of cellulose and polymers, four on X-ray diffraction (including one on the structures of CI_4 and CBr_4), and twelve review papers. In all Mark contributed to more than seventy papers. More impressive, he penned six books while at the University of Vienna. Two of his books, "Roentgenographic Untersuchung von Kristallen" coauthored by F. Halla (57) and "Hochpolymere Chemie" coauthored by K. H. Meyer (58), were particularly well received. A seventh book, "Physical Chemistry of High Polymeric Systems" (59) was written in Vienna but not published until 1940.

Mark's activities in Vienna were not limited to research and administration. During this period, he was appointed a member of the Austrian Ministry of Education, President of the Committee on Wood Utilization, and a consultant to the Ministries of Industry, Agriculture, and Foreign Relations. The appointments were made by Engelbert Dollfuss, a wartime comrade of his, who became first Chancellor in 1932 and later anti-Nazi strong man of Austria. In addition, he served on the local avalanche squad and wrote several articles on prediction of avalanches. He climbed glaciers searching for a new source of D_2O and kicked with the First Chemistry's football team.

Again, as in Germany, the changing political situation led to a turning point in his life. Dollfuss was assassinated in 1934, and the Nazis pushed in earnest for control. Mrs. Mark saw clearly what these events harbingered, but the effervescent, optimistic Mark did not. He continued his diverse activities undaunted, until late 1937. Finally, at the urging of Mrs. Mark he agreed to investigate opportunities outside of Austria.

One such contact was made in the fall when C. B. Thorne, Technical Director of the Canadian International Paper Co., visited Europe. Mark and Thorne at Thorne's invitation met in Dresden. They discussed Mark joining the company staff as Manager of Research. Thorne realized that the company laboratories had been outdated by the recent developments in polymer science. His hope was to obtain the necessary leadership to modernize the facilities by bringing Mark, a leader in those developments, to Ontario. Mark, hopeful of improvement in the political climate of Austria, neither accepted nor declined the offer. In a ploy for time, he suggested an alternative one year appointment to start the fall of 1938.

Mark returned to Vienna, and Thorne completed his European tour. Thorne returned to Canada convinced that Mark was the right man to lead Canadian International's modernization.

Unfortunately, the political developments in Vienna grew darker. Their effects could by then be felt by the whole scientific community. Until that time, the chemistry department was safely ensconced in a building a kilometer from the political forum at the center university. However, by December, 1937, the "arrival of Hitler" was no longer doubted by even the most optimistic. The brightest view was that the repression would not be as severe in Austria "since the Nazis were no longer threatened". Immediate effects on research were reduced funds and fewer students. By then, even the garrelous Mark grew reticent as he recognized the ramifications.

Mark knew in early 1938 that his emigration was necessary. At that time he began to delegate administrative duties to individuals at the Institute who would stay. This selfless duty, painful as it must have been, prepared the Institute for the difficulties ahead, but the preparation was successful. As one associate said, "Mark left a well appointed and orderly house".

Hopes and preparations almost delayed Mark's departure too long. German troops occupied Austria sooner than expected. They entered Vienna on March 15, 1938. Mark was arrested the following day, and retained in the Hotel Metropolitan. After four days of questioning, he was released without his passport and with a stern warning not to have contact with anyone Jewish. During the next eight weeks, he rarely slept at home in order to avoid the periodic night visits by the Gestapo.

Time, for Mark and many others faced with Nazi control, suddenly became priceless. Mark surmised he had to quickly solve three problems. They were: contact Thorne in Canada and secure the position he offered and his assistance in obtaining a Canadian visa, get his Austrian passport back, and withdraw as much money as possible without drawing adverse attention. Thorne quickly confirmed the offer and solved the visa problem by obtaining the signature of the Canadian ambassador. A Swiss visa came easily with the Canadian visa in hand.

Return of the passport was more difficult. Mark solved it by appealing to a Nazi lawyer (and former schoolmate) to petition for it for him. This step was necessary since Mark was not allowed to act on his own behalf. A liberal application of the third ingredient, money, brought the document home. The money problem was handled discretely by Mark, the scientist.

He bought over 1100g of platinum/iridium from a number of suppliers. The wire was bent into coat hangers.

After eight harrowing weeks, the Mark family--Herman, Mimi, their two boys, and a niece, strapped skis, picks, and ropes to the top of their car and hung a Nazi flag on the radiator. They drove into Switzerland on May 10, 1938.

> "There are no workers in the world (in polymer science) who at some point in their investigations are not indebted to his pioneering work."
>
> Eric Rideal

> "None of the big American universities at that time had a department for polymers. There was only Marvel at Illinois. It was a difficult beginning, but it was lucky for Speed and me. In America everything still depends on having something others don't have."
>
> H. F. Mark

When Mark and his family arrived in Zurich in May of 1939, they immediately notified the Canadian International Paper Company, Manager of Research, Sigmond Wang. Anxious to help, Mr. Wang arranged their passage to Canada in such detail that they were able to spend the summer visiting friends and touring laboratories. The Mark family passed May in Switzerland, June in France, and July-August in England. Near mid-September Mark embarked for a new position in the quieter surroundings of Hawkesbury, Ontario. Two months later in November, Mrs. Mark followed with their two sons and household goods which they "bribed" out of Austria.

In Hawkesbury, Mark found the state of the laboratory just as Thorne and Wang had described. The laboratory equipment was outdated, and the scientists were employing the older, conventional, empirical methods. Clearly, the laboratory was well behind in the principles and procedures advocated by the new schools of high molecular weight chemistry in Europe. Even Mark, who was very well known throughout Europe, required introduction. Yet, despite these problems, modernization of the laboratories was not difficult. Above it all, Mark was greeted by enthusiastic, inquisitive coworkers and sufficient capital to initiate the required changes.

On arrival, he found that his duties were not limited solely to modernization of the Hawkesbury laboratory. He was also charged with supervision of the pulp quality control laboratories scattered in a dozen paper mills throughout Canada. Always the opportunist, Mark used his frequent business travels to initiate contact with other scientists and present lectures.

Mark returned to research after the modernization program was organized and operating smoothly. Limited to cellulose chemistry by the necessity of commercial feasibility, he studied cellulose acetate and the effect of the degree of acetylation on product solubility. At the same time he worked closely on the development of suitable analytical methods to measure molecular weight distribution and the degree of functionality.

The most important problem to be solved in Hawkesbury was how to improve the quality of wood pulp so that it could be used in the production of rayon tire cords. The problem was particularly important since Canadian International Paper was the major supplier of cellulose to the largest rayon manufacturer in the hemisphere, E. I. duPont, Mark and his associates found that two conditions were crucial for strength and fatigue of the yarn, namely near complete absence of impurities (resin, lignin, ash) and a narrow molecular weight distribution of the cellulose. Under Mark's tutorage, the scientists in Hawkesbury expanded the classical methods for isolation of α-cellulose and introduced an improved xanthate-$Cu(NH_3)_4$ viscosity relationship. Mark's broad knowledge and uncanny ability to make the complicated or confused simple was in a large part responsible for their successes.

When he arrived in Canada in 1938, he recognized the acute need for a literature-books in particular-in the Americas. Seizing the opportunity he met with his former associate and publisher E. S. Proskauer in Montreal on a cold morning in January, 1939. Proskauer, who had fled Germany in 1937 and was living in New York, was the Chief Science Editor at Interscience Publishers. Walking several miles in the snow, Mark and Proskauer conceived the Interscience High Polymer Series. They agreed to initiate the series by publishing a collected volume of the papers of Wallace H. Carothers (60). The High Polymer Series grew from this conception to become one of the better known and more important polymer source works.

Despite Mark's diverse activites as a manager, lecturer, researcher, author, and editor, the isolation of living in Hawkesbury was too quiet for such a gregarious person. Restless, he investigated the possibility of obtaining a position in the United States at the 1939 fall meeting of the American Chemical Society in Baltimore. A scramble ensued for his services among several American firms. One firm, duPont, offered him a position as a consultant in its Rayon Department and agreed to assist him in obtaining an academic position and status as a permanent resident. A short time afterward, assured by duPont's offer of support, Dr. Harry Rogers, President of the Polytechnic Institute of Brooklyn (now called the Polytechnic Institute of New York), offered Mark a position as an Adjunct Professor. "Canadian International," Mark says, "was not particularly enthusiastic about the proposal, but condescended in view of their business relationship with du-Pont." Mark and his family left Hawkesbury late in May of 1940.

In 1940 most of the shellac imported into the United States came in through the docks of Brooklyn. It was, therefore, appropriate that W. H. Gardner had selected the Polytechnic as the site to establish a national testing laboratory a few years before. Research at the laboratory, called the Shellac Bureau, centered on evaluating the properties of this important natural resin. On assuming his position in Brooklyn, Mark was assigned to the Shellac Bureau.

Encouraged by Mark, Gardner quickly agreed that the research of the Shellac Bureau should be expanded to include a study of synthetic coatings. Students were brought in to work on these expanded projects. The presence of students convinced Mark that his most important single task was to establish a complete curriculum of study of polymer chemistry and technology. The teaching of students began that fall. The first course, taught by Mark, was "General Polymer Chemistry". Establishment of the program made "Brooklyn Poly" the first university in the United States to offer a degree in polymer science.

"The beginning was not easy," Mark explains, "but related experience in Vienna and the peerless support of Rogers and (Dean R. E.) Kirk made the Institute a quick success." The success was also in part brought about by the labors of a future "who's who" contingent of associates attracted by Mark. A few of the scientists who came to work in the converted razor blade factory which housed the Polytechnic laboratories are T. Alfrey, P. M. Doty, F. Eirich, I. Frankuchen, W. P. Hohenstein, H. Morawetz, C. G. Overberger, C. C. Price, R. Simha, A. V. Tobolsky, and B. H. Zimm.

Mark, realizing that the Polytechnic was surrounded by numerous industrial organizations, sought in those first years a method to encourage scientific and technical interchange between them and his growing department. He was successful in his effort on a scale beyond his hopes. This was accomplished by two innovations.

First, he established a series of regular symposia to be held each Saturday from October to May. Active areas of research and timely subjects were presented and discussed at these meetings. The series, called the "Saturday Symposia" as the years passed, were widely known and well attended. Additional prestige was given the series by the distinguished foreign visitors who, drawn by Mark's reputation, often served as chairmen, speakers, and discussion leaders. The net impact was to bring the exposure necessary to establish the department as the national leader in polymer science and give it international recognition.

The second vehicle for closer cooperation between the department and industry was his establishing "Special Summer Courses". The courses which were the result of a scheme by Mark and an associate, I. Fankuchen, served an ancillary purpose. The funds obtained by collecting fees from the attendees were placed in a special travel fund reserved for the faculty. A typical course consisted of a limited number of visiting scientists who remained at the Polytechnic for a period of one or two weeks. While there, they were given intensive theoretical and experimental instruction in a special area of the discipline. The first courses were taught in the summer of 1943 by Mark and a number of his associates including Fankuchen and H. S. Kaufman (X-ray crystallography), and T. Alfrey, P. M. Doty, and B. H. Zimm (molecular weight).

Mark was named Director of the Shellac Bureau in 1941 when W. H. Gardner was called to Washington, D.C. on government work. Under Mark's leadership research was expanded from shellac to synthetic coatings and eventually to the preparation, properties, and polymerization mechanism of a variety of synthetic polymers. Research by the end of World War II had far outgrown the Shellac Bureau. In 1946, Mark guided a reorganization of the polymer effort and thus was instrumental in creation of the now famous Polymer Research Institute. At the same time another long-time goal of Mark's was realized. Simultaneous with formation of the Institute, the Polytechnic initiated a course of study leading to a degree of Doctor of Philosophy in Polymer Chemistry.

Government spending during the Second World War and in the cold war years afterward provided a boom to Mark and the Institute. A number of large and important projects were sponsored at the Institute in the area of synthetic fibers, films, and rubber by the OSRD, ONR, and other U.S. Government agencies. The work at Polytechnic did not only contribute to the goal of the projects, but also to numerous publications in scientific journals. The combined output of the Polymer Research Institute and several other rapidly growing polymer research efforts such as that of C. S. Marvel in Urbana, IL soon caused a bottleneck of publications.

When Carothers began his studies in 1928 a great deal of the American chemical research was reported in the Journal of the American Chemical Society (JACS). The arrangement was satisfactory for more than a decade, but with the sudden increase in the number of publications, A. Noyes, editor of JACS, became leery of accepting too many polymer manuscripts.

Mark's reaction was to approach Noyes about the American Chemical Socity starting a journal on polymer chemistry. Noyes, however, rejected the idea since the number of papers, about a hundred per year, was regarded as too small to merit a separate journal.

Mark then turned to the ubiquious Interscience editor, Eric Proskauer. Together, they founded the Journal of Polymer Science. The journal was not in the early going as profitable as they would have preferred, but within a few years the volume of papers and subscribers increased a great deal. The journal became so active that in 1953 it was branched to include the Journal of Applied Polymer Science. Even this second journal was not enough. Within ten years the original publication was split into chemistry, physics, and letters sections. Mark, the founder, serves to this day as an editor of both journals.

Most of Mark's Brooklyn associates will readily testify that the guidance, teaching, and encouragement rendered by the chief was invaluable in their attaining personal and research goals. Their collective success and this compliment is not surprising when Mark's unique lifelong talent for leading is considered. The breadth of their research similarly reflects his diverse interests and abilities. Scientifically, the research at the Institute was a continuation of the Mark directed programs in Vienna and, to a lesser extent, in Hawkesbury. Yet the Brooklyn research was improved by better instrumentation, refinement in techniques, and a unique alliance of scientists with a broad range of experiences.

Since joining the Polytechnic, Mark has written over three hundred research papers, related articles and books. A list far too long to be reviewed in this section; however, a list should contain several outstanding items. Several books were written including "Principles of the Physical Chemistry of High Polymeric Systems" with A. V. Tobolsky (61) and "High Polymeric Reactions" with R. Raff (62). A detailed, systematic study of the influence of concentration, solvent, catalysts, and temperature on vinyl addition polymerization was conducted. Polymer fractionation techniques were improved and extensive use was made of osmometry, viscometry, and X-ray analysis. Related important research was initiated with Mark's assistance and encouragement. A list of these projects includes the use of radioactive bromine end group labels; hydrogen-deuterium exchange rate measurements; the rate of diffusion of vapors through polymer films; and the consequences of molecular orientation, crystallinity, plasticizer content, and temperature (this latter study led to the discovery of the "so-called Second Order Glass Transition Point"). Other projects of special interest to Mark included Alfrey and Goldfinger's development of a copolymer composition equation, and the Alfrey-Price Q-e scheme. A timely, comprehensive treatise of the copolymerization field was published in Volume VIII of the High Polymer Series as "copolymerization" by T. Alfrey, J. J. Bohrer, and H. Mark (63).

Throughout the years, it has been customary that Professors at the Institute pursue their own research projects with the assistance of the younger members and in close coordination with Mark. Among the projects conducted by these Professors under Mark's influence between 1940 and 1960 are T. Alfrey's study of mechanical properties, R. B. Mesrobian on graft copolymerization, M. Goodman on the stereochemistry of polymers, F. R. Eirich on rheology, G. Oster on photochemistry of polymers, H. P. Gregor on polyelectrolytes, H. Morawetz on solic state polymerization, and C. G. Overberger on the organic chemistry of monomers and polymers.

The presence of a world reknown figure such as Mark in a city like New York placed the Polymer Research Institute in an enviable position. Taking full advantage, Mark and, later, Overberger as head of the Institute always attempted to stay abreast of new developments by symposia, seminars, and the exchange of scholars. The academic events, for example, during G. Natta's first visit to the United States were organized at the Institute. A. Keller presented one of the first reports in the United States on his discovery of chain folding at "Poly", and Max Perutz (whom Mark placed with Bernal in 1936) visited after receiving the Nobel Prize and presented the results of his hemoglobin studies. C. V. Ramon came from India, L. Sakurada and S. Okamura from Japan, V. A. Kargin and A. Topchiev from Russia, K. Ziegler from Germany, the Bergmans and Katschalskys from Israel, Claesson and Rånby from Sweden and many other distinguished scientists from other countries. Even Hermann Staudinger visited the Institute in 1953. (When given the "royal" treatment by Mark, Staudinger was prompted to say, "Even Herman Mark now believes in polymers!").

Through his more than forty years in Brooklyn, Mark has been heaped with acclaim, medals, and degrees. He raised his family, and saw many sons both blood and academic go on to careers. He has witnessed an expansion in the science he helped create which mocks those who disagreed with the concepts he and a thin line advocated so many years ago. He has, as the years passed, also encouraged the formation of new centers of research which have also diluted the influence of "his" Institute. Perhaps the greatest compliment of all is to point out that through it all, it was his selfless encouragement of this growth, his refusal to protect personnel, his push to march away from Brooklyn and start new centers, his willingness to go out and lecture rather than "mind the store" which is no small part responsible for the growth.

In 1961, Mark was appointed Dean of the Faculty of the Polytechnic. After assisting in the reorganization of the structure of the Institute, he retired in 1964 and was appointed as a member of the Corporation of the Polytechnic Institute and received the title, Dean Emeritus of the Faculty. The title Mark prefers was not given by the administrators. On the occasion of his sixtieth birthday in 1955, his associates in Brooklyn bestowed the title of "Geheimrat" on him. "Geheimrat", a title given to haughty, pompous professors in the old German schools, seemed quite funny since it embodied so much of what Mark was not. Yet, the name stuck to the delight of all--especially "the Geheimrat".

Acknowledgements

The author gratefully acknowledges the cooperation of Drs. E. H. Immergut, H. Morawetz, C. G. Overberger, E. Proskauer, and especially the Geheimrat. They endured taped interviews and endless repetitive questions. Special thanks is due Dr. Claus Priesner for generously sharing his findings on the Staudinger-Meyer-Mark polemics. The information used by this author is presented in reference 27; however, Dr. Priesner's forthcoming book, "H. Staudinger, H. Mark, and K. H. Meyer -- Thesen zur Groese und Struktur der Makromolekule", is the most complete work on this subject.

Literature Cited

1. Mark, H. Z. Angew. Chem., 1927, 40, 16; ibid, 1927, 40, 645; and ibid, 1927, 40, 1497.
2. Kallman, H. and Mark, H. Z. Physik, 1926, 36, 120.
3. Mark, H. and Schochen, K. Naturwissenschaften, 1927, 15, 139.
4.. Mark, H. J. Soc. Glass Tech., 1926, 11 (41), 54.
5. Mark, H. Z. Metallkunde, 1928, 20, 342.
6. Mark, H. Naturwissenschaften, 1928, 16, 892.
7. Hengstenberg, J. and Mark, H. Z. Krist Miner., 1928, 69, 271.
8. Meyer, K. H. and Mark, H. Ber., 1928, 61B, 1932.
9. Meyer, K. H. and Mark, H. Ber., 1928, 61B, 1936.
10. Meyer, K. H. and Mark, H. Ber., 1928, 61B, 1939.
11. Meyer, K. H. and Mark, H. Ber., 1928, 61B, 593.
12. Mark, H. and Susich, G. V. Z. Physik, 1930, 65, 253.
13. Bergmann, E. and Mark, H. Ber., 1929, 62B, 750.
14. Hengstenberg, J. and Mark, H. Z. Krist., 1929, 70, 283.
15. Mark, H. and Wierl, R. Naturwissenschaften, 1928, 16, 725.
16. Mark, H. and Wierl, R. Z. Physik, 1929, 53, 526; ibid, 1929, 55, 156; and ibid, 1929, 57, 494.
17. Mark, H. and Wierl, R. Naturwissenschaften, 1930, 18, 205, 753, 778.
18. Mark, H. and Wierl, R. Z. Physik, 1930, 60, 741.
19. Mark, H. and Wierl, R. Z. Elektrochem., 1930, 36, 675.
20. Hengstenberg, J. Chemie Kunststoffe Aktwell, 1980, 34(2), 23.
21. Meyer, K. H. and Mark, H. "Aufbau der Hochpolymeren Substanzen"; Hirschwaldsthe Buchhandlung: Berlin, 1930; 240 pp..
22. Meyer, K. H. and Mark, H. "Der Aufbau der Hochpolymeren Organischen Naturstoffe"; Akad. Verlogsges: Leipzig, 1930; 264 pp..
23. Mark, H. and Wierl, R. "Die Experimentellen und Theoretischem Grundlagen der Elektronenbeugung"; Gebrueder Berntraeger: Berlin, 1930; 126 pp..

24. Mark, H. "Physik und Chemie der Cellulose" in Herzog, R. D. (Ed.) "Technologie der Textilfasern"; J. Springer: Berlin, 1932.
25. Olby R. "The Path to the Double Helix"; University of Washington Press: Seattle, WA, 1974; p. 6.
26. Staudinger, H. Ber., 1920, 53, 1073.
27. Priesner, C. Chemie in unserer Zeit., 1979 (2), 43.
28. Meyer, K. H. Z. Angew. Chem., 1938, 41, 935.
29. Staudinger, H. Ber., 1928, 61, 2427.
30. Meyer, K. H. Z. Angew. Chem., 1929, 42, 76.
31. Staudinger, H. Z. Angew. Chem., 1929, 42, 77.
32. Staudinger, H. Naturwissenschaften, 1929, 17, 141.
33. Mark, H. and Fikentscher Kolloid Z., 1929, 49, 135.
34. Staudinger, H. and Heuer, W. Ber., 1930, 63, 222.
35. Staudinger, H. and Nodzu, W. Ber., 1930, 63, 721.
36. Mark, H. Kolloid Z., 1930, 53, 32.
37. Guth, E. and Mark, H. Ergbnisse exakt. Naturw., 1933, 12, 115.
38. Guth, E. and Mark, H. Monatsh., 1934, 65, 93.
39. Kuhn, W. Kolloid Z., 1934, 68, 2.
40. Mark, H. Chimie and Industrie Spec. No. 788, 1934.
41. Meyer, K. H. and Mark, H. Z. Elektrochem., 1934, 40, 728.
42. Eirich, F. and Mark, H. Ergeb. exakt. Naturw., 1936, 15, 1.
43. Dostal, H. and Mark, H. Osterr. Chem. Ztg., 1937, 40, 25.
44. Mark, H. Der Feste Koerper, Leipzig, 1938, p. 103.
45. Houwink, R. J. prakt. Chem., 1940, 157, 15.
46. Dostal, H. and Mark, H. Z. Physik. Chem., 1935, B29, 299.
47. Dostal, H. and Mark, H. Trans. Fara. Soc., 1936, 32, 54.
48. Mark, H. Naturwissenschaften, 1937, 25, 753.
49. Mark, H. and Raff, R. Z. Physik. Chem., 1936, B31, 275.
50. Dostal, H., Mark, H., and Raff, R. Ind. Eng. Chem., 1937, 29, 595.
51. Dostal, H. and Mark, H. Angew. Chem., 1937, 50, 348.
52. Dostal, H. and Mark, H. Oesterr. Chem. Zgt., 1937, 40, 50.
53. Dostal, H. and Mark, H. Trans. Fara. Soc., 1937, 33, 350.
54. Guth, E. and Mark, H. Oesterr, Chem. Ztg., 1937, 40, 304.
55. Guth, E. and Mark, H. Naturwissenschaften, 1937, 25, 353.
56. Guth, E. and Mark, H. Z. Elektrochem., 1937, 43, 683.
57. Halla, F. and Mark, H. "Roentgenographische Untersuchung von Kristallen"; J. A. Barth: Leipzig, 1937.
58. Meyer, K. H. and Mark, H. "Hochpolymere Chemie"; Akademische VerPagsgesellschaft: Leipzig, 1937; 345 pp.
59. Mark, H. "High Polymers, Vol. II, Physical Chemistry of High Polymeric Systems:; Interscience Publishers: New York, 1940; 345 pp..
60. Mark, H. and Whitby, G. S. "High Polymers, Vol. I, The Collected Papers of Wallace H. Carothers"; Interscience Publishers: New York, 1940; 459 pp..

61. Mark, H. and Tobolsky, A. V. "High Polymers, Vol. II, 2nd Ed., Principles of the Physical Chemistry of High Polymeric Systems"; Interscience Publishers: New York, 1950; 506 pp..
62. Mark, H. and Raff, R. "High Polymers, Vol. III, High Polymeric Reactions"; Interscience Publishers: New York, 1941; 476 pp..
63. Alfrey, T., Bohrer, J. J. and Mark, H. "High Polymers, Vol. VIII, Copolymerization"; Interscience Publishers: New York, 1952; 269 pp..

RECEIVED August 10, 1981.

Herman F. Mark: The Man

HANS MARK

Department of the Air Force, Las Vegas, NV 89114

Ladies and Gentlemen. It is an honor and a privilege - as well as a source of profound joy - for me to be here this morning and to talk about the distinguished member of your Society whom we are honoring here today. I am supposed to talk about the man. You must understand that this is a very difficult thing for me to do. Herman Mark has loomed so large in my life for over half a century that I simply do not know how to separate him from my world.

Perhaps I can best do what you have asked me to by talking about some memories that stand out vividly in my own mind. As long as I can remember, there was never any doubt that he is a very great man. One of my earliest recollections is of his fortieth birthday party in 1935. We lived in a large house in a Vienna suburb at the time and I remember that we had a very large backyard. There were a great many people in the yard at the party - more than I had ever seen before and I remember thinking that my father must indeed be a very important man if so many people would come to celebrate his birthday.

Even though I knew that he was very important, it was also true that he paid quite a bit of attention to my brother and to me. He was always calm and very serene but also capable of some startling abberations from this general attitude. I remember once when I was about ten or eleven years old, my brother and I were going somewhere with my father and he was driving our car. My mother was not with us so there was no one to keep the peace between my brother and me. We were sitting in the back seat and having an argument over something or other that eventually became at least partly violent. Finally, it got so bad that my father stopped the car, turned around to face us and in ominous tones threatened to have a nervous breakdown if we did not stop. Well, we stopped because we were so absolutely shocked by the threat. A nervous breakdown? Impossible! This serene and rock-like person we both knew so well; could he really have a nervous breakdown? No! While my brother and I were considering the enormity of this

0097-6156/81/0175-0089$5.00/0

possibility, my father started the car and drove to our destination. Needless to say, we arrived there before my brother and I even considered the prospect of starting another argument.

My father's fiftieth birthday party, which was a great milestone, is also a vivid memory. We had been living in this country for a few years by then and I was just finishing high school. I was also developing a strong interest in science myself at the time so I spent quite a bit of time at the laboratories of what was then the Polytechnic Institute of Brooklyn. The Polymer Research Bureau had just been started and I remember that it was at that time located in the old South Building of the Institute on Livingston Street. I remember being involved in a number of projects during that time - the most famous one being the "Icecraft Carrier". This was the brainchild of one Geoffrey Pyke who was an English inventor close to Mr. Churchill. The idea was that we would put together some large ice flows into a large artificial island which could be used as a landing field by airplanes guarding the North Atlantic convoy routes. There is no doubt that ultimately airplanes made the difference in suppressing the U-boat threat in the North Atlantic so that any idea to get them there had to be investigated. It was soon discovered that pure ice was not a good surface for landing fields since it tended to shatter upon the shock of an airplane landing. The job that was taken on by my father's laboratory at Poly was how to make ice as tough as concrete - we were going to call it "Pykecrete" after Geoffrey Pyke, the originator of the idea. My job at that time was to make mixtures of water and sand, water and sawdust, water and cotton batten and similar things, freeze them in quart ice cream containers and then take them over to the Civil Engineering Department and see if I could shatter the resulting ice block in their hydraulic press. Well, we succeeded in developing "Pykecrete" which was a mixture of ice and something I don't remember but which had the properties of concrete as far as aircraft landings were concerned. By then, however, Henry Kaiser had solved the problem by building his famous jeep carriers and "Pykecrete" was no longer needed. I tell this story only to illustrate the breadth of the work going on at Poly at the time and the atmosphere of excitement which it generated. I was perhaps 14 or 15 years old when I worked on the "Pykecrete" project but it made a very lasting impression on me, since it demonstrated clearly to me the power of applied science. My father's fiftieth birthday party came at the end of this period and it was a great affair. I remember all the colleagues at the Polytechnic Institute joining around the punchbowl and singing happy birthday. I believe that this was the first time that I had some "hard" liquor because I felt I was now one of the Poly family. You could not possibly be a member of that family without drinking!

My father's sixtieth birthday in 1955 is another vivid memory. By that time I was already living in California but I did come back for this event. It was on that day that we bestowed the title "Geheimrat" on him. It was, of course, a joke. As you all know, the title Geheimrat was bestowed by the Emperors of Germany or Austria on distinguished academic people during the period of the empires. Now, normally, German or Austrian academics tend to be stuffed shirts and so the association of the name Geheimrat with someone usually means that he is a stuffed shirt. Our Geheimrat is, of course, just the opposite and this is why we thought that bestowal of the title was so funny. To our great surprise, the title stuck and I think it has been a great thing. The Geheimrat, of course, has turned around our original joke and has given his own special meaning to the old title.

What sort of a man is the Geheimrat? Some things are well known about him. His astonishing breadth of interest and his acute historical perspective make him a superb research planner. He is insistent that things be done right, and he does not get involved in hair splitting over what "right" means. He simply knows and, if you work with him, you will also know. Furthermore, he believes that high standards are important and that life is nothing without them. He is a great teacher and he has influenced thousands of people in developing standards of their own. He is a liberal in the best sense of the word not in the political sense that is usually used today. He is generous, he accepts people for what they are, he has no prejudices and he helps people if he feels they are worth it. He is a genuine liberal in the old European sense.

I cannot speak about the Geheimrat without mentioning some of his scientific work and his achievements. Even though science has an objective and a deductive facet, it does, just as any other human endeavor, reflect the personality of the practicioner. In that sense, Polymer Science is just the kind of thing in which an eclectic individual like the Geheimrat would find sufficient elbow room for the exercise of his talents. Maybe I have even said this backwards: Polymer Science is what it is today because an eclectic individual like the Geheimrat was one of the leaders who created it. Either way, the statement is true! The Geheimrat started his scientific career as a "classical" organic chemist, then switched to physics for a while by mastering the experimental techniques of the new field of X-ray physics. During that period, he did some work in basic physics such as measuring the polarization of Compton-scattered X-rays. This was a difficult experiment that contributed significantly to the understanding of X-ray scattering. He also was among the first (along with Wirl) to demonstrate the phenomenon of electron diffraction. He made some important contributions to the field of X-ray crystallography and in doing so helped a number brilliant students, among them being Eugene Wigner, start their careers in Science. He also found time to write a beautiful little book for the "Sammlung Goschen" called "Teilchenstrahlen" or, in English, "Particle Beams" which is one of the first scientific books that I read really carefully. My own early interest in nuclear science and particle physics was stimulated by this book.

What is most important about the Geheimrat's scientific work is that he turned what he learned about organic chemistry as a student in Vienna and what he did in physics during the 1920's in Berlin toward understanding the new field of Polymers. This choice had far-reaching consequences and the state of Polymer science and technology today is ample testimony of the wisdom of this decision. The astonishing diversity of things that have come from the basic work in polymers that he helped to lead, the fabrics, the plastics, the petrochemicals and the vast number of products have had economic impact beyond measure. But this is not all. The study of macro-molecules and polymers has led to the beginning of our understanding of the basic process of life itself. Moreover, it was the same application of physical techniques such as X-ray diffraction and electrophoresis that were decisive in establishing this new science as well. All of this is a great tribute to his foresight.

What is the Geheimrat's most outstanding quality? I think that it is his unquenchable optimism. This is not a foolish optimism, like that of Candide and Dr. Pangloss, but rather a kind of optimism that is both sophisticated and enduring because it is based in a profound knowledge of people and of life. It is made up of a combination of Viennese fatalism and the American belief that the next day will be better than the last. In that respect, he is, and always has been, a quintessential American.

Four hundred years ago, an English school teacher, Robert Whittington, wrote a passage about a very great man who was very different from the Geheimrat, but I do believe that the passage also applies to the Geheimrat. What Whittington said was that "he is a man of angel's wit and singular learning; for where is the man of that gentleness and affability? And as time requireth, a man of marvelous mirth and pasttimes; and sometimes of as sad a gravity; a man for all seasons."

Ladies and Gentlemen, even though these words were written about Sir Thomas More, I think they describe the Geheimrat just as well. He is truly "A man for all seasons", someone who has seen everything life has to offer and to take away, and who has through intelligence, courage and indomitable will, conquered.

Thank you very much.

RECEIVED February 5, 1981.

8

Herman F. Mark and the Structure of Crystals

LINUS PAULING

Linus Pauling Institute of Science and Medicine, 2700 Sand Hill Road, Menlo Park, CA 94025

Herman Mark is famous for his tremendous contributions to the field of polymer science. He is not so well known for his early work in the field of crystal structure. I think that it was his experience in the crystal-structure field that gave him the background of knowledge that permitted him to make his important contributions to the understanding of polymers. He was, in fact, one of the leading investigators in the field of the use of x-ray diffraction for the determination of the structure of crystals in the years 1923 to 1928, and it was through this work that he developed the feeling for atoms and their interaction with one another that permitted him, later on, to make an effective attack on the problem of the structure and properties of macromolecules.

The first volume of the Strukturbericht, written by P. P. Ewald and C. Hermann, covered the period 1913 to 1928. Herman Mark ties for third in the number of page references on crystal structures reviewed in this compendium. First place is held by the mineralogist Viktor M. Goldschmidt, with 89 references, many of them, however, to his great multivolume work Geochemische Verteilungesetze der Elemente. Second place, with 53 references, is held by Wheeler P. Davey, who was diligent in applying the powder technique that had been invented by Hull (independently of Debye) to a large number of elements and simple compounds. The tie for third place, 50 references, is between Mark and R. W. G. Wyckoff.

In his x-ray work Mark showed a greater range of interests than the other investigators in the field. Mark and his collaborators studied elements, both metallic and nonmetallic, minerals, inorganic compounds, simple organic compounds, condensed gases, and macromolecular substances, and in addition they studied the physics of x-rays and of the diffraction phenomenon.

Most of Mark's work done during this period was done at the Kaiser Wilhelm Institute of the Chemistry of Fibers or the Kaiser Wilhelm Institute for Physical Chemistry and Electrochemistry, Berlin-Dahlen. His collaborators during this six-year period

0097-6156/81/0175-0093$5.00/0

included W. Basche, R. Brill, W. Ehrenberg, K. W. Gonell, C. Gottfried, O. Hassel, E. A. Hauser, J. Hengstenberg, H. Hoffmann, H. Kallmann, J. R. Katz, H. Mehner, K. H. Meyer, W. Noethling, E. Pohland, M. Polanyi, P. Rosbaud, J. Steinbach, G. von Susich, Leo Szilard, S. Tolksdorf, K. Weissenberg, and E. Wiegner.

At the time that Mark began x-ray work the crystal structure had not yet been determined for any organic compound. At that time the x-ray techniques could be applied with the greatest prospect of success in determining the complete structure to crystals with high symmetry, especially cubic crystals. It is not surprising that Roscoe G. Dickinson, who had begun crystal structure work in 1917 in the California Institute of Technology and who was hoping to be the first to determine the structure of an organic crystal, should have selected hexamethylenetetramine, $C_6H_{12}N_4$, one of the few organic compounds that forms cubic crystals, for his investigation, and that Mark himself should have selected the same substance. In January 1923 Dickinson and a student of his, Albert Raymond, published their determination of the structure, and eight months later Mark and Gonell published their determination, with essentially the same results. Dickinson soon went into another field of research, but Mark continued for some years to investigate crystals of organic compounds. In most cases he found it impossible to make a complete determination of the positions of the atoms, but he usually succeeded in drawing some conclusions about the structure of the organic molecules from the x-ray results. In 1923, with Eisenberg, he published an account of a preliminary study of urea, followed by studies of other organic compounds, usually made in collaboration with von Susich, Mehner, Hassel, Hengstenberg, or Noethling. The crystals investigated include carbon tetraiodide, tetramethylmethane, tetranitromethane, pentaerythritol, oxalic acid, metaldehyde, biphenyl, stilbene, triphenylcarbonyl, triphenylmethylbromide, fluorene, phenanthrene, and (in 1929), D-glucose, D-fructose, and D-cellobiose.

In 1925 he reported on his work on condensed gases, CO_2, B_2H_6, NH_3, and CS_2, carried out together with Pohland. His structure for carbon dioxide is the correct one, except that the value of the parameter determining the carbon-oxygen distance is incorrect. He and Pohland reported this bond length to be 1.6 Å, rather than 1.16 Å. The boron-boron bond length found in their study of diborane, 1.8 to 1.9 Å, is close to the correct value, 1.77 Å. His first study of an element was carried out in 1923, with Weissenberg, Gonell, and Wiegner. They reinvestigated orthorhombic sulfer, which had been reported by W. H. Bragg to have a unit cell containing sixteen sulfer atoms. Mark and his collaborators found that each of the edges of the unit had to be multiplied by 2, to give a cell containing 128 atoms. They were not able to evaluate the parameters determining the positions of the atoms. Mark and Hassel in 1924 reported on their reinvestigation of graphite, which they found to have the structure assigned to it by Hull in 1917, rather than an alternative structure suggested by Debye. They evaluated the parameter as 0 ± 0.10. J. D. Bernal in England was investigating graphite at the same time, and in the same year he reported similar results: the Hull structure, with the parameter equal to 0 ± 0.06.

The structure of white tin had been subjected to investigation first by Bijl and Kolkmeijer in 1918, who reported an incorrect structure. In 1923 Mark and Polanyi carried out a second investigation, and found the correct structure for this tetragonal crystal, a structure involving no variable parameters. In 1924 Mark and Hassel reported the results of their reinvestigation of the structure of bismuth, whose structure had been determined in 1921 by R. W. James, in Manchester, who assigned the value 0.232 ± 0.004 to the variable parameter. Hassel and Mark verified the James structure, with the parameter equal to 0.236 ± 0.003. The presently accepted value for the parameter is 0.2339.

Mark also extended his interests in 1923 to inorganic crystals, beginning with calomel, mercurous chloride. Calomel had constituted a puzzle for inorganic chemists in the 19th century, in that the position of mercury in the periodic table is such as to lead strongly to the conclusion that mercury is bivalent, as in mercuric chloride, $HgCl_2$. Calomel has the composition that would permit the formula HgCl to be written for it, suggesting univalence. Chemists discovered, however, that calomel and similar compounds of mercury (mercurous mercury) ionize in solution to produce the diatomic cation Hg_2^{++}. Mark and his collaborator Weissenberg in 1923 determined the structure of the crystal, and found it to contain linear molecules Cl-Hg-Hg-Cl, with the distances indicating that the molecule contains a mercury-mercury covalent bond as well as mercury-chlorine covalent bonds. This was an important contribution to structural inorganic chemistry.

Other inorganic crystals studied by Mark and his collaborators, sometimes leading to complete structure determinations, include strontium chloride, zinc hydroxide, tin tetraiodide, potassium chlorate, potassium permanganage, and ammonium ferrocyanide. Minerals investigated by them include $CaSO_4$ (anhydrite), $BaSO_4$ (barite), $PbSO_4$, Fe_2TiO_5 (pseudobrookite), and three forms of Al_2SiO_5 (cyanite, andalusite, and sillimanite).

Mark's x-ray work on fibrous macromolecular substances began in 1925, with his publication, together with Katz, of a paper on cellulose. He continued the work on cellulose with Meyer and Susich (1929). In 1926 he and Hauser published a report of their studies of rubber. He had developed excellent ideas about the nature of rubber and the explanation of its extensibility and elasticity. I remember that when I visited him in Ludwigshafen in the summer of 1930 both he and I took pleasure in a demonstration that he showed me. He took a large piece of unvulcanized rubber, and stretched it to twice its length. When it was released, it contracted to its original length. He then stretched the rubber and held it under a cold-water faucet, so that it was cooled in the stream of water. On being released, it remained in the stretched form, which had crystallized. He discussed the part played by the increased entropy of the contracted form in the extensibility of rubber, much to my edification.

During this period he had already developed an interest in the fibrous proteins. R. O. Herzog and W. Jancke had made moderately good x-ray fiber diagrams of silk fibroin in 1920, and Brill in 1923 had assigned indices to about twenty spots in the fiber diagram in terms of an orthogonal unit with identity distances of 7.0Å along the fiber axis and 9.3 and 10.4Å perpendicular to it. Meyer and Mark in 1932 then discussed the structure more definitely, on the basis of Brill's x-ray data. They suggested, as Brill had earlier, that the polypeptide chains extend along the fiber axis, and contain glycine residues alternating with amino-acid residues with larger side chains, principally alanine and serine. Four polypeptide chains, extending along the fiber axis, would pass through the unit cell, with one glycine residue and one alanine residue of each chain in the cell. They also suggested that the peptide chains are strongly attracted to one another by forces between the CO groups and the NH groups of adjacent chains. None of their conclusions could be drawn from the analysis of the x-ray diagrams alone; instead, they were the result primarily of considerations of the length of chemical bonds, the bond angles, and the nature of intermolecular forces. Although the nature of the hydrogen bond was pretty well understood by 1932, twelve years after the first important paper on the hydrogen bond, by W. M. Latimer and W. H. Rodebush of the University of California in Berkeley, had been published, Mark, in common with most other European scientists, still had little knowledge about this important structural feature, and accordingly he and Meyer could not make their suggestions about the interaction of the CO and NH groups precise.

In fact, even after the later refinement of experimental techniques, including the preparation of doubly-oriented fibers of silk fibroin by the stretching and rolling of silkworn gut and the discovery of the Patterson diagram and other increasingly powerful methods of interpreting x-ray data, it was still not possible to derive the structure of silk fibroin except through the detailed application of principles of structural chemistry. By 1948 enough detailed x-ray analyses of the structure of crystals of amino acids, simple peptides, and other simple substances related to polypeptide chains, all carried out by Robert B. Corey and his associates in the California Institute of Technology, had provided precise information from which the interatomic distances and bond angles in polypeptide chains could be predicted with high reliability, about 0.01Å in bond lengths and 1° in bond angles. In particular, these observations verified the predicted planarity of the peptide group and the importance of the formation of the maximum possible number of N-H...O=C hydrogen bonds. The application of these structural principles and the use of accurate values for interatomic distances and bond angles permitted the exact description of several possible configurations of the polypeptide chain, the alpha helix and the two pleated sheets. In particular, it was found that acceptable sheet structures of polypeptide chains could not be formed by fully extended polypeptide chains; instead, the chains need to be contracted somewhat, and staggered in the direction perpendicular to the fiber axis and the lateral hydrogen bonds. The predicted length of the two-residue unit of a completely extended polypeptide chain is 7.23 Å, that for the antiparallel-chain pleated sheet is 7.00 Å, and that for the parallel-chain pleated sheet is 6.6 Å. The close agreement of the predicted value 7.00 Å and the experimental value 6.97 Å for the fiber-axis unit length in silk fibroin is strong indication that the structure is based upon antiparallel-chain pleated sheets. Moreover, the length of the a axis is calculated to be 9.5 Å for the antiparallel-chain pleated sheet, in good agreement with the observed value 9.4 Å. In 1955 R. E. Marsh, R. B. Corey, and L. Pauling showed that the antiparallel-chain pleated-sheet structure accounts satisfactorily for the intensities of the x-ray diffraction maxima of silk fibroin. The nature of the structure of silk fibroin suggested by Meyer and Mark in 1932 was thus completely substantiated, and somewhat refined, twenty-three years later.

In the meantime, as is well known, Herman Mark had continued to apply his understanding of the basic structure of fibrous macromolecules in many important ways.

Work on the structure of crystals and fibers was not the only way in which Mark made use of x-rays. With several collaborators, he reported the results of a number of significant investigations of the physics of x-rays in 1926 and 1927. With Ehrenberg he reported studies of the index of refraction of x-rays, and with Leo Szilard studies verifying the linear polarization of x-rays scattered from electrons at 90°. An investigation of the width of x-ray lines was carried out by Mark and Ehrenberg, and Mark and Kallmann reported work on the properties of Compton-scattered x-radiation and on the theory of the dispersion and scattering of x-rays.

I cannot conclude this account without mentioning another contribution to science by Herman Mark, not involving x-rays. I believe that this contribution, the discovery of the technique of determining the structure of gas molecules by the diffraction of electrons, constitutes Mark's most important contribution to structural chemistry, one which, moreover, was of great significance in my own development. In 1930, when I visited Herman Mark in Ludwigshafen, I learned that he and his young associate R. Wierl, had constructed an apparatus for scattering a beam of electrons from gas molecules and had determined the interatomic distances in carbon tetrachloride and a number of other molecules by analysis of the diffraction pattern (Die Naturwissenschaften 18, 205, 778, 1930, and later papers by Wierl in Phs. Zeit. and Ann. Phys. in 1931 and 1932). The equations describing the diffraction pattern produced by a wave (x-rays or electrons) scattered by a molecule had been derived independently by P. Ehrenfest and P. Debye in 1915. The electron-diffraction pattern from a molecule such as carbon tetrachloride showed a series of concentric rings, with different intensities, and with the radii of the rings and their intensities depending upon the interatomic distances and the scattering power of the atoms in the molecule. I was overwhelmed by my immediate realization of the significance of this discovery. For several years I, in common with other x-ray crystallographers, had been disappointed by the repeated failures to determine the structures of crystals by application of the known procedures. What was expected to be a simple structure determination often turned out to be impossibly complex. For example, in 1922 the crystal $K_2Ni_2(SO_4)_3$ which I had made and examined was found to have a structure determined by nineteen parameters, locating the atoms in the cubic cell, and it was not possible even in 1930 to determine more than perhaps half a dozen parameters from the x-ray intensities. In fact, despite the interest of many x-ray crystallographers, such as J. D. Bernal in London, in amino acids and simple peptides because of their significance for the protein problem, even as late as 1937 no one had succeeded in locating the atoms in any amino-acid crystal or simple peptide. The simplest amino acid, glycine, would probably require in its structure determination the evaluation of ten parameters, two for each of the five heavy atoms, assuming, on the most optimistic assumption, that the molecule has a plane of symmetry, and the methods had not yet been discovered for solving such a problem. Whereas the investigation of any crystal by x-ray diffraction was a gamble, in that a simple molecule might interact with its neighbors in the crystal in such a way as to make the structure complex, no such complicating effect was possible in a gas. For example, Dickinson in 1923 had found that the unit of structure of tin tetraiodide is a cube containing eight molecules, with the atomic positions determined by five parameters, which he succeeded in evaluating. But the SnI_4 molecule is tetrahedral, with its structure determined by a single parameter, so that one could predict with confidence that the investigation of the vapor by the electron diffraction method would surely permit the verification of the tetrahedral structure and the determination of the value of the one parameter, the tin-iodine bond length, without trouble. As the impact of the significance of this discovery burst upon me I could not contain my enthusiasm, which I expressed to Mark - my feeling that it should be possible in a rather short time, perhaps ten years, to obtain a great amount of information about bond lengths and bond angles in many different molecules. I asked Mark if he and Wierl were planning to continue with such a program, and he said that they were not. He added that if I were interested in building an electron-diffraction apparatus he would be glad to help, and in fact he gave me the plans of their apparatus. On my return to Pasadena in September I talked with a new graduate student in the California Institute of Technology, Lawrence Brockway, about this project, and he agreed to undertake the construction of the apparatus (with the help and advice of my colleague Professor Richard M. Badger). During the following twenty-five years the structures of molecules of 225 different substances were determined by the electron-diffraction method in the California Institute of Technology, through the efforts of 56 graduate students and post-doctoral fellows. These studies led to the discovery of several valuable principles of structural chemistry. I continue to have a feeling of gratitude to Herman Mark for his discovery of this important technique and for his generosity to me in connection with it.

Herman Mark is thought of by most chemists as a pioneer in polymer science. I think of him, with affection and admiration, as a pioneer in modern structural chemistry and an important early contributor to its development.

RECEIVED February 5, 1981.

9

Salute to Herman F. Mark

MAURICE MORTON

University of Akron, Akron, OH 44325

When I was asked to contribute to this volume on the lifetime accomplishments of my good friend Herman, I seized the opportunity with enthusiasm and alacrity. Here was my chance to put on the record how I have always felt and still feel about this remarkable and unique individual. I use the term "unique" advisedly, since I feel that Herman F. Mark has indeed made a unique contribution to the world of polymers, and that he stands alone in that regard.

Since its early beginnings more than a half-century ago, the field of polymers has attracted a number of brilliant minds and has benefitted from the contributions of many bright minds. The scientific researches of Herman Mark, especially his pioneering work on polymer chain structure by X-ray diffraction, are part of such contributions. But that is not where his uniqueness lies. His special and unique contribution has been and still is in "spreading the gospel" of polymers, to borrow the phrase from great spiritual movements. But unlike the proselytes of such movements, who seek to "convert" people, Mark's activities have been aimed at disseminating knowledge about polymers, and he has the unique ability to do so in a fascinating, stimulating and charming manner. As everyone in our field knows, these activities have contributed no small part in the enormous growth of polymer science in our lifetime, and have also "popularized" polymers even among the general public. Small wonder that he is the only one in this field who has merited inclusion in the "Profiles" which appear from time to time in the New Yorker magazine. Significantly his profile appeared (New Yorker, Sept. 20, 1958) under the heading "Polymers Everywhere"!

Perhaps the most important influence that Herman Mark has had was to stimulate a lifelong interest in polymers in young scientists and engineers, who were thus attracted to this field. It was my personal experience to come under his influence in this way, and I would like to share this experience with you.

0097-6156/81/0175-0101$5.00/0

Meeting Herman Mark in 1943

At the time of the outbreak of World War II, I had spent a number of years as a chemist in the building products industry in Canada, and had decided to return to McGill University for graduate studies. My experience in industry had been with a variety of organic and inorganic materials, ranging from bitumens and rubber to minerals and cement. The early plastics had already appeared (vinylites, bakelite, etc.) but were technically ill-defined. Upon returning to the university in 1942, however, I noticed that a growing interest was being shown in the macromolecular aspects of these plastics. The current interest at that time was in learning how to measure molecular weights and their distribution. Names like Flory, Huggins, Kraemer, Schulz, Fuoss, etc., were becoming familiar to us. Many discussions were held about the significance of the various molecular weight averages. The fascinating world of polymers was opening up.

I, too, was caught up in the wave of enthusiasm for this new science which had the lofty goal of relating the properties of materials to their molecular structure, and, in the end, to "tailor-making" molecules for specific properties. Since one of the big developments at that time was the newly-started synthetic rubber programs of the American and Canadian governments, I chose the topic of the emulsion copolymerization of butadiene-styrene as the subject of my doctoral dissertation.

As it turned out, the local chemical society in Montreal had scheduled Herman Mark as one of its monthly speakers, and there was a large turnout to hear him, both from industry and academe. Most of us had already heard about this "refugee" from war-torn Europe who had recently come to the New World, first to Canada and then to settle in Brooklyn, and we were very curious to meet him.

His effect on the audience could almost be described as "electric". Here was this smartly-dressed handsome man, in his forties, looking more like a business executive than a professor, explaining in a quiet, yet enthusiastic, manner what macromolecules were, and holding his audience enthralled. His lucid presentation, his charming humour, and his delightful Viennese-accented English made a profound impression on his listeners. As I recall, he described, at that time, the general mechanism of free-radical polymerization of vinyl monomers, emphasizing the factors which control the molecular weight and its distribution, and such processes as copolymerization and chain transfer. He made a special point of predicting that polymer research would eventually lead to the synthesis of the "tailor-made" macromolecule, of predictable structure and properties.

I remember that many of my friends later remarked how much they enjoyed his lecture and how clearly he had explained the "mysteries" of the new plastics, which they were now beginning to understand for the first time. As for myself, I found his talk exciting and stimulating, and it made me proud to have chosen to enter the field of polymers. That was my first encounter with Herman Mark, and I marvel that he has never lost his touch, and still delivers the "goods" in the same way at all his lectures.

Subsequently, we arranged to have him visit our laboratories in the Chemistry Department at McGill (how simple and unsophisticated were our methods and equipment then!). Suffice it to say that we had a stimulating session with him, discussing our research on emulsion polymerization and obtaining many valuable suggestions for future work. His special contribution then, as is the case even now, was that he was aware of the wide spectrum of polymer research which was being carried out in all parts of the world, and could immediately refer you to the literature or to people who could answer your questions.

The Intervening Years

It is now over thirty years since I came to the University of Akron and have been part of the polymer scene in this active area. The Akron Polymer Lecture Group dates from that time, too, and has been one of the prime catalysts in keeping the Akron community abreast of new developments by means of its annual series of monthly lectures by invited speakers. Here again Mark figured prominently as the one person who could keep us aware of new developments, especially when these represented a real "breakthrough" in polymers. No wonder that he received more repeat invitations to Akron than anyone else. Let me quote just a few examples of the more important polymer topics which Herman Mark first told us about.

a) Graft and block copolymers

We first heard this interesting idea explained by Mark in early 1953 (Akron Polymer Group Lecture, Jan. 9, 1953). Compared to today's elegant methods based on anionic polymerization, the early work involved relatively crude free radical polymerization, but the idea itself (and the concomitant nomenclature) represented a real advance in polymer synthesis.

b) Stereospecific polymerization

When this giant breakthrough occurred, in the early 1950's, its reverberations soon reached us at Akron, and this immediately prompted another invitation to Mark to "tell us all about it". He spoke to a packed house, again giving a lucid presentation not only of the remarkable results of these Ziegler-Natta polymerization systems, but even stimulating many thoughts about possible theories to explain these results.

c) Heterophase polymers, blends and composites

In more recent years, Herman Mark has, as we all know, concentrated more on the effects of heterophase morphology of polymers on their mechanical properties. This has enabled him to set up a useful classification system of the various types of heterogeneities which can occur in polymers, e.g., crystallinity, incompatibility, particulate and fibrous inclusions, etc. and to discuss these in the context of their effect on the mechanical properties. Such an "overview" has again kept Mark in great demand as a speaker.

In addition to Mark's role as a harbinger of new developments in polymer science, he also played the leading role in reporting on the growth of polymers in different parts of the world. Thus, in the early 1960's, with the first thaw in the cold war, he was the first to bring back a report on polymer research in the USSR (since he was the first to be invited there!). Then, again, in the early 1970's, he was the first to go to China (after Nixon, of course!) and bring back his usual fascinating report on polymers in that emerging nation. It would be an interesting statistic to know how many miles of travel this perennial traveller has logged to date!

Conclusion

I have restricted my discussion to those activities of Professor Mark which have impacted directly on me, i.e., his role as a popularizer and teacher of polymer science. His other accomplishments in nurturing education, research and publication in polymer science will undoubtedly be more appropriately described by others more directly involved. As far as I am concerned, however, I shall always remember Herman Mark for his outstanding ability to present his inimitable lectures on a variety of polymer topics in a lucid, fascinating and charming manner. That is what helped to bring me into the field of polymers, and has undoubtedly done the same for countless others.

RECEIVED September 2, 1981.

10

Herman F. Mark: The Continuing Invasion

G. ALLAN STAHL

Phillips Petroleum Company, Research and Development, Bartlesville, OK 74004

> "When he (Mark) left the Institute you could tell the weather by his face: a low - he was going to the Dean's office, a high - he was on his way to the football field or to see a billiards game at the cafe Josefinum. But a smiling chief meant he was engaging his students in a discussion. Yes, one should have such teachers!"
>
> H. Tschamler

A popular encyclopedia defines a teacher as one who promotes the student's gain of knowledge, skills, and ideals by taking an interest in the student, speaking and writing effectively, and being knowledgeable on his subject matter. If taken literally, Herman Mark is the quintessential teacher. An outgoing, personable man who strives to make each contact meaningful, comfortable, and rewarding, he is a man with a predilection to conversation, photographic memory, and nearly all-encompassing knowledge of polymer science. It is then only natural that he teach, or in his own words, "talk, talk, and talk" about chemistry. But talking is not enough. Mark's genuine interest in people and near perfect Jeffersonian ability to communicate at any level make him one of the premier science educators in the world today.

It has been said "that for the price of the bus ticket to the auditorium Herman Mark can be bought for an evening." If this is true, it is a nonpareil bargain. There are many examples of his ability to hold an audience of any background. One was when he addressed a diverse assembly of students, faculty, industrial scientists, and businessmen at Kent State University in the honor of the inventor and educator Waldo L. Semon (1). Speaking slowly, gesticulating, and smiling he

0097-6156/81/0175-0105$5.00/0

charmed the audience--leaving the impression with all that he had prepared his remarks with each group's level of interest and background in mind. The reaction to his address in Kent was not an exception; rather, it was the same reaction he evoked in Brooklyn, Vienna, Ludwigschafen, Berlin, and throughout the world. Without further elucidation, it is safe to say that Mark's greatest skill is his ability to communicate, and his greatest contribution is teaching enhanced by intelligent application of this ability.

When Mark accepted a position at the University of Vienna in 1932 he was at last formally answering the call of teaching. From the time he left Vienna in 1921 until his return, the years at the Kaiser Wilhelm Institute and I. G. Farben, he remained interested and in close contact with the education programs of young scientists. At Ludwigschafen for example, he was simultaneously an industrial scientist and manager, and an Associate Professor at the technische hochschule at Karlsruhe.

His lectures at Vienna, as well as those at Brooklyn years later, have been variously described by such adjectives as brilliant, inspirational, and informative. The impact being greater than these adjectives or the gentleness of his delivery might suggest, but adjectives do not convey the simplicity of a concept when it is explained by Mark. "His greatest skill," says E. H. Immergut, "is in making the complicated and dry seem simple and interesting." E. Katchalski-Katzir said of him, "We were deeply impressed by his informality, by his wealth of knowledge, his elucidity of thought, his constructive comments, and his encouragement and fatherly attitude . . ." Mark, the teacher, clearly satisfies the three conditions--interest in the student, communicability and knowledge of the subject--set forth in the definition of a teacher.

In Vienna, a former associate recalled, Mark usually had a broad smile, strickly parted hair (with strands of white even then), and a fashionable suit. Walking briskly into the lecture room as his weight shifted from one foot to the other, he would pause, smile, and begin lecturing. His style was slow, careful, and pleasing. If visitors were present who spoke a language other than German, he would pause periodically and outline his concepts in their language. He spoke English, French, and Italian in addition to his native German. His lectures were clear, cleverly constructed, enthusiastically delivered, and laced with samples and experiments. Nearly fifty years later at the American Chemical Society Symposium in honor of his 85th birthday, Mark stepped before several hundred attendees and employed the same mannerisms and once again charmed the audience.

Writing and speaking excellence are not the only interactions this teacher has had with his students. Mark, over the years, has been a close associate and "hands-on" experimenter. He has always worked closely with those in his department,

Figure 1. Professor Herman Mark at his desk at the Polymer Research Institute.

personally advocated an "open door", and held "round table" discussions. He has also been a leader in providing the necessary medium for spreading findings and theories. His many books, especially the High Polymer monographs, Encyclopedia of Chemical Technology, Journal of Polymer Science, and Journal of Applied Polymer Science, are examples. Other devices initiated by Mark are the "Saturday Seminars" and "Summer Courses" at Brooklyn Polytechnic, and subconferences that he and Milton Harris initiated at the Gibson Island Research Conferences. This latter series of conferences are still being held, and attended by Mark, but are now named after their founder, the former Johns-Hopkins Professor Neil Gordon.

Herman Mark is a natural teacher who has practiced the art in all the positions he has held. A review of his accomplishments in education is, therefore, a review of his life. Many agree with J. Hengstenberg's high tribute. Hengstenberg said, "Professor Mark can tally many valuable successes outside of his scientific and technical accomplishments." One of the greatest is that "many of his younger coworkers have been molded by the example of his personality. His dealings with coworkers and attitudes toward work have remained, for many of us, an example suitable for emulation."

> "There were so many practical jokes
> and so much cheerful mischief
> that the Master (Mark) on his 80th
> birthday asked himself, 'How is it
> that in spite of it all, something
> worthwhile came out of it'."
>
> J. Hengstenberg

In the preceding essays the contributions of Herman F. Mark have been reviewed. Yet in them very little has been said about his personality and ways. Certainly he is capable of inspirational leadership and possesses keen insite. Equally true, he works hard and has produced a prodigious amount of research. What then beneath these traits is this remarkable person like? Besides his obvious intelligence and organizational skill, Mark is a relaxed, informal man with a broad sense of humor. Certainly his manners are commensurate with those required of one who has attained the zenith of scientific endeavor, yet Mark's informality has never allowed protocol to interfere with expedience.

In the late 1920's, when Mark was a manager at the I.G. laboratory in Ludwigschafen, managers of research were traditionally esteemed, but distant, unapproachable personalities. Mark's personality and style of management, however did not fit in this mold. Discussions with his various work groups were held when necessary, on the spot. He frequently discussed the day's problems with coworkers during his morning shave in his office, walking on the laboratory grounds, or, should a particularly enticing topic surface, in one of the region's local wine cellars. This unorthodox style quickly drew comparison, and Mark and his staff were labeled "somewhat disorderly". The record, however, indicates that its effectiveness was unquestionable.

Memorable to his associates was the absence of tempermental outbursts by Mark during times of stress. In a period when these things were expected from management, Mark's demeanor was especially appreciated. "Difficulties and malfunctions," said one associate, "were mastered by charm and humor."

The "somewhat disorderly" Mark was known as "undignified" a few years later when he made his daily appearances at the Institute in Vienna. Each morning, he would run up the steps of the main staircase to his office, pausing only to check the bulletin board on the half-way landing. Undaunted by this criticism, Mark founded and fitted out a football team for the First Chemical Institute. Then, he kicked with them. Mark, a former player for the Vienna Sports Club, played center-half for his team when it met the Second Chemical Institute in a big athletic showdown. The game was held at the Elektraplatz on Engerth Street, and although few recall who won, those who attended do remember that Professor Spaeth, the Director of the Second Institute, observed Mark's exertions indignantly.

Mark is also known for his love of the practical joke. Clearly he has taken advantage of his respected experience and sunny disposition, for the merriment of (almost) all. Otto Kratky tells of the time in Dahlem when a quartz vacuum pump used for evacuating X-ray tubes ceased to function. After failing to find the reason, Kratky took the problem to Mark. With a straight face, Mark replied, "It probably has a crack; put it in water and pressurize." Kratky responded, "It might break." Mark then sealed the pump's fate by saying, "Not the total pressure, open it slowly; otherwise it will serve it right!" Kratky says he had deep doubts as he "murdered" the pump.

Kratky reported that a short time after his experience, he observed Mark in action again. A physicist's, H. Kallman, experiment required he make a hole in a glass plate. Knowing Mark's reputation as a successful experimenter, he asked Mark how he might accomplish the task. Mark said, "Do you know what will happen if you shoot a glass plate with a bullet? It will make a hole because the blow is so quick. You need to get a pointed object and experiment with a hammer in order to give it quick blows." Kallman, reports Kratky, "practiced and practiced" producing a "mountain of broken plates" until the Director, Professor Haber, passed by and put an end to the joke.

Years have not changed Mark. On a European tour in the late 1950's, he visited a particularly attractive campus in France.

Figure 2. Professor Herman Mark receiving an Honorary Degree at Lowell Technological Institute. Beside Dr. Mark at the far right is President John Kennedy.

hosts took special pleasure in showing him the Institute's spectacular architecture. Tongue-in-cheek, he responded by telling them of the beauty of his own campus at Brooklyn Polytechnic. Brooklyn Poly is of course located in the serviceable, but not "ivy walled" downtown of Brooklyn, not far from New York's East River.

This episode does not end here. A few months later the Frenchmen toured the U.S., and naturally came to Brooklyn to see Professor Mark's campus. In the tradition of his Geheimrat, Murray Goodman, then a young Professor at Poly, greeted the visitors, and saved honor by convincing them that the New York Supreme Court Building, a few blocks away, was the building Dr. Mark had described.

Mark's informality and sense of homor often carries him to the point of near irreverence toward the conventional symbols of scholarly achievement. "I am sorry to say that I haven't read your latest work yet, but I have weighed it, and it is clearly of considerable importance." Many of Mark's associates have found themselves the victim of this old gag. His generous use of the titles "Professor" and "Doctor", regardless of station, has been carried to extremes. One classic story is his introduction of a young student to Poly's Dean Raymond Kirk. "Ah, Dean Kirk, please let me present to you Dr. Smathers, who is coming to us to earn his Master's degree."

His jokes and irreverance have not been given maliciously. In fact it is this geniality, his ability to divert his mind and energy, which allows him to maintain the pace he does without wearing himself down. His temperment is known in his native tongue as "Gemutlichkeit". The person so named could be expected to be easy going and good natured, have a capacity for hard work, a sense of duty, and take delight in literature and science. An apt description of Mark.

His friend, E. S. Proskauer, says of Mark, "The secret, I believe, is that he is such a skillful juggler of projects that he can keep a dozen going at once without ever seeming to bear the full weight of any one of them. He manages to keep from getting emotionally committed to any one part of his work, so if things start to go wrong, he doesn't have to throw himself into a death struggle to save something from the wreckage." Proskauer says this trait of character--"an eagerness to be busy and helpful, but with reservations that spare the usual feelings of distress or guilt about errors and failures"--is Goethe's "die Frohnatur", the joyous disposition.

Herbert Morawetz, whom Mark hired at Brooklyn Poly in 1951, says that he believes that Mark is incapable of negative emotion, "or at least they are buried beyond retrieval". As an example, it is well known that Hermann Staudinger disliked Mark as a result of the events about the establishment of the macromolecular concept. His dislike bordered on open contempt, but Mark steadfastly refused to openly criticize Staudinger. In fact, Mark was first to suggest that Staudinger be invited to the IUPAC Conferences after the Second World War and he and his wife, Mimi, spared nothing in hosting Staudinger and Hopff during their visit to the U.S. in 1953.

Mark's open feelings about those who drove him out of Austria in 1938 are curiously free of contempt. He describes the Nazis as "misguided", and those scientists who supported them as "unfortunate". Before emmigrating from Austria he attended a conference in Mainz where a Professor Stuart sneered, "what are YOU doing here?" Yet, Mark has never criticized his antagonist. After the Second World War, Mark was, in fact, very active in the reestablishment German and Austrian scientists to the World scientific community. His first action on returning to Vienna in 1947 was to call on his indirect successor at the First Chemical Institute, Professor L. Ebert, and reassure him that he would "never attempt to drive him out of the position he filled in such an excellent manner."

Those meeting Mark for the first time are usually impressed by his amiability, utter lack of pretentiousness, and overflowing cheerfulness. Additional contacts usually reinforce the extent of his genuineness, but his overpowering personality, optimism, and diverse interests prevent all but the more persistent from knowing him too well. Close associates have often been surprised to learn of particulars of his life, his war record for example. One coworker of more than thirty years recently commented regarding Mark and his family's flight from Austria in 1938, "I knew he left, but I never realized the circumstances. I guess I always assumed they (the circumstances) were the best possible since the Geheimrat has never complained. He doesn't seem bitter, yet he has a right to be."

Mark's response to "how do you relate to your associates" gives insite to his enigmas. He says that he is "everything to everybody". Considering the pervasiveness of his personality and interests, this comment is not to be taken lightly. Further, it is not given egotistically for in many ways it is completely accurate.

As we have discussed in the preceeding chapters, Mark rose rapidly to positions of high responsibility at the Kaiser Wilhelm Institute and I.G. Farben. A young, gregarious leader among the autocratic leaders of German science and industry, he developed a style which was unique at that time and place. He was simultaneously companion, collaborator, and authoritative director. He was called on, and considering the testimony of his former associates, was successful at handling a variety of challenges not limited to research problems. Those successes permitted him to move to even higher levels of responsibility. At some point about the time of his Directorship of the First Chemical Institute in Vienna, he assumed the role he has even today, that of a benevolent father to those below his position in the discipline.

Figure 3. *Herman Mark and his family at their Lake Peekskill summer home. (left–right, wife Mimi, Mark, sons Peter and Hans, and daughter-in-law Marion.)*

In this lofty position, he has for nearly fifty years followed and often guided the research efforts of so many workers that his label, "everything to everybody" takes on real meaning. With his breadth of contacts and genuine interest in people and his ability to communicate with interested audiences of almost any level of knowledge, he has been and remains one of the most liked, visible, important leaders in science.

> "In 1934 it was known that D_2O melted at 4^oC instead of 0°. So we thought, why not take a mass of crushed ice and melt out the D_2O? How? Why not on a glacier? So we went to the Jungfrau Glacier and later to the Bezinghi Glacier. Enrichment of several times was found, and we got to climb a mountain too!"
>
> H. F. Mark

Herman F. Mark and his wife, Mimi, true to their Viennese heritage, love winter, snow, and especially the sports of winter. From boyhood until his seventies when his doctors made him stop, he has skied, hiked, and climbed with a ferver matched only by his activities in the physical sciences. Just as an example, his love of these activities influenced his decision to enlist in the Kaiserschutzen regiment, an elite Alpine infantry unit, in 1913.

Between 1921 and 1932 in Germany, he kept in form by taking every opportunity to get on the cliffs and slopes. These outings ranged from gentle skiing to arduous vertical scaling. They were, of course, conducted with characteristic Markian zest. On one trip in the Engadine region of Switzerland, Mark's enthusiasm was nearly overdone. Shortly after leaving the Tschieva Glacier with M. von Laue and H. Pelzer, he suffered a dangerous and painful fall off a cliff. Laue and Pelzer were shaken after extracting him from the precipice, but with his usual tenacity, Mark lead the party home with a hurt leg and one ski. There were many advertures, but for an aficionado, the opportunities came far too infrequently.

His return to Vienna with his wife and their young sons in 1932 was thus a return to the city of his birth and childhood, and most of all, some say, to the mountains he knew so well. Unable to resist their Loreli call he reimmersed himself in their activities. Within a few months, he was skiing regularly, serving on the local avalanche rescue squad, and conducting a personal, yet thorough, study of the causes of avalanches. Throughout the winters of 1932 and 1935 he dashed to the scenes of avalanches to question inhabitants about the conditions--weather, noise, stacking--just before the snow started to slide.

The Ministry of the Interior soon recognized his growing expertise. In the winters of 1933-35, he was asked to serve on a committee charged with recommending road sites. The committee was made up of a mountain guide, an Army officer, several Alpine troops, and Mark, who served as the scientific advisor. The results of their study was used by the Ministry and Army when constructing new roads. Independently, Mark published the findings as part of a community report in an Austrian journal, thus establishing himself as an expert in the inexact art of avalanche prediction. A few years later at the outbreak of World War II, the U.S. government called on his prognostic abilities and assigned him to the testing program of the U.S. Army-Studebaker Snow Weasel.

In 1934, Mark published a short but interesting book titled, "Das Schwere Wasser" (2). In the book, he suggested that deuterium enriched water could be obtained by using the higher melting temperature of D_2O. D_2O melts at about $4^{\circ}C$, while H_2O melts at $0^{\circ}C$. Mark reasoned that in the laboratory ice could be maintained at $4^{\circ}C$, and the enriched water slowly collected. A more practical confirmation, he suggested, was to study the D_2O levels in a glacier. In a glacier the time frame of the experiment would be significantly expanded.

Shortly afterward in 1934, F. Eirich was dispatched to Switzerland to collect samples on the Jungfrau Glacier, but the results were inconclusive. An older, more obscure glacier was needed. Mark organized and led an expedition to explore the Bezinghi Glacier in the central part of the Greater Caucasius Mountains. The region is a desolate, wind-swept range between the Black and Caspian Seas, inhabited by only a few nomadic Kirghiz and Kalmuck shepards. The expedition was jointly financed by the Academy of Science in Vienna and the Russian Academy of Science.

In the company of E. Baroni, J. Schintlmeister, and three Russian physicists, Mark journeyed by train, muleback, and finally skis to the glacier. The results warranted their sacrifices. They found the concentration of D_2O at the tongue of the Bezinghi several times that of ordinary water. Before returning to Vienna, Mark celebrated the outcome by climbing a few sixteen-thousand foot mountains.

Mark's preoccupation with ice was the reason for his involvement in another project. In 1942, German submarines were very active in the North Atlantic and threatened the continuity of Allied shipping. Further, the only effective protection, airplanes, were unable to fly over the ships in midocean because of their limited flight range. The situation was critical. One solution, suggested by John D. Bernal, was to build unsinkable airstrips on icebergs in the North Atlantic. The difficulty was that natural ice shatters far too easily to withstand aircraft landings, much less torpedos or bombs. Mark, the famous Viennese ice, avalanche, and glacier expert, was called in as a consultant.

The Allied effort to make an "icecraft carrier" was known as Operation Habakkuk and headed by a British officer, Geoffery Pyke. Assisted by Pyke and W. P. Hohenstein, a former Viennese associate, Mark rented a Manhatten warehouse, and began preparation and testing of ice reinforced by various additives. The optimum mixture, ice and sawdust, was identified and then named "Pykrete". After the laboratory scale testing, the project was moved to Lake Jasper in Alberta where a 100' prototype was constructed by the Canadian Army. Although capable of withstanding an artillery shelling, a Pykrete carrier was never constructed. The technique has, however, found use in both arctic and antartic scientific stations.

> A student once asked Mark, "How do you manage at your age to keep up the pace? You get up early, stay up late, and fly between Europe and America as often as we would take a street car." Mark replied, "You know, this is true. I guess I have no time to get old."

When retracing Herman Mark's footsteps once quickly learns that his career, from meteoric rise to prominence in Germany to founding of institutes in Vienna and Brooklyn, has truly been international. Indicative is the fact that over the years he has had contact and often close interaction with a large percentage of the notable scientists of this century. Among them are Haber, Planck, Nernst, Einstein, Perutz, Bernal, Staudinger, Pauling, and Flory. A man molded by the largeness of his actions, accomplishments, and contacts, he has very little regard for parochial interests. These interests in his mind include national boundries. As a result he has shared his time and knowledge, as well as, the time and experience of those under his leadership, so freely that there are polymer science establishments in nearly all the industrialized nations which in some way or another owe a part of their existence to him.

Mark's international prestige was assured before he immigrated to Brooklyn in 1940, but it was his second career, that of world traveler and "Johnny Appleseed" of polymer science, which has made him the patriarch of the discipline. This new, yet not so different, Mark was created by the increased managerial and editorial responsibilities he voluntarily assumed after the Second World War.

In the time after the War, his role in science, which some mourn as abandonment of experimental chemistry and others praise as attending of larger duties, has taken him all over the globe. His travels began in earnest in 1946 when he and many of his American associates started regular commutes to Europe using the Military Air Transportation System (MATS) flights. The resulting liaisons were so successful that in a short time Mark was recognized as an active and effective protagonist of European science. To the delight and relief of many, he was also acknowledged as one who assisted without regard for creed or nationality.

Among many activities, he played a major part in the establishment of the Polymer Section of the International Union of Pure and Applied Chemistry (IUPAC) in 1946, and served as the section's first chairman. One of his first actions as chairman was to request that representatives be accepted from Germany. The request is still remarkable considering the plight of Mark and many other scientists in the years preceeding 1946.

Mark, however, sincerely and persistently sought true international membership for the organization, despite opposition so intense that his commitment was tested on many occasions. One scientist, for example, remarked sarcastically at the height of the controversy, "You came from over there (United States), you don't know what happened here." This observer was wrong for Mark did know and understand. Besides his own emigration from Austria, he lost many members of his family, during the war and his own mother in the battle for Vienna in April, 1945. The result of the struggle which lasted for four years was that H. Staudinger and G. V. Schulz of Germany were invited to participate in the IUPAC in 1950.

During the same period Mark was active on behalf of German scientists, he was nominated by Chaim Weizmann to head the Scientific Planning Committee of the newly established Weizmann Institute. Throwing himself into the job with unusual zest, he visited Rehovot the summer of 1946 despite the threat of terrorist attacks, and ascertained first hand the requirements of the Institure. Working closely with David and Ernst Bergmann, Aharon and Ephraim Katchalski, and C. Pekeris in Rehovot, and I. Fankuchen, K. Stern, and W. P. Hohenstein in Brooklyn, he organized many of the operational functions of the Institute and in 1947 procured its first sophisticated laboratory equipment.

Those were times of great turmoil in the region. Terrorist attacks could be expected with alarming frequency as the British prepared to withdraw. Because of the unrest, Mark and the Planning Committee set-up and operated the Institute's instruments in Brooklyn until the political situation stabilized.

Professor E. Katchalski-Katzir of the Weizmann Institute and newly named Herman F. Mark Professor of Polymer Science at the Polytechnic in Brooklyn points out that Mark was important to the Weizmann Institute in another way. "Mark", he says, "was the first scientist to establish ties between Israeli scientists and the international community." In the years since 1947, Mark has served as Vice-President in Charge of Project Research of the Board of Governors of the Weizmann Institute, and in 1975 he was elected as an Honorary Chairman of its Scientific Advisory Council.

At the request of the Indian government and representatives of the Indian textile industry, Mark toured India in 1948 and established a close, working relationship which still exists even today. One result was that Mark, and W. P. Hohenstein cofounded the Indian Textile Institute in Ahmedahad. A few years later Mark cofounded India's National Chemical Laboratory in Poona which S. L. Kapul directed and Mark served as Polymer Division Chairman. In 1956, Mark and S. Palit cofounded the Polymer Division of the Indian Institute for the Cultivation of Science in Calcutta.

In 1950, Mark became the first western scientist to contact counterparts in the post-war Soviet Union. Renewing old friendships dating back to Vienna on the first visit, he has been treated to a number of special tours of the chemistry establishments of Russia since then. In return he and Mimi have hosted a number of Soviet scientists in New York and at their summer home at Lake Peekskill, NY. Even today he maintains close contact with many of them. One of his major contributions to science in the Soviet Union was his able service on the organizing committee of the Polymer Institute at Lomonosov State University in Moscow in 1963.

He was also one of the first U.S. scientists to visit post-war Japan. On his first trip in 1956, he was specially honored by a personal audience with Emperor Hirohito. To his hosts' delight, he demonstrated with the able assistance of Mimi the "Nylon Rope Trick". In 1957, he returned and was active in the establishment of a polymer section at Professor Sakurada's Institute of Radiation Chemistry near Kyoto.

A special recognition came in 1966, when U.S. Secretary of State Dean Rusk asked Mark to use his contacts and arrange a good-will tour of chemistry departments in South America. The trip was, of course, sponsored by the State Department. The arrangements were easy for the Polytechnic family had several alumni in South America including Eloisa Mano in Rio de Janiero and Jose Swarts in Buenos Aires. Mark and his companion, S. Atlas of the Polytechnic, traveled with their wives through Venezuela, Brazil, Uruguay, Chili, Peru and Argentina. His many lectures and experiments assured a successful tour.

Possibly his greatest international adventure began also in 1966. While attending an IUPAC meeting in Tokyo, he met Liu Dagang, a Vice-President of the Chinese Academy of Science. The two became overnight friends, and before leaving Tokyo exchanged invitations to visit their respective institutes. Although it seemed unlikely in 1966 that their invitations would be honored, U.S.-Chinese relations warmed immediately after President Richard Nixon's famous trip in early 1972. As a result of his own "opening to China", Mark was the first western chemist to visit colleagues in the Peoples Republic. This historic visit was made in June, 1972.

Figure 4. *Dr. Mark and associates during his 1972 tour of Japan.*

Although he only visited China once more, his initiative has led to an exchange of U.S. and Chinese scientists. Mark's views of chemistry in China was one of the earliest published reports (3).

Since 1965, Mark has also been very active in his native country, Austria. He has returned nearly every year to teach, lecture, maintain connections, and through these efforts encourage science and, in particular, polymer science. In recognition of his many contributions, he has been named Honorary President of the Chemisches Forschungsinstitut der Wirtschaft Osterreichs and the Director of the Science Advisory Committee of the Osterreichische Kunststoffinstitut. A special tribute was given him on the occasion of his Eightieth Birthday in 1970. The H. F. Mark-Medaille was presented for the first time. The medal is given annually to an Austrian scientist who has gained special merit in polymer research.

> "After nearly sixty years he resists slackening either his interest or his effort. He youthfully persists in his continuing invasion of the world of science."
>
> H. Ringsdorf

Over the years Herman Mark has been known as polymer science's advocate, early explorer, spokesman, representative, teacher, and senior citizen. Starting, as we have seen, as a young man when the concept of high molecular weight was not accepted, Dr. Mark and his many associates confirmed the structure of polymers and helped open whole new areas of scientific research.

In the ensuing years he led an eniment industrial research team to the creation of a number of commercially valuable polymers and established an internationally famous Polymer Research Institute. His principal research has been on X-rays and electrons for the study of the structure of matter, and on the synthesis, characterization, reactions, and properties of natural and synthetic polymers. Yet his contributions include clarification and dissemination of basic principles, encouragement of academic-industrial interactions through research and education programs, and the founding of several key aspects of the polymer literature including the Journal of Polymer Science, the Encyclopedia of Polymer Science and Technology, and the High Polymer monographs. It is, strictly speaking, not possible to single out any single achievement as his magnum opus.

Mark's energy, insite, leadership, and communicability have all received considerable attention, but his greatest strength is probably his ability to move from one position or place to another without looking back. He has wasted little time in retrospect because of his flexibility, rather, he, in his own words, has pushed ahead in "a broad offensive". "Moving slowly, systematically, and irresistibly to probe new terrain", Mark has always willingly gone into new areas, developed new skills, and by recognition of the importance of interdisciplinary research, remained creative.

He has earned a lasting place in the history of polymer science through his research contributions, the successes of his students, his organizational genius, and his tireless promotion of the science and its students. It is entirely accurate to say that Herman Mark found polymers a curiosity and made them a science.

Acknowledgements

The author wishes to make special note of the use of a special issue of Chemie Kunststoffe Aktuell (4) and two taped interview sessions with Professor Mark (5, 6). A special thanks is given to Ms. Jeanne Strauss, Phillips Petroleum Co., for her patient clarification of a great deal of German.

Literature Cited

1. Mark, H. F. "Waldo Semon Lectureship Series"; Kent State University: Kent, Ohio, November 3, 1978.
2. Mark, H. "Das Schwere Wasser"; F. Deuticke: Leipzig, 1934; 32 pp..
3. Mark, H. F. Naturwissenschaften, 1974, 61, 157.
4. Chemie Kunststoffe Aktwell, 1980, 34(2), 1-36.
5. Stahl, G. A. "Taped Interview with H. Mark", October 21, 1979.
6. Stahl, G. A. "Taped Interview with H. Mark", February 16-17, 1980.

RECEIVED August 21, 1981.

Polymer Education and the Mark Connection

CHARLES E. CARRAHER, JR.

Wright State University, Department of Chemistry, Dayton, OH 45435

Educators traditionally are expected to contribute to the areas of teaching, service and research - the academic trinity. Those kinder administrative souls have noted that contributions in two of these areas are acceptable. Even so, few make significant contributions to even one of these areas and only a small handful make significant contributions to all three areas. Professor Doctor Herman Francis Mark is one of these exceptional educators who has and continues to contribute significantly to all three areas. He continues to actively review research areas and to consult (research), to speak to learned and popular audiences (teaching) and to edit the Journal of Polymer Science and associated journals (service).

The development of polymer education in the USA will be traced using specific examples to illustrate general trends, developing briefly Professor Mark's many contributions to the growth of polymer education. Many others have made significant contributions to polymer education, but in this brief paper only a few of these individuals or centers of learning can be noted.

Polymers are integral parts of mankind as building blocks of himself (protein, nucleic acids), his clothing (wool, cotton, polyester), building materials (wood, polystyrene), in the accumulation of knowledge (paper, ink), etc. The polymer industry today is at parity with the total metal-based industry and employs more chemists than all the other areas of chemistry combined.

0097-6156/81/0175-0123$5.00/0

Tomorrow offers new frontiers in all major areas of macromolecular science with new building materials, automobiles and other vehicles consisting mostly of polymeric materials, polymeric space ship wrappings, new synthetic organs, further unraveling of biomacromolecules, etc. A greater focus will be placed on the use of natural, regenerable polymers as stockpiles of nonregenerable feedstocks diminish.

Polymer chemistry is also important because (a) it serves as a basis of many related industries such as ceramics, textiles and leather, (b) its principles are applicable to materials in general, and (c) its principles are vital to biochemistry and medicine.

Polymers are large molecules, often called macromolecules, bound by covalent bonds. The transmittal of both the general and specific properties that contribute especially to the understanding of large molecules and their synthesis is polymer education. Informally, education occurs mainly through the written word-journals, books; visually-laboratory demonstrations, movies, slide presentations; and the spoken work-classroom presentations, short courses, seminars, lecture series, discussion groups. Education occurs both informally by discussion of research problems or solutions with fellow workers, reading the latest monograph and formally by attending classes, lecture series, etc. While elements of all of the above will be considered, emphasis will be placed on the academic aspects of polymer education.

Finally, the importance of polymer education cannot be separated from its importance in industry and the "real world" - yet unfortunately this is what has and often continues to be done.

Prior to 1940 there was little or no education directed toward polymer chemistry. With the contributions of Carothers and others that transformed the practice of polymers into a science, and with the leadership of Herman Mark and others (informal graduate and postgraduate), polymer education took root but rapid expansion and general acceptance by more traditionally oriented academic institutions was slowed by many barriers. Only since 1974 has there been a concerted effort to include education in the basic concepts of polymer chemistry into the undergraduate curriculum.

The development of polymer education will be divided into three time frames. First, Age of Empiricism - to 1938, where the key word might be art. Second, The Mark Connection - 1940 to 1974, where the key word is science. Third, Recognition - 1974 to present.

The Age of Empiricism - to 1938

The key word in the early years of polymer science is empiricism. Thus, Simon in 1839 made polystyrene but it was not until about 1937 with the recognition that high purity monomers were necessary to achieve desired polymer strength that the synthesis of industrially "satisfactory" polystyrene was achieved.

Much of the early development of science, including polymer chemistry in the USA, focused on "application" of natural materials to our needs - cotton, wood, flax, rubber, cottonseed oil, linseed oil, and wool.

Devoe and Raynolds (now a division of the Celanese Corporation) has been making paint in New York since about 1754, the decades prior to our independence. It may have been the oldest corporation in the USA.

The first successful factory was a cotton mill, under Samuel Slater, which began to produce on December 21, 1790. In 1814 the world's first factory to convert raw cotton into cloth by power machinery within the walls of one building was erected in Waltham, Massachusetts by the Boston Manufacturing Company. Lowell, Massachusetts, now the home of Lowell University which is a center of polymer education, was the first American mill city employing the Pawtucket Falls of the Merrimack River for power. Charles Goodyear's discovery of the vulcanization of rubber in 1839 was one of the most perplexing and exasperating, yet important, tales in U.S. history.

The first synthetic polymer utilized on a large commercial scale was Bakelite produced by Leo Baekeland, one of many foreign-born scientists who helped create American technology. By 1909 he had demonstrated a number of Bakelite articles to members of the recently formed Chemists Club of New York.

The earliest courses which considered polymeric materials dealt with them on an empirical basis, with recipes directing additions of specified amounts or treatments of materials being typical rather than the exception. These early courses concentrated on adhesives, oils and coatings, resins, textiles, paper and pulp and natural rubber. Specific examples are given in the following sections.

The Mark Connection - 1930's to 1974

The vast majority of polymer education focused on graduate and postgraduate education from the late 1930's to 1974. Early progress was delayed because of World War II.

For many chemists experimenting in such diverse areas as perfumes, natural products, drugs, steroids, dyestuffs, the end product included some waxy, gooey "glop". Without recognizing the potential of this "find", such scientists often mumbled vigorously four-lettered words which did not include "like" or "love". Even today this "passion play" is reenacted in laboratories throughout the "learned" countries. Yet this "glop" and "gunk" has often been transformed into today's plastics, elastomers and fibers - the basis for the largest, both money- and manpower-wise, subdiscipline of science, typically with a ladle turned by the Der Geheimrat, Professor Doctor Herman Mark, stuck within the "brimming soup of discovery and application".

Many of us date Professor Mark's contributions to polymer education and science as beginning in 1940 with his arrival at Brooklyn Polytechnic. This is only because we have not recognized the pages of scientific discovery prior to the coming of the Geheimrat to the USA.

After reading several accounts of Professor Mark's life I am tempted to tackle a biographical sketch of his life along with negotiating for him, with the Geheimrat's permission, of course, a contract to have it made into a motion picture. (I highly recommend to you the two articles written by Morton M. Hunt for the New Yorker in 1958. These are fast-moving, delightful and even true.) Even so, my task permits me to briefly recall only glimpses of his "candle which continues to burn so brightly on the stage of life". He was born Herman Francis Mark in Vienna, Austria on May 3, 1895, the oldest of three children. His mother was a Viennese Lutheran and his father a physician of the old family-doctor variety. Even as a child his zeal for life and zest for living life to its fullest was evident. He played tennis and soccer as often and long as his playmates could endure and then ran 2 to 3 miles about a cinder track and then two miles to reach home in time for supper.

He became infected with the love of chemistry through a visit to the chemistry laboratories at the University of Vienna when he visited it with a close friend, Gerhard Kirsch, whose father was a professor of chemical technology at the University. Gerhard and Herman assembled a fairly respectable assembly of glassware, chemicals, Bunsen burners and chemistry books during the next few years. At eighteen, Mark decided to get his compulsory military service term over before attending college. He joined the elite Alpine infantry unit, the Kaiserschutzen Regiment and in 1914 participated in the bitter battles of Lvov and Przemysl where he became highly decorated for bravery including a shrapnel wound in the ankle which required two months' traction. He then rejoined the Kaiserschutzen and served for three and a half years on the Italian front, rising through the ranks to lieutenant. In 1918 he was captured, leading to a "stranger-than-fiction" escape in 1919, coming home to his Vienna to learn that he was the most decorated company-grade officer in the Austrian Army.

In a little less than two years he completed a typically four-year task of achieving his doctorate with Wilhelm Schlenk, summa cum laude in 1921, while simultaneously courting a tall, blond young woman named Maria Schramek, who became in 1922 Mrs. Herman Francis Mark. He was instructor at the University of Berlin in 1921 and one year later joined the Kaiser Wilhelm Institut fuer Faserstaff-chemie in Berlin-Dahlem. During this time he met and became friends with von Laue who had discovered that when a narrow beam of X-rays was passed through a crystalline material, geometric patterns formed on the X-ray films. His first two of well over 600 publications appeared in 1922 in Berichte der Deutschen Chemischen Gesellschaft on "Nature of the Carbon Bond", "Free Pentaphenylethyl" and "Analogs of Pentaphenylethyl", both coauthored with Schlenk.

In 1923 one of Mark's superiors handed him a piece of hemp and asked that he study its molecular structure. Hemp was one of the many natural materials we now recognize as macromolecules which to then defied structural analysis. After much thought and work, also including X-ray studies, he concluded that the hemp fiber contained crystals. On examination, so also did cotton, wood and flax.

The acceptance of macromolecules was still hotly debated even though the work of Hermann Staudinger in the 1910's and the 1920's showed, at least to some, that macromolecules did indeed exist.

Gradually Mark and his colleagues determined that the cellulose molecules were, in fact, long chains of thousands of atoms making up hundreds of glucose units connected to form single polymeric molecules of cellulose. The significance of this work is not to be overlooked. This was one of only a few basic cornerstones upon which the giant of polymer science now rests.

In January 1927, the Marks were lured by Kurt Meyer, director of research for I. G. Farben, to be both a professor and head of a laboratory staffed by 20 scientists. He became a highly popular lecturer throughout Europe. While this popularity delighted Farben directors, his research efforts had not filled the company's tills. Within Mark's mind were pictures of branched, elongated, "zig-zagged", interlocked chairs and the persistent taunting regarding property-structure relationships.

During this time, publication number 50 came forth, his first book "Die Verwendung der Roentgenstrahlen in Chemie und Technik" was published. According to Maurice Huggins, "Mark contributed importantly to my own education (in X-ray science and in the German language) in the mid-1920's through this excellent book". Thus, Mark began early to contribute within all three areas of polymer education-teaching, research and service, the latter including publication of significant works.

Using insights contributed to by Mark, his laboratory group was steadily concocting new polymers, turning them over to the development department. Toward the end of 1929, word was rushed about the plant that one of these showed great promise. This polymer was formed from the common liquid styrene and could be transformed into a clear plastic from which cigar holders, toys, handles, etc. could be constructed. About a month later the first order for polystyrene was placed. The crisis was past.

During the next four years many additional new polymers were synthesized. Most proved of little commercial value, but the list includes three of the top-ranked popular plastic families - the polyvinyls used for phonograph records and floor tiles; polyacrylics (such as Lucite) used in paints, airplane windows and buna N and buna S, two versions of synthetic rubber. Thus few households are not affected by even his early contributions.

In October of 1932 the Mark family, which now included Hans, then three, and Peter, then six months old, returned to Vienna to escape the growing Nazi influence in Germany. Herman Mark took the post of director of the First Chemical Institute at the University of Vienna. With the pressures of filling the Farben coffers removed, he began to assess the science behind the polymers previously synthesized. During this time he also exhibited a common Mark trait of juggling numbers of different topics and interests seemingly simultaneously. He collaborated with Eugene Guth on a basic study of the nature of rubber elasticity, resulting in a book; made analyses for the university's medical school of natural polymers created by various infections; served as a member of Austria's Board of Education; and served as a member of the Agriculture Ministry's avalanche-warning commission, a post that permitted him to keep his muscles in top form. He also discovered that the glaciers contained higher percentages of D_2O than did newly-formed ice.

In March 1938 he was arrested by the Gestapo and, after a night of questioning, freed without his passport. After bribing a Nazi lawyer to retrieve his passport and having it inscribed with an authorization for him to visit Zurich, he began liquidating his savings by buying, piece-meal, strips of platinum wire which he bent into coat hangers. Shortly thereafter he, with his family, escaped carrying with them some family belongings including several dresses neatly hung from hangers.

Now in Zurich, he began contacting various chemical concerns eventually accepting a position with Industrial Cellulose Research, Ltd., Hawkesbury, Canada (60 miles west of Montreal) as research manager. He continued to publish (now numbering over 200) including 8 books and 18 patents through 1938. DuPont was instrumental in bringing Mark to the USA by offering him a liberal consultantship, promising to get him permanent residence and an adjunct professorship at Brooklyn Polytechnic. In May, 1940 he and his family settled into a roomy apartment in Brooklyn. It was from this home base that Herman Mark launched his attacks on behalf of polymer science.

As a side note, his contributions to society have been quite broad. During WWII, at the request of the Office of Scientific Research and Development, he assisted the Army in developing its DUKW, a sea-going truck, and Weasel, a snowmobile. He also assisted in the development of an unsinkable aircraft carrier made of ice which could heal torpedo wounds. He has been chairman of the wood-utilization committee for the UN and has been involved with numerous other socially-related programs, always bringing along his great energies, keen intellect, creativity and administrative ability.

Shortly after his arrival at Brooklyn Polytechnic, he and a physicist on the staff, Isidor Frankuchen, quietly hatched, unbeknownst to the higher echelons, a scheme to bring in money to the Polytechnic to assist faculty to travel, etc. They prepared an invitation to many of the chemical industrial personnel to attend, on the Brooklyn Polytechnic campus, an intensive two-week summer course on polymer science and X-ray diffraction analysis, unhindered by the usual formalities of prerequisites and credentials. Twenty scientists enrolled, leaving the two entrepreneurs with a profit which was utilized to help other faculty to travel to scholarly meetings. Each summer the enrollment grew until, by 1945, their impromptu school was netting several thousands of dollars. The higher administration was probably aware of the goings-on from the beginning, but by 1945 President Harry Rogers summoned Mark and Frankuchen into his office to discuss this venture. They talked of the traveling fund operation and Rogers warmed to the idea. He not only encouraged further development of the operation, but relieved them of the routine details incident to operation of the summer program, making it a part of the normal University operation.

During this time the developing team of polymer scientists at "Brooklyn Poly" began giving weekly seminars which attracted industrial and academic chemists from the surrounding greater New York City Metropolitan area. The lectures were given by graduate students, faculty and visiting scientists from many different coun tries. The Saturday seminars were a welcome change of pace since, because of the war effort, most chemists had to work on Saturdays with the seminars being a recognized legitimate alternative by the industrial superiors. Also they were free and "after speech" interchanges were encouraged. A Thursday night series also emerged. Thus the critical bridge between academics and industry was being built - a bridge which has continued to grow but which is currently in great need of repair and enlarging to support the needed ongoing revitalization in the growth of polymer science.

Mark was also proving to be quite a magnet as an academic research adviser, adding five professors and 60 graduate students to the Polytech fold.

To Mark, it became plain that the polymer activity at Brooklyn Poly needed to be highlighted to gain the support and recognition it deserved. This time he approached President Rogers with his plan. On paper the establishment of an institute to study polymers would cost about two hundred thousand dollars annually, but Mark eagerly contended that the staff, equipment, etc. were already present so the "start up" costs would be limited to ordering new letterheads. Rogers was reported to have muttered for several days, at intervals, "Genius! Sheer genius!" Thus the Polymer Research Institute officially came into being with the printing of the letterheads.

Along with the need to "spread the news" of polymer science was the increasing need to develop an adequate vehicle which would focus on the dissemination of generated experimental results in polymer chemistry. In 1940 Drs. Maurits Dekker and E.S. Proskauer of Interscience Publishers consented to begin the publication of a series of Monographs on polymer chemistry, "High Polymers", with H. Mark and G.S. Whitby as editors. The response was so positive that in 1946 the publication of the Journal of Polymer Science under the editorship of H. Mark and P.M. Doty began. The contribution by Herman Mark to polymer education is clearly seen in the growth of both literary series as the backbones of the current polymer literature. The Journal of Polymer Science is the most popular journal of polymer science today. It has subdivided many times and "fathered" such Journals as the Journal of Applied Polymer Science and Polymer Letters. (Today there exist two additional primary US based polymer journal groupings - Macromolecules - ACS and the Journal of Macromolecular Science and associated journals - M. Dekker and two major ACS divisionally sponsored volumes - Polymer Preprints and Organic Coatings and Plastics Chemistry. Additionally there are numerous other polymer related journals such as Polymer News, Plastics Technology, Plastics, Plastics Industry, Textiles, Modern Coatings, Polymer Engineering, SPE, Plastics World, Modern Plastics, etc. which also contribute to a healthy contribution by the USA to the polymer literature.)

It is of note that the arrival of Mark into the Du Pont sphere of influence coincided with the emergence of a midwestern bred and trained chemist, Wallace H. Carothers, as director of Du Pont's polymer research. The work associated with Mark and Carothers signaled the break from the empirical practice of polymer chemistry and the birth of the science of polymers. Carothers directed the research group which on October 27, 1938 publicly announced the synthesis of a synthetic polymer which, for the first time in history, had properties superior to natural fibers. The polymer was nylon.

The science of polymer chemistry was born, an infant, when most of its brother and sister scientific disciplines were already adults.

The beginnings of academic polymer education in the USA are unclear at best because, as with most sciences, polymer chemistry emerged, grew from a number of sciences - science and art associated with fabrics and textiles (making of boat sails, sheets, clothing), coatings (oil based), adhesives (rosins, glues) and rubbers. For instance a course in paint technology was started in 1906 at North Dakota State University. Other courses in textiles, coatings and cellulosic materials were also being taught prior to 1920. These early courses were quite applied and might be referred to as courses in technology and not science if such a distinction can in fact be made (technology is applied science). Considerable subject matter now termed polymer chemistry was contained in a course designated as "Colloid Chemistry" at the University of Delaware by E.O. Kraemer as early as 1941. This was followed by the first courses taught under titles involving polymer chemistry taught in 1944 by G.E. Landt in the chemical department. John Delmonte taught courses in plastics at the Armour Institute of Technology in 1937-39 and expanded these efforts to courses in the Plastics Industries Technical Institute (1940-50). Paul O. Powers taught courses in plastics under the auspices of Pennsylvania

State College of Franklin and Marshall College (1939-44) and published the first book for use in colleges (Synthetic Resins and Rubbers, John Wiley, 1943).

The early focus of polymer related education was then also largely empirical. The first formal course in polymer chemistry which balanced theory with practice was probably one associated with Herman Mark given at the Polytechnic shortly after his arrival. Ray Seymour offered such a course at the University of Chattanooga in 1945.

By the middle 1940's the Polytechnic had attracted a number of outstanding "polymer" chemists (Herbert Morawitz, Turner Alfrey, Paul Doty, Bruno Zimm, Kurt Stern, A. McLaren, Fred Eirich and Charles Overberger) under the directorship of Herman Mark. A program of graduate study leading to a M.S. and Ph.D. degree with a major in polymer chemistry was developed, undoubtedly the first such program in the USA. Course offerings were:

Introduction of Polymer Chemistry (Mark)
Polymerization Kinetics (Alfrey)
Solution Properties of High Polymers (Elrich)
Organic Polymer Chemistry (Overberger) and
Chemistry of Proteins (McLaren)

Compared to more classical areas of chemistry there was little formal education and academic based research related to polymer chemistry through the late 1950's although there existed a small number of individuals and groups who were conducting important pioneering work including Carl Marvel and F.T. Wall at the University of Illinois, Peter Debye and Paul Flory at Cornell, Ray Seymour at the University of Houston and elsewhere, Malcolm Dole at Northwestern University, James Long, Ray Myers and Al Zettlemoyer at Lehigh University, Walter Stockmayer at M.I.T., R. Guoss at Yale University, A.V. Tobolsky in Princeton; F. d'Alelio and E. Guthin at Notre Dame, Richard Stein at the University of Massachusetts, Charles Price, University of Pennsylvania; John Ferry, University of Wisconsin; George Butler, University of Florida; Leo Mandelkern, Florida State University, etc.

In the mid to late 1940's polymer chemistry was "raised" to the status of optional material for certain academic courses in chemistry. Polymer chemistry was generally considered to be within the realm of organic chemistry and only a few ventured the inclusion of polymer topics in physical chemistry and fewer still considered it beyond these two areas.

A few "staunch believers" did give polymer chemistry a realistic portion of class time and these "adventurers" were generally well rewarded with a stream of good, interested graduate students. Carl Marvel relates, "At Illinois I did work in polymer chemistry and I always spent approximately four to six weeks on synthetic polymer chemistry as part of the second semester of organic chemistry which I taught, and that was started along about 1940..."

Industry, seeing a need for training those employed by them in the art and science of polymer chemistry, began to experiment. Certain groups with common research interests and needs joined together to form societies which in turn offered short courses typically in areas of very specialized applications. Societies active in these ventures include SPE, PRI, ACS, APS and SPI. Several companies such as Hercules, Monsanto, and DuPont sponsored or permitted informal courses in polymer related topics to be presented during company time and/or using company facilities. The "teachers" or discussion leaders could be either one of the participants or an invited academic or industrial "expert". These "in-house" courses have increased in popularity and today are part of many companies' schedules of activity.

These "sorties" into polymer chemistry continue today with a greater number of academic institutions acting as nucleating agents around which series of short courses on related topics are offered.

While a number of introductory or comprehensive texts dealing with polymer chemistry were written, the most influential was probably Paul J. Flory's textbook "Principles of Polymer Chemistry", published in 1954. No prior knowledge of polymers was assumed with particular chapters directed at the beginner. It also contained much information useful to the experienced investigator. A wealth of experimental data was included to illustrate the applicability of the presented concepts and conclusions. Admittedly missing are topics related to the mechanical properties of polymers and to the application of polymers in industry - i.e. fabrication, synthesis, etc. Even so Flory's text is a landmark book in science.

A dominant textbook of the 1960's, and one which even today is possibly the most utilized introductory polymer chemistry textbook, was Fred Billmeyer's "Textbook of Polymer Chemistry" (the 1961 and subsequent editions were named "Textbook of Polymer Science"). Billmeyer's text contains portions on most aspects of polymer chemistry from synthesis, to characterization, to fabrication. Other significant texts include "Introduction to Polymer Science" by Raymond Seymour, 1971 (McGraw-Hill) which was superseded by "Polymer Chemistry" by Ray Seymour and Charles Carraher, 1980 (Marcel Dekker), "Polymer Chemistry" by Bruno Vollmert, 1973 (Springer-Veriag); "Introduction to Macromolecules" by Leo Mandelkern, 1972 (English Universities Press); "Introduction to Polymer Chemistry", John Stille, 1962 (John Wiley); "Organic Polymers" by Turner Alfrey and Edward Gurnee, 1967 (Prentice-Hall); "Organic Chemistry of Synthetic High Polymers" by Robert Lenz, 1967 (John Wiley); "Polymer Chemistry" by Malcolm Stevens, 1975 (Addison-Wesley); "Principles of Polymer Systems", by F. Rodriguez, 1970 (McGraw-Hill); Polymers: "Chemistry and Physics of Modern Materials", by J.M. Cowle 1974 (In Text); and "Principles of Polymerization" by G. Odian, 1970 (McGraw-Hill).

Much informal postgraduate education involved reading books written by those within a particular research area. Such early (pre-1953) books include T. Alfrey, J. Bohrer and H. Mark, "Copolymerization", Interscience, 1952; C. Ellis, "The Chemistry of Synthetic Resins"; T. Alfrey, "Mechanical Behavior of High Polymers", Interscience, 1948; R. Boundy and R. Boyer "Styrene-Its Polymers, Copolymers, and Derivatives", Reinhold, 1952; H. Mark and H. Tobolsky, "Physical Chemistry of High Polymeric Systems", Interscience, 1950; C. Bawn, "The Chemistry of High Polymers", Interscience, 1948; G. d'Alelio, "Fundamental Principles of Polymerization", John Wiley, 1952; R. Houwink (Ed), "Elastomers and Plastomers", Elsevier, 1948; R. Burk and O. Grummitt (Eds), "High Molecular Weight Organic Compounds", Interscience, 1949; and A. Schmidt and C. Marlies, "Principles of High-Polymer Theory and Practice", McGraw-Hill, 1948.

Many of the editors and authors of these books and numerous research and review journal articles were from industry and represent a continued, critical commitment and investment by industry in postgraduate education. Others are from outside the USA emphasizing the necessary cross-fertilization of scientist and scientific knowledge critical in the development of a science.

The American Chemical Society has had a great deal to do with fostering polymer education through several of its Divisions and the Journal of Chemical Education. In 1919 the Division of Rubber Chemistry (then classified as a Section) was organized as the ACS's ninth Division. In the early 1920's the Scientific Section of the Paint Manufacturers of the United States and the National Varnish Manufacturers Association provided the impetus for bringing paint chemists together to discuss mutual problems and to provide a forum where papers could be presented and discussed. The first of these meetings occurred in June 1922 in Washington, D.C. and was organized by Henry Gardner. After a second meeting (1923) the subject of affiliation with the ACS was considered with H.A. Gardner appointed as a committee of one to confer with ACS officers. On October 20, 1923, W.T. Pearce sent to the ACS Secretary Parsons a petition with about 100 signatures. On December 21, 1923 Pearce was notified that he had been appointed secretary and Gardner was chairman of the Section, eventually called the Paint and Varnish Division, currently known as the Division of Organic Coatings and Plastics Chemistry. Other polymer related Divisions formed including the Divisions of Colloid and Surface Chemistry; Biological Chemistry; and Cellulose, Paper and Textile.

The ACS High Polymer Forum, similar to today's Macromolecular Secretariat, with the help of C.S. Fuller and A.C. Elm, began meeting at the 1946 Atlantic City ACS meeting. Through the efforts of the Polymer Forum, the Division of Paint, Varnish and Plastics Chemistry and others in 1948 a group of university and industry chemists petitioned the American Chemical Society to establish a separate division of Polymer Chemistry. After a probation period of three years the Division of Polymer Chemistry was formed, which today is the second largest division in the ACS.

Through 1959 there existed only one large concentration of polymer chemists in academia (at the Polytechnic Institute of Brooklyn) with several smaller academic centers featuring concentrations in textiles, celluloses and wood, rubber and coatings. Series of short courses, seminars, journals, monograms and textbooks had been established and were to grow through the 60's to the present.

The 1960's saw an increase in the number of formally designated polymer or macromolecular institutions. What I believe to be an unhealthy trend of including polymer chemists and polymer chemistry in engineering colleges and departments of chemical engineering rather than integrating within departments of chemistry had begun. For example, the comprehensive programs at Case-Western Reserve, The University of New York at Buffalo, Washington State Universiy, University of Massachusetts, North Carolina State University all began in the 1960's and all were headquartered within engineering colleges. The danger is that if the academic institutions sense they can hide, segregate or ignore polymer chemistry with selected token institutions, then it will be far more difficult to integrate polymer science into the curriculum of our academically trained chemists - most of whom will eventually end up in polymer related research.

Some of the older polymer institutions grew, such as the rubber and polymer based institute at the University of Akron, which was formed in the 1940's, and where Maurice Morton took the Directorship in 1954. Others remained essentially the same and still others slipped. "Brooklyn Poly", the center of academic polymer chemistry in the USA in the 1940's and 1950's remained as such but became less dominant because a number of their excellent staff left to found other institutes. For instance, Charles Overberger, who rose to Dean of Science and Director of the Polymer Research Institute, left in 1967 to take the chairmanship of the Department of Chemistry at the University of Michigan where he helped found their macromolecular based institute.

The beginning of the 1970's saw a continuation of the academic apathy towards formal training in polymers which marked the 1950's and 1960's. This topic was hotly debated by ACS presidents at ACS meetings (for instance 1-3), within workshops (for instance 4) and within the inner circle of the ACS (for instance 5). Still significant steps towards actually achieving advances in the formal inclusion of polymers in the academic training of chemists had to wait for a beginning until 1974.

In summary, the period from the 1940's through 1973 saw a steady, healthy increase in postgraduate education spearheaded by academics, industry and government and a slower, steady increase in the number of schools offering graduate education in polymer chemistry, but the latter was clearly undernourished in comparison with the relative importance of polymer chemistry in industry. By the 1950's, polymer chemistry had largely made the transition from a largely empirically-based branch of endeavor to a strong, well-developed scientifically-based major branch of science.

Recognition - 1974 to Present

While previous ACS presidents had called for greater emphasis in the academic training of chemists, it was during William Bailey's presidency that the call was heeded. Significant inner committee assignments were being filled by those sympathetic with the cause. Further, the Polymer Division's Education and Public Relations Committees were primed for action.

The formulation of an overall plan to achieve the needed greater emphasis of polymer education was begun at a meeting at Atlantic City of the ACS in 1974 at the direction of Eli Pearce and Joe Salamone who called myself and others under the auspices of the Division of Polymer Chemistry's Committees on Education and Public Relations. It was agreed that not only was graduate education in polymer chemistry lacking the emphasis it deserved and that previous attempts aimed at increasing the emphasis of polymers in graduate education had received only minimal success, but that there was almost no emphasis on undergraduate education related to polymer chemistry. Rather than repeating the same programs that had received minimal success aimed at graduate education, we were convinced that the solution to the problem of polymer education throughout the educational levels would better be solved if we focused our efforts on undergraduate education which would result in a much easier upward flow of interest in polymer chemistry if we were successful at the undergraduate level. Further, we realized that we needed a large-scale, well-orchestrated approach with the cooperation of all of the sub-polymer disciplines and their many members. That week I contacted Lieng-Huang (Sam) Lee, Clara Carver and George Brewer concerning our proposed emphasis. Their support was quite positive asking me to serve as Education Liaison and to compose an Education Committee for the Division of Organic Coatings and Plastics Chemistry.

Of note was (and continues to be) the efforts of the Joint Education Committees of the Divisions of Polymer Chemistry (chaired by Eli Pearce; presently chaired by Guy Donaruma) and Organic Coatings and Plastics Chemistry (chaired by Charles Carraher). Most of the accomplishments noted following could not have been accomplished without the joint cooperation of these two committees which acted as one "super" committee.

The period between the Atlantic City meeting and the Philadelphia meeting was very active and directions for achieving our goal were cemented. The major steps were (a) preliminary investigation aimed at obtaining NSF support for summer short courses in polymer chemistry for college teachers of undergraduate general, organic, analytical and physical chemistry; (b) generation of a laboratory manual for use in undergraduate courses of general, organic and physical chemistry -using polymer exercises to illustrate chemical principles usually illustrated utilizing non-polymers, (c) contacting Stanley Kirschner and the ACS Committee on Education concerning support for the greater emphasis of polymer chemistry in undergraduate curriculum, (d) contacting Theo Ashford and the Examinations Committee offering the services of polymer chemists to serve on examination committees (Les Sperling has served admirably as our Liaison in this behalf), (e) working with John Howard and the Committee on Professional Training (ACS) to include a course in introductory polymer chemistry as a recommended advanced course for the ACS - Approved Major, and (f) working with "popular" chemistry journals toward greater inclusion of polymer related topics.

A preliminary proposal related to the summer short courses was drafted by Eli Pearce and myself. This was submitted to the NSF Education Office of Experimental Projects and Programs in November of 1974. Papers were also prepared and presented to members of the ACS Committee on Professional Training and the Committee on Education. During this time, Bill Bailey and others were active in our support. In December 1974, we received from NSF a request for a formal proposal.

The Council Committee on Chemical Education went on record at the Philadelphia meeting favoring greater emphasis of polymer chemistry in the undergraduate training of chemists. Through the efforts of many, the Committee on Professional Training voted to include a course in polymer chemistry in the ACS approved major in the 1978 publication. This topic will be considered later in greater detail.

At the Philadelphia meeting, Ray Seymour and I met with the ACS Examination's Committee. We were asked to generate answers to a number of questions in preparation for a decision by them to support or not support an ACS standardized test in polymer chemistry. To gather information needed for this and other projects, Rudy Deanin surveyed over 110 departments offering courses in polymer chemistry assessing the potential need and use of a standardized test in polymer chemistry and evaluating topics which should be included within an introductory polymer course. Shalaby Shalaby and Eli Pearce were asked to edit an issue of CHEMISTRY devoted to polymer chemistry.

During this time the first ACS Interaction Series - "Polymer Science and Technology - An Interdisciplinary Approach" - under the direction of Eli Pearce, Shalaby Shalaby and Garth Wilkes was under preparation.

I met with the ACS Examinations Committee at the Chicago ACS meeting and presented the results of Deanin's survey. I was asked to formulate and chair a testing committee for the purpose of generating a standardized test.

The results of the Deanin-Carraher survey are of interest because they were used to help "drive" the proportion of questions within given polymer areas. They also show a fairly good commonality of desired course content for an introductory course in polymer chemistry, something that was hoped for but something that was not known.

The Ninth Biennial Education Conference of the ACS met on October 16-18, 1975 and strongly endorsed the need for increased emphasis of polymer education.

At the ACS meeting in New York City, two pretests were constructed for use. This was done with the considerable help of Charles Gebelein, Les Sperling and Angelo Volpe utilizing Deanin's survey of topics as a guide. The ACS, through the active support of the Divisions of Polymer Chemistry and Organic Coatings and Plastics Chemistry, helped in the generation of this examination. Over 60 contributors from schools and industry directly helped in such areas as pretesting and submission of questions. Thus, the generation of the first ACS Standardized Examination in polymer chemistry was accomplished with help from a wide base. The examination has sold about 2,000 copies during the first year since its publication in 1978. Considerable interest has been shown by chemical societies from other nations in the examination and the ACS Committee on Examinations is currently considering permitting its use outside of the USA. This examination should assist in generating additional commonality in introductory courses. Some diversity is important to reflect the breadth of polymer science - textiles, natural materials, rubbers, plastics, materials engineering and processing, etc. Thus, while there is an aim towards standardization of content and topics, it is generally accepted that the amount of this "standardization" should only be in the 70% to 80% range.

The NSF proposed Short Course in Polymer Chemistry died a death of old age since by the time it emerged from CEPACC in 1977, the climate at NSF had changed so that teacher retraining programs were no longer in fashion even though a strong case was made that this was of such a strong need and of sufficient worth to the USA that an exception should be made. The Divisions of Polymer Chemistry and Organic Coatings and Plastics Chemistry still believe in this program and are currently looking towards industry to help finance several mini-short course projects. Two short courses on Polymer Crystallinity and Rubber Elasticity will be offered at the New York ACS in 1981. The short course on Rubber Elasticity will be held just prior to a symposium covering the same topic. This is directed by Frank Harris and is free. In the fall of 1979, the Columbus Section of the ACS offered a short course series, similar to that outlined in our proposal, illustrating that such programs can be conducted locally, using local "talent".

Probably the most significant single event in polymer education occurrred in 1978. The latest edition of "Undergraduate Professional Education in Chemistry: Criteria and Evaluation Procedures" by the ACS Committee on Professional Training contains "In view of the current importance of inorganic chemistry, biochemistry, and polymer chemistry, advanced courses in these areas are especiallly recommended and students should be strongly encouraged to take one or more of them. Furthermore, the basic aspects of these three important areas should be included at some place in the core material." After almost thirty years as a king of the sciences, polymer chemistry has been recognized as essential core material in the training of all ACS accredited undergraduate majors.

The full impact of these new provisions is yet to be fully recognized. What is being advocated is both that polymer chemistry is being recommended as advanced work but possibly of greater importance, that "basic aspects" of polymer chemistry be included in the core material. The education committees of a number of divisions and societies associated with polymer science are working towards adopting recommendations involved with these two major related points.

The establishing of a suggested syllabus has been undertaken with the leadership of Ray Seymour. The form is to note the topic and suggest percentages of course time suggested for this broad topic.

The content, level and duration of the introductory course is actively being evolved and evolution will continue, hopefully with the active assistance of those teaching introductory courses. There is a strong feeling among those in many of the education committees that the introductory course should contain a laboratory portion which may include both synthesis and characterization of a polymer. This laboratory portion should emphasize topics covered in the lecture portion giving students first-hand experience in the "Joy and trauma" of polymer science (such as polymers typically taking several days before becoming dissolved).

Seven committees have been named to (1) establish topics which would be useful, applicable and suitable for introduction within each of the traditional undergraduate core courses of General Chemistry (Art Campbell), Inorganic Chemistry (Norm Miller), Organic (Charles Gebelein), Analytical (Roger Hester), Biochemistry (Raphael Ottenbrite), Physical (Les Sperling), and Chemical Engineering (F. Rodriguez); (2) establishing general guidelines as to the level and depth of coverage of these topics; (3) generation of specific illustrations which could be directly used by teachers of a specific course; (4) suggesting general, broad guidelines regarding proportions of time to be spent in polymer related topics within a given course. To be emphasized is "substitutive" type of information - i.e. replacement of examples, etc. which typically utilize nonpolymer related material with polymer related material which will "enhance" the "covered" topic. Results of these studies are being coordinated through a central committee made up of the core course committee chairmen, Guy Donaruma, Eli Pearce and Charles Carraher. Results should become available in 1982.

Thus every effort is being made to permit the fulfillment of the newest ACS guidelines with regard to the upgrading of polymer topics to core material within the academic training of chemists in ACS accredited programs.

A biyearly polymer newsletter which will go to every academic college within the USA will shortly be published under the editorship of Charles Carraher, Shalaby Shalaby, Guy Donaruma and Eli Pearce.

A number of timely surveys have been completed. The Industrial Relations Committee under Charles Gebelein recently completed a survey of 60 of the top (president, chairman of the board, etc.) officials within chemical related industries regarding what courses they would like to see undergraduate students take above the present core courses. The survey showed a surprisingly strong trend toward the feeling that more polymer science should be included in the academic training of chemists and chemical engineers. Twenty courses were listed on the original survey along with the category "other". Essentially all the respondents (59 out of 60) recommended that students should take a general course in polymer chemistry. Six of the top seven top rated courses dealt with polymers (Polymer Laboratory - 72%; Polymer Processing - 72%; Engineering Polymers - 60%) with only Advanced Organic (63%) breaking into the top 5. The second five were Emulsion Polymerization - 50%, Elastomer Chemistry - 48%, Advanced Physical - 45%, Instrumental Analysis - 42% and Paints & Coatings - 38%.

A second question asked in the survey measured possible ways the administrators believed their companies could best promote closer academic-industrial relationships. Following are results relating topic, %-useful and, lastly, %-which would actively support that mode: Industrial Speakers for Academic Seminars - 88%, 45%; Academic Speakers for Industrial Seminars - 53%, 28%; Industrial-Academic Workshops - 40%, 12%; Additional ACS Symposia -22%, 8%; Short Courses in Industrial Location - 43%, 15%; Short Courses at a College - 42%, 17%; Short Courses at National ACS Meeting - 22%, 5%; Sabbatical and/or Summer Research Programs -38%, 18%; Joint Research Projects - 43%, 18%; Grants or Contracts for Applied Research - 35%, 13% and Grants or Contracts for Basic Research - 27%, 10%.

An updating of the Salamone, Deanin, Young and Pearce 1972(6) survey to determine trends with academics regarding polymer training has recently been completed under Rudy Deanin. The previous study showed about 110 schools offering some polymer science. The recent survey shows about 135 now offering polymer science, an increase of about 25% over eight years. Further, the trend towards offering polymer science has greatly increased since 1978 and it is expected that a doubling to about 350 schools will occur by 1985.

Thus the work begun by Herman Mark and others in the late 1930's and early 1940's is greatly accelerating and every indication is that this growth toward almost universal acceptance of polymer topics within the core academic training of chemists and chemical engineering will continue, but even at a faster pace.

Following are other current and recent-past ventures related to polymer education. A series of experiments appeared in the 1978 Journal of Chemical Education which is aimed at introducing polymer related experiments into traditional courses (7). More of this type of presentation must be forthcoming.

A visitation program to help support visits by those already involved in polymer education to colleges that are not involved but which desire to learn about polymer education related programs has begun. Lon Mathias chairs this committee.

NSF funded polymer science modular experiments are now available from Eli Pearce. There are 15 of these modules for use with an introductory polymer science course or for use with a separate beginning polymer laboratory course.

There is an increased awareness of the place in history of science, particularly polymer science where many of those who contributed the basic building blocks are still alive. Ray Seymour is doing a series on the Pioneers of Polymer Science for Polymer News. Fred Eirich has been asked by the Journal of Chemical Education to write a history of the development of the "macromolecular concept." (As a side comment, those associated with the Journal of Chemical Education such as Tom Lippincott, have been "friendly" towards polymer chemistry and have assisted in the attempt to bring polymer chemistry into the mainstream of material presented to both the teaching staff and chemistry students alike).

Recently, the Education Committees of the Polymer and Organic Coatings and Plastics Chemistry Divisions initiated a tape cassette program asking a number of pioneers of polymer science to recite their thoughts, philosophies, recollections, etc. not to form an oral history, but rather so in years to come teachers can play portions of the tapes to their students and say, "This is what Carl Marvel says about . . . in his own words." This program is ongoing. V. Harry DuBois has also compiled a series of cassettes by plastics pioneers and these are displayed in the Smithsonian Institute in Washington.

Polymer News, edited by Gerry Kirshenbaum has as a long-standing series featuring academic polymer centers. Those associated with the particular institution are asked to present their programs, aspirations, intents, etc. These are typically well done. Polymer News also features a column on Polymer Education which offers a wide variety of information including booklets, tapes and films related to polymer science, special education (academic and inservice) experiments, polymer science education outside of the USA, recent activity in education committees, featured polymer chemists in academics, reports on education related conferences, etc.

Under the leadership of Eli Pearce and Shalaby Shalaby, and the sponsorship of the Divisions of Polymer Chemistry and Organic Coatings and Plastics Chemistry, an entire issue of CHEMISTRY was devoted to polymer chemistry (Vol. 51, No. 5, June of 1978). Several thousand extra copies were purchased for a wide variety of uses by academic and industrial organizations for distribution to youth and the general public.

Two national ACS related symposia with polymer education were conducted in 1979. Both were co-chaired by Eli Pearce and Charles Carraher. The first occurred during the Hawaii ACS meeting and was initiated at the invitation of Robert Brasted on behalf of the Division of Chemical Education and jointly sponsored by the Divisions of Organic Coatings and Plastics Chemistry and Polymer Chemistry. Topics were centered about helps and introductory course contents. The second symposium occurred at the Washington, D.C. meeting co-sponsored by the same three Divisions. This was a full symposium with about 20 talks given varying from how to introduce polymer topics into undergraduate courses of inorganic, organic and physical chemistry, to those noting the importance of polymers to society and to the proper education of chemists and to special projects related to polymer education.

A summer scholarship program has begun under the sponsorship of Proctor & Gamble. The program resulted in about 500 inquiries with eventually six summer scholarships awarded to support work on polymer related research projects. Flyers were mailed to about 1200 departments of chemistry. Such propagation is worthwhile in advertising the prospects that await those who desire to enter industry in polymer-related ventures. It is hoped that this program will grow.

Several awards related to polymer education are in the "wings".

Thus the present activity within the various polymer education committees is broad, far-reaching, timely and progressive.

In summary, the 1970's began with the same academic passiveness toward polymer education that marked the 1950's and 1960's, but in 1974 major developments aimed at the greater inclusion of polymer chemistry into the curriculum of chemists began and for the first time emphasized the undergraduate training of chemists. The most significant single event was the ACS pronouncement that aspects of polymer chemistry were to be core material in the training of ACS accredited chemistry majors.

Summary

Polymer education in the USA can be divided into three periods - Empirical, Mark Connection and Recognition with the latter period still in effect. We must continue to be active at promoting the greater inclusion of polymer topics, illustrations and concepts into the formal education of our chemists and chemical engineers not just because we are polymer chemists but because it is critical to the growth of our industries, our nation, as well as many related branches of science, health and technology.

The Geheimrat continues to be active in all the aspects of an educator - teaching, service and research. He has been a constant source of help and support during the present period, working behind the scenes in our behalf, counseling and prodding.

Polymer education has grown from a seed at "Brooklyn Poly" to a young giant redwood which is taking its rightful place as a giant among giants. This growth has been catalyzed by the Geheimrat, not only from his ivy-covered walls in Brooklyn, but from a constant preaching, as a possessed evangelist of the "polymer gospel" at almost any school or institution which would invite him. In the earlier years he would arrive by bus, deliver his message of "hope and light" and after an appropriate visitation with those present, leave - his apppearance costing no more than bus fare. His guidance continues as he continues his "camp meetings" trying to "save the lost" and "comfort the saved" often speaking of future needs, future solutions - the future - indeed a man for ALL seasons. A teacher, servant of mankind, editor, chemist, student, father, husband, author, administrator, organizer - a futurist, helper and friend.

We thank you for your help, guidance and friendship.

Acknowledgments

I thank the following for their direct contributions in the preparation of this paper: Charles Overberger, Ray Seymour, Paul Flory, James Scott Long, Zeno Wicks, Maurice Huggins, Al Zettlemoyer, Walter Stockmayer, Fred Elrich, Fred Billmeyer, Malcolm Renfrew, Carl Marvel and Malcolm Dole; and Edgar Hardy for suggestions as to how to develop the presented material.

I particularly thank Eli Pearce for being my "comrade-in-arms" for the past six years without whose energies much of what has been accomplished within those six years in polymer education would not have been accomplished.

Finally, I thank those many pioneers of polymer science on whose shoulders we proudly and firmly stand and those currently involved in polymer education for closing ranks and offering the needed solidified stance for the recognition and development of polymer education.

Literature Cited

1. J. Chemical Education, 45(8), 498-506 (1968).
2. J. Chemical Education, 42(1), 2-27 (1965).
3. M. Morton, J. Chemical Education, 50(11), 731 (1973).
4. F. Eirich and M. Williams, Eds., "Polymer Engineering and Its Relevance to National Materials Development, A National Science Foundation Workshop," Washington, D.C., Sec. 5, 6, 1972; Salt Lake City, University of Utah Press Page 1x.
5. Chemical and Engineering News, August 14, 37(1972).
6. J. Salamone, R. Deanin, M. Young and E. Pearce, J. Chemical Education, 50(11), 768(1973).
7. C. Carraher, J. Chemical Education, 55, 51, 269, 473 and 668(1978).

RECEIVED February 5, 1981.

12

The Interfacial Macromolecule

F. R. EIRICH

Polytechnic Institute of New York, Brooklyn, NY 11201

The now so popular retelling of developments of ideas which led to a discovery or new understanding, may not serve a wider purpose than that of the historian, or our general interest in the drama of the growth of an idea and the dynamics of its formation. Still, it is quite possible that the account of the many false starts, the many blind alleys or alternative routes to a solution, in the unpredictable ways of intellectual connections, might stimulate some readers to take-off on a trip of their own. The following, then, is the story of a development starting from an afternoon conversation with H. Mark and still winding its way today through the minds of many investigators.

During the first years of the 1950's at the Polymer Institute of the Polytechnic Institute, Brooklyn, N.Y., and following earlier work on the viscosity of colloids and macromolecular solutions, one of my interests centered on the magnitude of the volume affected by the macromolecular coils in solution as expressed by their viscosity. W. Kuhn (1) almost twenty years earlier had shown that the "flow-through" resistance of the solvent is so high that the fluid will move around rather than through a coil. Later we had shown with models how the viscosity of a suspension containing rigid rod-like particles is lowered when the particles become flexible (2). The "flow-through" resistance had just been calculated for a special model by Debye and Bueche(3), while Scheraga and Cerf (4) had

0097-6156/81/0175-0143$5.25/0

pointed out that part of the viscosity and flow birefringence of macromolecular solutions was not only due to orientational, but even more to deformational contributions from the coil, a point later elaborated by Peterlin(5).

When we compared the viscosities of solutions of natural rubber and of guttapercha and of other elastomers and later of polyethylene vs.(poly)cis-butadiene, with such bulk properties as moduli, densities, X-ray structures, and adhesiveness, we were greatly helped in understanding these behavioral differences by the studies of Wood (6) on the temperature and stress dependent,melting and freezing,hysteresis of natural rubber, and by the work of Treloar (7) and of Flory (8) on the elasticity and crystallinity of elastomers on stretching. Molecular symmetry and stiffness among closely similar chemical structures, as they affect the enthalpy, the entropy, and phase transitions (perhaps best expressed by $T_m = \frac{\Delta H_m}{\Delta S_m}$, and by Clapeyron's equation), were obviously as powerfully behavior determining factors in macro - as in smaller molecules, for which such steric variants as succinic, and maleic and fumaric, or stearic versus oleic and elaidic acids have been school book examples for a long time.

Working with less dilute solutions of elastomers one cannot fail to notice the influence (the stiffer the greater the effect) of molecular structure on the onset and course of non-Newtonion flow, on gelation and swelling, and the influence of the solvent as expressing itself by virial coefficients, molecular dimensions in solution, spinnability, and film forming. The sensitivity with which the tack of adhesives, demonstrated by pressure sensitive tapes which at that time reached the market, depends on the structure and composition of the elastomer was similarly striking and raised the question, which molecular structure or state was best suited to exhibit tacky adhesion, or adhesion per se.

From my work in surface and colloid chemistry, I was aware of the energetics and mechanics of wetting and spreading. It followed that cis-compounds with their greater entropy of fusion and flexibility must always spread better and stay tackier on the surface under a given set of conditions. I discussed this with Dr. Mark during an afternoon stroll in 1951. He agreed and also pointed out that cis-chains with their more convoluted coils would form thicker surface layers than the stiff rods of the trans forms and would not contain their ordered shear planes. At that time also data just published by Jenkel and Rumbach (9), and some earlier work of other authors (10), showed that polymers were adsorbed from dilute solutions in quantities amounting to 10 to 20 monolayers of monomeric units which indicated either a dense, semicrystalline packing or, more likely, long segments or loops dangling from the adsorption sites. When these macromolecular coils collapse on removal of the solvent, or are being applied in bulk (melt) to the surface, the wetting should then occur by partially adsorbed molecules which extend into the adjacent phase. Surface layers thus structured should assist adhesion. In view of the great importance of adsorption and adhesion for a wide range of natural phenomena, including biological, pharmaceutical and industrial processes, I undertook a study of adsorption that was to continue for over 20 years (II). Exploring the adsorption of macromolecules and its underlying mechanism has since, because of its implications (see Table I), attracted many other workers and become a major field of theoretical and experimental study.

I first approached my theoretition friend and co-worker of many years, R. Simha, for statistical-mechanical assistance, and we obtained further the cooperation of H. Frisch, then just completing his Ph.D. at the Polytechnic Institute. The model we evolved was that of a macromolecule in solution colliding first with one of its segments with a solvent-solid interface, becoming adsorbed when a complicated set of energetics becomes negative. This set included the free energies of desolvation of the adsorption sites and adsorbed segment, the free energy of the segmental adsorption step, and the loss of conformational entropy when the rest of the coil becomes restrained in its number of available conformations. Once one segment is held, further segments will at first be more easily adsorbed, followed later by a greater resistance against adsorption when the restraints on the chain conformations become severe.

The extent of equilibrium adsorption of a given chain, thus, would depend on the counteraction between the three enthalpic and entropic components, plus the negative entropic contribution of the chain restraint. The rate of adsorption should be fairly fast, in fact, accelerated after the first segment was held, while the rate of desorption requiring simultaneous multiple desorption steps should be very slow, so slow indeed as to be barely measurable except when assisted by a displacing species.

Based on this, the theory developed by Frisch and Simha (12) accounted for the already known temperature and solvent dependencies and also the number and length of the attachments. It gave further the not yet observed, and not readily evident, result that the length of the unattached loops would be proportional to the macromolecular weight (MW) to the 1/2 power; the amount adsorbed should show only a weak dependence on the macromolecular weight. For weakly adsorbed macromolecules, that is usually for physical adsorption in the absence of strong surface-segment interaction such as acid base reactions, salt-or complex-formation, or of porous surfaces with much entrapment,or of surface condensation reactions etc., the extension of partially held coils into the solvent may be a large fraction of the dimension of the free coils in solution. Over the years, this theory has been much debated, has been improved, alternative theories developed by many other authors (13), special features and limiting cases analyzed; much experimental work was stimulated (14) which added a great deal of information but also introduced new complications for our understanding.

Before our theory was fully developed, extensive work by J. Koral in cooperation with R. Ullman (15) confirmed in detail and with considerable accuracy all previously known features. They ascertained, in addition, the particulars of the adsorption isotherms for a number of polymers and dispersed adsorbates and established the remarkable degree to which most isotherms could be approximated by 2-parameter equations, like Langmuir's isotherm for monolayers of small molecules. They found the dependence of the adsorption on MW to be weak and determined the area per adsorbed molecule. The concept of intermolecular repulsion and compression within the adsorbed monolayer and between the surface layers of 2 dispersed particles was also invoked.

However, the most essential information for a physical picture of the adsorbed layers (Fig. 1), that of their dimensions perpendicular to the adsorbing face, e.g., the length of the loops, was lacking up to that time. Experiments yielding such information are difficult to design. Eventually, I decided on hydrodynamic determinations which, while not yielding the actual, but only the effective, thicknesses of the adsorbed layers, are sensitive to the MW and the solvent. The work of Oehrn (17), and of Tijnman and Hermans (17) showed that there was promise in obtaining the needed data from flow measurements before and after macromolecular adsorption onto the capillary walls. The drawback was that adsorption isotherms in individual capillaries, needed to obtain the necessary information, are most difficult to establish. Thus, we began to work with suspensions of nearly isodiametric pigment particles (TiO_2, $CaCO_3$, SiO_2, carbon black) for which the relation between their original dimensions and average volumes and their sedimentation and diffusion rates, and intrinsic viscosities were well known. The adsorption isotherms for polymer fractions on dispersions of the previously defined particles were then established. Unfortunately, the settling rates of pure and coated particles could not be used because boundaries of settling or rising swarms are beset by convection instabilities. Thus we determined the viscosity of pigment dispersions, of predeterminated Einstein factors and of known surface areas, stabilized first with the aid of small soap molecules. The same pigment dispersions were measured after being stably dispersed by adsorbed polymers (18). From the increase in viscosity with increasing MW of the polymer, the thicknesses of the adsorbed layers were calculated and found to be approximately equal to the coil diameters in solution. By knowing the surface areas, the layer densities, or the degree of surface coverage respectively, could be determined and, in particular, the dependence on MW which indeed turned out to be close to $MW^{1/2}$; even more interestingly, the layer thicknesses were exactly proportional to the intrinsic viscosities of fractions of the adsorbed polymer. This proved not only that the coils were held by relatively few segments, but also that the conformations dangling or looping into the solvent were affected by MW, solvent, or temperature with respect to their dimensions in the same way, as if they were free in solution. It should be noted that, although the theory of adsorption has been developed

TABLE I

Adsorption of (or onto) polymers plays a key role in:

Adhesion	Corrosion
Coatings	Aging of composites
Paints	Crack resistance
Lamination	Drag reduction
Reinforcements	Textile finishing
Emulsions	Flocculants
Suspensions	Chromatography
Detergent action	Soil structure
Flotation	Films and membranes
Drilling and cutting	Biological agglutination
Solid lubricants	Immune reactions
Crystallization	Cell recognition
Precipitation	Drug direction
Agglomeration	Genetic reproduction

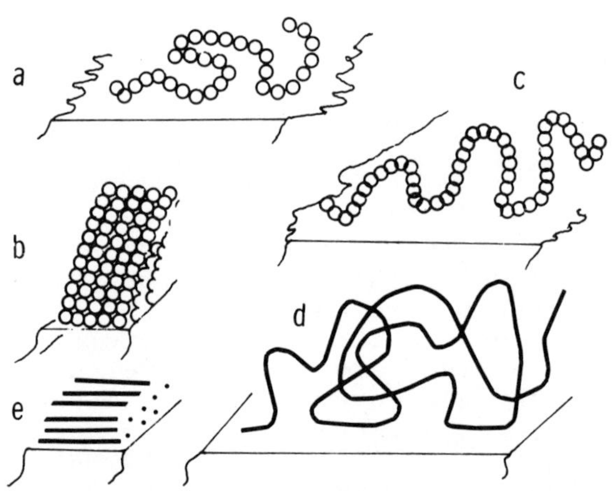

Figure 1. Hypothetical conformations of chain molecules (16) *adsorbed on solid–liquid interfaces:* (a) *lying on surface;* (b) *standing on end,* (c) *looping,* (d) *coiled,* (e) *flat multilayer.* a,c,d *single chains adsorbed with decreasing affinity;* b,e, *condensed surface layers.*

for flat surfaces, the curvature of smaller particles offers no difficulties in principle. Particles smaller than the coils become engulfed within the latter. Only when the particles sizes become comparable to those of the adsorbed trains, will steric hinderances affect the latter so that fewer segments and molecules adsorb per area; perhaps, it would now be more correct to say that the particles become adsorbed on the chains.

We have little information on the way low molecular weight molecules and oligomers adsorb (19). Apparently below DP's of about 100 they lie flat on the surface for concentrations up to a monolayer of segments, then seem to form thicker islands of smectic or nematic structure. Ordered condensed mono, -di, -or multi-layers are primarily the arrangements of smaller, especially amphipatic molecules on liquid-liquid interfaces. Polymers are too large to adsorb, in the ordinary sense, on micelles but segments of linear polymers may act as nucleation centers for micelles of small molecules which probably is one of the mechanisms for the lipid-, or detergent-, polymer interaction.

These principal results were confirmed by further work of Rowland (19), Wadman (20), Chough (21), Chao (22) and Kudish (23) employing a capillary (fritted glass or sintered steel wool filter) flow method. Their results showed again that for MW's sufficiently high to follow Flory's viscosity relation (24), the loops responded to changes in temperature, solvent, and additives in practically the same way as if they were free. This result means further that for so-called physical adsorption (non-specific interaction) the dangling chain portions are long enough and the restraints on them weak enough, to allow them to behave like random coils. For stronger interactions, e.g., between charged surfaces and oppositely charged polyelectrolytes, the situation may be very different as indicated first by the work on such isotherms of Lopatin (25) and Lauria (26), to be discussed further below.

The final proof of the physical reality of our model of macromolecular adsorption was provided by simultaneous independent work, by an entirely different method, at the National Bureau of Standards (27). Stromberg et al., allowed polystyrene fractions, besides other polymers, to become adsorbed on ferrochrome plates and determined the thicknesses of the adsorbed layers by ellipsometry with the help of assumptions on the segment density distribution. They found the thicknesses again to be proportional to $MW^{1/2}$, except for the very highest MW's, i.e., above 10^6 (see Fig. 2). They also observed fast adsorption, very slow desorption and proved the important fact, by working with tagged polymers, that adsorbate exchange occurred regularly at the interface, i.e. that this type of adsorption was truly reversible.

The substantial thickness of adsorbed polymer layers has important consequences for many systems that contain macromolecular components. When adsorbed, e.g., on the particles of fluid dis-

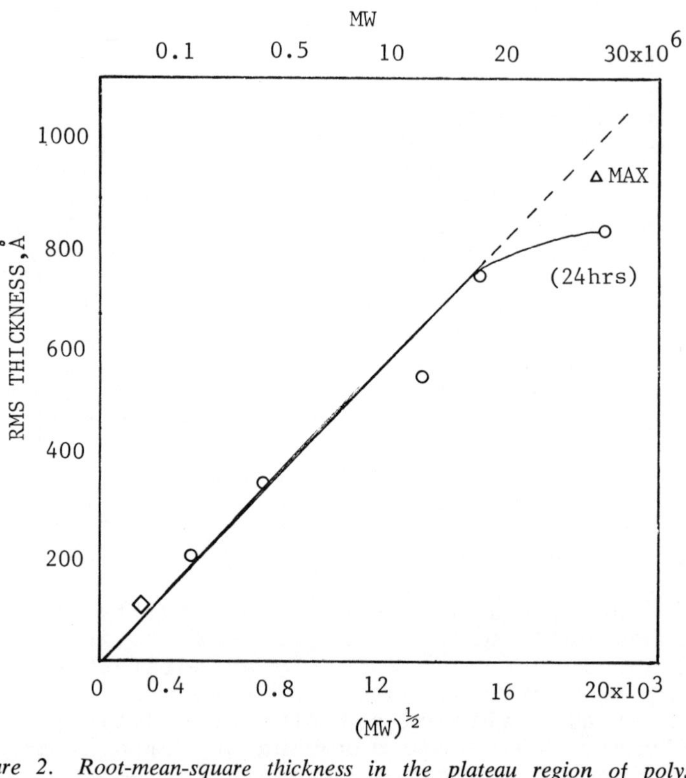

Figure 2. Root-mean-square thickness in the plateau region of polystyrene, adsorbed on chrome ferrotype as a function of the square root of the molecular weight, in cyclohexane, 36.4°C.

persions so as to cover most of the particle surfaces macromolecules will resist close inter-particle approach and contacts because this would increase segmental density. Resistance to reduction of conformation surface gives rise to dispersion stability. In other words, the "staying"character of the dangling portions is the basis for dispersion and emulsion stabilization, and of macromolecular detergent action. If the polymers were polyelectrolytes, stabilization would further be due to the repulsion between their electrical double layers. On the other hand, if the dangling chains would carry 2 types of, locally separated, groups which are capable of associating with their counterparts, this could lead to mutual particle (cell) recognition and association, or to targeted adsorption of the macromolecules on specific cells, or cell sites. Bridging, i.e., simultaneous adsorption of one long chain on 2 or more particles, leads to easier particle filterability or flocculation; if hydrophobic side groups, or blocks of hydrophobic segments, are part of the macromolecules, they will adsorb preferentially, by hydrophobic bonding on less hydrophilic surfaces or surface sites, exactly analogous to detergent action: the polymer will function now as a long chain, ionic or non-ionic, soap. If the hydrophobic segment adsorption occurs on water-air interfaces, such as of airbubbles, it will allow flotation, whereas joint adsorption on particles and large liquid-solid interfaces is the basis for pigment coating. Blockpolymers, made up of 2 sections, one of a cohesive density close to that of the dispersed phase and lying mostly on the interface, the other extending into the continuous solvent phase, will be exceedingly good polymeric soaps or compatibilizers (28).

The parallelism between δ and $[\eta]$ does not hold when $[\eta]$ grows either much faster, or slower, than α $MW^{0.5}$. The former happens for rather extended molecules of low affinity, or for polyelectrolytes of low charge attraction, the latter for strong affinities, e.g., due to molecular group reactions between adsorbate and adsorbant surface, such as polyethers on polycationic particles, polyacrylamide on strongly hydrogen bonding, or PVP on negative, surfaces and, in particular, for the adsorption of polyelectrolytes on surfaces of opposite net charge. Examples for the latter cases of "binding" by ion pairing are: Poly(sulfonates) and poly(acrylates) on calcium carbonate (29), or collagen on hydroxy apatite (30).

Interesting observations are made, if a polymeric adsorbate is capable of more than one mechanism of adsorption. If poly(styrene sulfonate) (PSS) is adsorbed on carbon black which contains multivalent cations on its surface, the adsorption depends strongly on the nature of the cations and on pH, with a maximum near pH 7. At very alkaline pH's, the OH^- ions compete with the sulfates for the cations and the PSS adsorption drops to the low values of hydrophobic bonding between the carbon itself and the PSS backbone. The latter low values are also obtained, if PSS is adsorbed on the same carbon after purification, i.e. when oxygen and/or the cations were removed. This adsorption of PSS is now pH independent (29).

The fact that the conformations of the free, solvent immersed, segments remains the same if, instead of a few adsorbed segment trains, one or a few segments are strongly, e.g. covalently, bonded to the interface, is the basis for the uninhibited function of enzymes, and other polymeric reagents, when bound to solid supports. This has been important for many forms of solid support syntheses and enzyme engineering. On the other hand, polymers that grow to very high MW, or stay adsorbed on the solid, or supported, catalyst after they have grown and become detached from a Ziegler-Natta site, inhibit monomer diffusion to still active sites or other growing chains, and thus reduce catalyst efficiency.

As the concentration of the macromolecules in equilibrium with the interface increases, so that the coil population becomes crowded, a sudden reaching of a very gently upward sloping plateau is observed, but hardly ever any multilayer formation, since no proper outer boundary of the adsorbed layer of coils develops (see below). Instead, as more coils compete for the same interfacial area, adsorption of further molecules occurs as interpenetration is still resisted, at the expense of the site areas held per molecule, i.e., the adsorbed trains become shorter, the loops longer, and the area under the coils smaller: the coils become sideways and upwards compressed, (15) (16), Fig. 3. Experimentally,

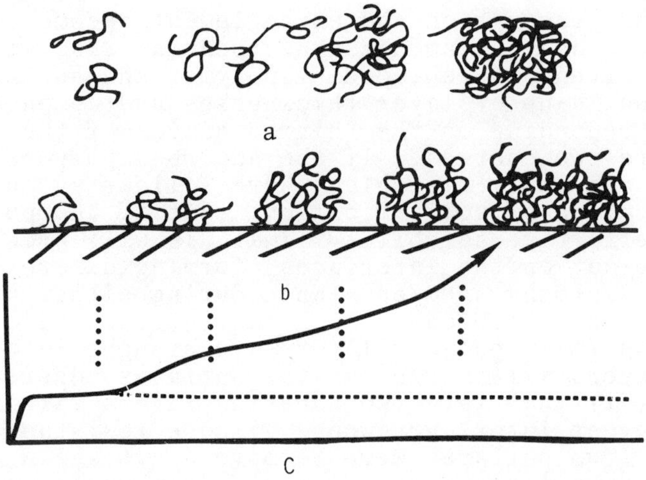

Figure 3. (a) *Conformation and coil interpenetration of polymer chains at the interface as a function of polymer solution concentration, or from melt (basic conformation corresponds to* d *of Figure 1).* (b) *Corresponding adsorption isotherms.*

this is seen as a slow thickening of the adsorbed layer in conjunction with gradually rising adsorption isotherms, until the number of segments held amounts to up to 10 to 12 monolayers. Such layers of adsorbed very high MW polymers can reach thicknesses of several thousand nonameters and thus impede the flow through capillaries (22) and porous media. This fact is of greatest commercial importance during tertiary oil recover, when the simultaneous parallel flow through water bearing rock beds is retarded by water soluble polymers, but not the flow of oil(23).

At a stage at which the density of the segments per coil volume reaches the density of a theta state, coil interpenetration and entanglement become extensive; from there on, the amounts adsorbed may rise steeply, as the solvent is replaced by polymer and an indeterminate build-up of layer thicknesses and compaction proceeds. Depending on the thermodynamics of the case, there are two courses. If the adsorbed polymer and the solvent are fully miscible, there follows a continuous densification or solidification. But if the polymer has an effective solubility limit, it will precipitate or plate-out on the interfaces, forming discrete patches of the adsorbing species. When during either process two surfaces which carry such concentrated adsorbed and/or built-up polymer layers are brought into contact, these layers will merge and the surfaces adhere by autohesion.(31).That is, two adhering layers will become one coherent interlayer whose viscose resistance (32), if those two surfaces were to move apart again, against flowing or splitting,makes itself felt as adhesion. In a dispersed system, colliding particles at this adsorbent layer density will aggregate instead of being repulsed and the particles become eventually imbedded in the polymer as a continuous phase: a filled reinforced composite system has been formed. On complete removal of the solvent from the polymer, the remaining interlayer whether rubbergy, glassy, or partly crystalline constitutes a truly adhesive bond, whose viscoelastic character allows for internal stress relaxation and energy dissipation, crazing,and crack formation or healing.

These latter mechanism are well understood for polymers in bulk and may now be seen on the molecular level by analogy to the nature of tack (33) referred to above. As the plates joined by an adhering macromolecular interlayer separate in the direction normal to their planes, they experience the opposing tensile force against the separation due the viscous resistance against the inward flow of the fluid. If the motion becomes so fast that the tensile viscous stresses reach the cavitation stress in the liquid, voids will form. For narrow separations and high viscosities this may happen even at low separation rates. If the liquid is purely viscous, especially when it is shear thinning, the voids will rapidly merge and the resistance break down at a low level of stress. If the fluid is shear thickening, e.g., solutions above the macromolecular entanglement point, or turning more glassy or crystalline under shear, and becomes strong relative to the surface tension, such fluids are spinnable and, instead of, Fig. 4, breaking early, will form strong threads which remain attached to the surfaces and are being spun out further as the plates separate (38). Such "tack" can be substantial. Adhesion proper will be felt,either if the viscosity and the cavitation stresses are high enough to prevent noticeable flow,or stop the fluid from breaking under the applied forces for the applied time,or if the fluid has solidified and the rupture and dewetting (desorption)stresses are higher than the ones applied.

Even minor debonding, i.e., a partial desorption due to volume contraction, by creating flaws, frozen strains and stress concentrations along the interfacial contact will weaken the macroscopic adhesion in a major way. Solid bodies do not wet. Thus, the essence of good adhesion resides in the close adherence of the molecular boundary layer to the adjoining surfaces on account of multiple molecular adsorption i.e., wetting which resists rapid dewetting while other sections of the same molecules extend into and are entangled with molecules of the overlying, also macromolecular, adhesive. The adhesive must be spinnable, or be crosslinked to form a fluid network, or be partially a strongly cohesive viscoelastic solid facing the interfaces with ductile domains.

Actually, the strength of all tough materials, including composites, derives from an analogous behavior, the ability to reduce imposed stresses by strain relaxation, particularly in front of cracks. There, the leverage action by the moment of:(crack depth x applied force),that is the stress concentration, is the most frequent cause of material failure. As the crack advances by converting excess free energy of deformation into surface free energy, the mode of forming of new surfaces is analogous to the separation of two adhesively connected surfaces. If the material is a macromolecularly entangled viscoselastic polymer (34), molecules must become strung out for the crack to advance. Mechanical energy will be consumed in this molecular movement, while any potential alignment and crystallization tend to blunt the crack and to raise the modulus ahead of it, as the material changes morphologically from isotropy to anisotropy oriented perpendicular to the crack plane. Cracks advance therefore during bulk cohesive failure by exactly the same mechanism as in adhesive failure without debonding. Even for this, Fig. 5., latter case Williams has shown (35) that the separation of two phases along the interface can be treated by regular fracture mechanics. The fact, then, that all good adhesives are polymeric derives from their viscoelastic nature as well as from the nature of their adsorption. The process of macromolecular adsorption improves the bonding, while where bonding fails, viscoelasticity will tend to change the failure mode from an interfacial to the bulk type.

One can expand on this by saying that the statements: "an adhesive bond is as strong as the adhesive", or "....is only as strong as the interface", are trivial only as long as one remains ignorant of the mechanism of how interfacial bonds are set up. If one subscribes to the generally accepted notion that physical adsorption (absence of special bonds), like mixing, occurs (36) only when the sum: a^2+b^2-ab, becomes negative, where a and be are the energies of cohesion within the materials a and b in bulk, and ab represents the interaction energy, then the interfacial layer, ab, should always be stronger than the weaker component, and the break always occur within the weaker material, a or b. However,interfacial contact, that is wetting, is never ideal or perfect. Any flaw within the zone of a-b contacts acts as a stress raiser from which cracks will start and either leap-frog through neighboring flaws causing bonding failure, or propagate into the weaker medium (cohesive failure) whichever is energetically more favorable. Macromolecular adsorption offers many composition and design options by allowing to vary interfacial vs. interlayer structure and strength. Other advantages of polymeric adhesiveness over that by small molecules, at the same degree of wetting, are the substantial fraction of covalent bonds which lie in the immediate first plane adjacent to the interface, allowing to bridge small flaws by chains, and encouraging stress dissipation and transmission through their loops away from the interface.

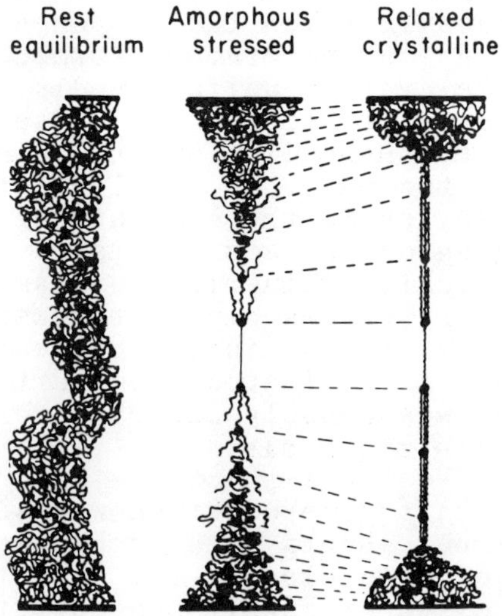

Figure 4. Nonuniform (nonaffine) straining due to network inhomogeneties, for example, of short network chains, causing local crystallization and relaxation (38).

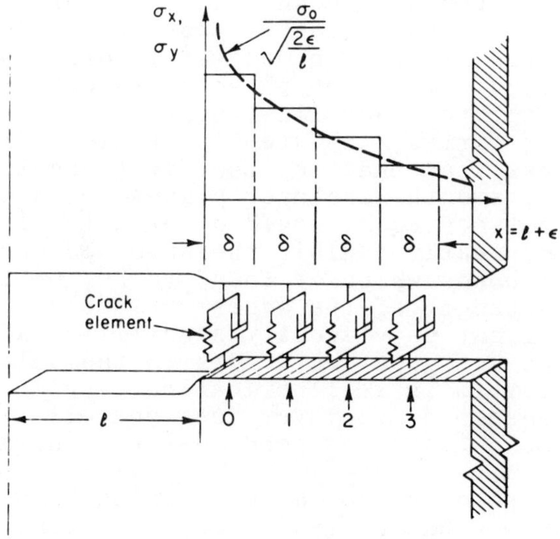

Figure 5. Representation of stress at advancing crack tip, and below, the state of viscoelastic strain indicated by Voigt models (35).

A specially interesting case of interparticle adhesion is presented by blockpolymers in bulk. They are solid dispersions in which the component of higher melting or glasstransion temperature formed the first discontinuous phase, rather uniformly distributed in the sea of the interpenetrating dangling sections of the second component. Depending on the volume ratios of the components, the major one will eventually be the continuous phase, while the segments of the minor component, depending on the amount, may arrange themselves to become spherical, cylindrical or lamellar islands. Such blockpolymer systems amount to dispersions with sections dangling from a single covalently bonded site. Compound systems precipitated from particles on which another polymer was densely adsorbed, are structurally similar composite materials.

One might raise the question why we did not consider, or observe, sheets of adsorbed monolayers of uniformly folded molecules as the structure of the severely compressed or highly concentrated layers towards the end of monolayer adsorption, that is a kind of epitactic, smectic, condensation. We found for rodlike molecules, e.g. prolin I in water (22), indications of a nematic array of the adsorbed molecules, but never any evidence for planes of folds when the polymer molecules were in their random coil form. Out of hundreds of cases observed we encountered only three instances of a step-up in the isotherms above the first plateau to be expected if an even moderately ordered layer of chains existed; all these three cases were H-bonding polymers (acetates, acrylics, peptides). When the second layer formed, the step in the isotherms was invariably less than that for the first monolayer laid down. Also, smooth monolayer planes, even with tie molecules between them, would offer natural shear planes and thus not nearly exhibit the observed shear strength of adhesive bonds which is found to be, or to exceed, the shear strength of the adhesive itself. Finally, though excellent as viscosity stabilizing additives, polymers are not good lubricants in themselves (except as Beilby layers in selflubrication), again a sign that the first adsorbed monolayer does not offer a smooth plane to the outside, as good adsorbed lubricants would do.

I have reported so far only on one non-specific adsorption, that of random coils. Thus, ordered monolayers of polymers were rare and randomly packed first layers the rule. There are of course many other types of macromolecules and their adsorption, and an infinite variety when it comes to the adsorption of more than one species onto nonhomogeneous, or patterned, surfaces. I want to refer here only to one system, that of polyelectrolytes and ampholytes on charged and uncharged surfaces, noting our main observations: Polyelectrolytes adsorb only if their net charge is opposite to the net surface charge; and compensation affects only the net charges, so that oppositely charged interfaces and polyelectrolytes interact like any two polyelectrolytes, at first by an exact compensation, so precise that net surface charges can be titrated to normalities of 10^{-5}, or less, by oppositely charged polyelectrolytes, using dispersion stability, conductivity, or potentiometric minima as indicators. Counter-charged polyelectrolytes lie flat at the interface, but may form loops, if they carry excess charges. For very weakly charged interfaces and polytelectrolytes of the same sign a van der Waals or hydrophobic, bonding for the corresponding sections of surface and polyelectrolyte is possible, while any charged sections of the polymer stay in loops off the carrier surface. Unbonded segments of polyampholytes form electrostatically determined loops whose extensions from the surface respond closely to the pH and salt content of the solution. Again, these loops parallel the dimensions of the free ampholytic coils in solution, showing for instance, a minimum for the isoelectric point (23).

The perhaps most important case of the effect of the conformaitonal state, or its changes, on the activities of adsorbed layers is that of the globular proteins, in vivo and in vitro. The native conformations of these proteins are partly helical, α-helical, β-sheeted or randomly coiled, fixed internally by S-S, H-bonding, and hydrophobic attraction. On becoming adsorbed they may or may not stay conformationally constant. The former is the case for serum albumin whose lining of blood vessels assists the endothelial prostacyclin in keeping the other blood constituents from adhering to the walls. Adsorbed on implanted parts, serum albumins reduce blood clotting. If they, or other plasma proteins, become denatured as a result of some damage to the organism, so that their native conformation opens up they, like synthetic linear polyelectrolytes become more extensively adsorbed on the walls of the blood vessels,or other formed elements and thus, become adsorptive for, and trap, blood platelets which then open up and trigger plaque formation. It requires the co-adsorption of strongly negative ions like citrate, or of polymers like heparin, to offset the adsorptive affinity of denatured proteins by overlaying them with a repulsive, stabilizing negative charge.

Among other methods to render surfaces of implanted materials bio-compatible, in particular to prevent them from initiating blood clotting, one may activate the surfaces by radiation and then graft water soluble, or negatively charged, polymers onto the active sites. These grafted polymers, though anchored covalently at one end, act like loosely adsorbed polymers of the same nature, except that their number of contact points per molecule will be small and the chains extend further in to the fluid.

I have tried to recount as a matter of personal recollection the course over two decades of a concept, godfathered by Dr. Mark and including the input from many co-workers and colleagues;also to describe the basic ideas and development and ramifications of the results. What started as a very limited inquiry into the role of macromolecules for adhesion, became one into their state at interfaces. Since it developed independently that macromolecules and interfaces are at the core of much of material science, engineering and biology, the answers acquired a broad significance. Macromolecules, partly fixed and partly reaching out from an interface act like tentacles, either collecting, holding, or keeping apart, the domains of dispersed multi-phased or cellular systems. Remaining responsive to the environment, they may become the mediators of interdomain mechanics in the chemistry of solids, of complex reaction sequences, and of stabilizing structures in biology.

I have been partial in emphasizing where we believe to have useful answers rather than where we miss them. Inevitably, every research opens more questions than it answers. Thus, we have no good way of finding experimentally the density distribution of the molecular segments away from the surface in a given case, whether they extend into a mobile or into a mixed solid bulk phase. We do not know why, during the increase of the surface population, there seem to be no critical points at which density or phase changes occur discontinuously, as one finds for the surface isotherms of small molecules. Macromolecules at interfaces are in a sense cross-linked by way of their adhering segments, but these links will travel dynamically along the chains. Will this be a dominant feature of "good" adhesiveness, as well as of the mechanical advantage of multi-phase systems with macromolecular interphases over the performance of homogeneous systems? May there be merit in studying other systems which are also cross-linked in a gliding way, e.g., is this in essence just another case of entangled molecules? Besides the case described here, to which extent is the structure of the first monolayer, ab, affected by specific molecular interactions of types a and b explored by so many studies (36) (37)? How do polymers adsorb during crystallization on seed surfaces, as random coils which order themselves, or do they fold on arrival? When macromolecular coils, instead on interfaces, attach themselves to each other, do they interpenetrate or fold, or order in some fashion? Would the latter process of macromolecular adsorption akin to that of macromolecular aggregations, including the reassembling of fused helically intertwining macromolecules? How do surface features influence the conformations of the loops and,thereby,is an amount of "information" already noticable some distance away from the interface? How is the chemistry of reactive groups, already greatly modified and usually more specific and catalystically more active when tied to polymers than when they are free, further changed by adherence of the chain to an interface? Could we design a catalytic system such that the active site (or enzyme) is adsorbed on, or fixed to, an interface which also adsorbs and concentrates the catalysts' substrate? If adsorbed coils billow above a porous layer, how can they snake their way through channels only wide enough for one segment? If a chain passes through a number of low molecular micelles, is it stabilizing or destabilizing the latter, or may it act as a core of a cylindrical micelle?

When proteins are absorbed into, or together with, lipids and phospho-lipids to form cell membranes, are the dimensions of the amphoteric domains sensitive enough to pH and ion concentration, so that by swelling or shrinking with changing gradients, they may act as gates or regulators, of ionic flows? How is collagen adsorbed on hydroxy apatite, or vice versa? How does the state of adsorption between the domains of a composite material affect its strength and other mechanical-chemical properties?

It will be seen that starting even from our limited insights into the state of adsorbed macromolecules, a large number of inquiries may be undertaken towards a better understanding of an extremely wide range of natural and technologically posed phenomena. Who would have expected, when we first asked why flexible materials are the better adhesives, that we would answer some so very different questions and raise so many others?

References

1. Kuhn, W., Z. physik. Chem., A161. 1 (1932)
2. Sverak, J. and Eirich F., Trans. Faraday Soc.,43, 57 (1946)
3. Debye, P. and Bueche, A.M., J. Chem. Phys. 16, 573 (1948)
4. Cerf, R. and Scheraga, A., Chem. Rev. 51, 185(1952)
5. Peterlin, A., Adv. Macromol. Chem., 1, 225 (1968)
6. Wood,L.A., Adv. Colloid Sci., 2, (1946); Interscience, N.Y.
7. Treloar, L.R.G., "The Physics of Rubber Elasticity" Clarendon Press, Oxford (1958)
8. Flory, P.J., J. Chem. Phys., 15, 397 (1947)
9. Jenkel, E., and Rumbach, B., Z. Elektro. chem., 55 612 (1951)
10. Claesson, I., and Claesson, S., Arkiv Kemi Mineral. Geol. 19A, No. 5 (1945)
    Heller, W., and Tanaka, W., Phys. Rev., 82, 302 (1951)
11. Eirich, F.R., J. Coll. and Surface Sci., 58, 423 (1977)
12. Simha, R., Frisch, H.L., and Eirich, F.R., J. Phys. Chem., 57, 584 (1953)
    Frisch, H.L., and Simha, R., in "Rheology", Vol. I (F.R.Eirich, Ed.) p. 525, Acad. Pr. (1956)
    Frisch, H.L., and Simha, R., J. Chem. Phys. 27,702 (1957)
13. Silberberg, A., J. Phys. Chem., 66, 1872 (1962); Silberberg, A., J. Chem. Phys., 46, 1105 (1967) ibid. 48, 2835 (1968);
    Silberberg, A., Faraday Disc. No. 59, p.203 (1975)
    Hoeve, C., J. Chem. Phys., 44, 1505 (1966)
    Hoeve, C., DiMarcio, E., and Peyser, P., J. Chem. Phys. 42, 2558 (1965)
    Roe, R., J. Chem. Phys., 43, 1951 (1965)
    DiMarcio, E., and Rubin, R., Amer. Chem. Soc.Polym. Preprints,11, 1239 (1970)
14. Stromberg, R.R., in "Treatise on Adhesion and Adhesives," I, (R.Patrick, Ed.) Dekker, N.Y. 1967
    Lipatov, Yu. S., and Sergeeva, L.M., "Adsorption of Polymers", Wiley, N.Y. 1974
15. Koral, L., Ullman, R., and Eirich, F.R., J. Phys. Chem. 62, 5 41 (1958)

16. Eirich, F.R., in "Interface Conversion for Polymer Coatings" (Phil. Weiss, Ed.), Amer. Elsevier Co. 1969
17. Oehrn, O.E., J. Polym. Sci., 17, 137 (1955)
Tuijnman, C.A.F., and Hermans, J.J., J. Polym. Sci. 25, 385
18. Rowland, F., Bulas, R., Rothstein, E., and Eirich, F.R., Ind. Engin. Chem., 57, 46 (1965)
19. Rowland, F., and Eirich, F.R., J. Polym. Sci., A1, 4, 2033, 2401 (1966)
20. Wadman, J.W., Dissertation, Polytech, Inst., Bklyn, 1966
21. Chough, E., Thesis, Polytechnic Inst., Bklyn, 1968
22. Chao, H.Y., Dissertation,Polytech. Inst., Bklyn, 1974
23. Kudish, A., Dissertation,Polytech. Inst., Bklyn, 1972
Chao, H.Y., Chough, E., Earl, Ch., Lopatin, G., Wadman, J., and Eirich, F.R., preprints, 48th Natl. Colloid Symp., Austin, Tex., 1974, p. 165
24. Flory, P.J., "Principles of Polymer Chemistry", Cornell Univ. Press, 1953, p. 602 ff
25. Lopatin, G., and Eirich, F.R., Proceed. 2nd. Inter. Congr. Surface Activity", B, p 97, 1961
26. Lauria, R., Dissertation, Polytechnic Inst.,Bklyn, 1962
27. Stromberg, R.R., Tutas, D.J., and Passaglia, E., J. Phys. Chem. 69, 3955 (1965)
28. Molau, G.E., "Block Polymers" (S.L. Aggarwal, Ed.), Plenum Pr., N.Y. 1970 p. 79
29. Earl, Ch., Dissertation, Polytechnic Inst., Bklyn, 1970, see also Ref. 18
30. Chen, T.M., Report to NIH on Train. Grant DE00099, 1973
Eirich, F.R., in "Dental Adhesive Materials", (Moskowitz, Ward, Woolridge, Eds.), Brookdale Dental Ctr. Symp., New York, 1973, p. 53
31. Kaelble, D., "Physical Chemistry of Adhesion", Wiley-Interscience, 1971; p.397
32. Eirich, F.R., and Tabor, D.,Proceed. Cambr. Philosoph. Soc., 44, 566 (1948)
33. Beatty, J.R., Rubber Chem., Technol., 42, 1040 (1969)
34. Eirich, F.R., Trans. Intern. Conf. Mechan. Behavior Materials, Vol. III, 405, Kyoto (1971)
Eirich, F.R., Pure and Appl. Chem., 46, 115 (1976)

35. Williams, M.L., Proc. 5th U.S. Natl. Congr. Appl. Mech., ASME, (1966); Williams, M.L., in "Fracture of Solids" (D.Drucker and J. Gilman, Eds.), p.157, Wiley-Interscience, N.Y. 1963
36. Kaelble, D., Ref. 31, p. 489
37. Mark, H.F., Adhesive Age, 22, 35, 45 (1979) Mark, H.F., Dechema Monograph, 51, 59 (1964) Adamson, A.W., "Physical Chem. of Surfaces", 2nd Ed., Interscience-Wiley, 1967
38. Van der Hoff, B.M.E. and Buckler, E.J., J. Macro. Sci., 1, 747 (1967)

RECEIVED March 16, 1981.

13

Polymers in the World of Tomorrow

W. O. BAKER

Bell Telephone Laboratories, Murray Hill, NJ 07974

As we have said in reporting the establishment of the national materials program two decades ago, the ages of civilization have long been named according to the materials used in the technology of those times (Stone - Tin - Bronze - Iron - Steel). In the world of today synthetics - plastics - fibers play a large part in civilization. So do electronic and optic materials, the semiconductors and magnetics and now lasers and light guides to facilitate communications and information, on which order in modern life so much depends.

Communication and information handling may be dominant modes of easing the strenuous demands for food, shelter, energy transport and security, as well as health and education. And the former remain also primary demands for raw materials in a crowded planet. So in looking toward tomorrow, the role of polymers would extend beyond the major part they play already in the conventional functions of society which we have noted.

We should emphasize as well, however, that those conventional functions are also ever expanding. They extend in the case of synthetic rubber to more than 2-1/2 million metric tons/year in the United States alone. Likewise for plastics, annual usage of 30 billion pounds of thermoplastics and about 5 billions of thermosetting systems added to 10-1/2 billion pounds of fibers (of which only about a billion are cellulose derived), all indicate the magnitude of service already provided by macromolecules. In Western Europe, currently about 28 billion pounds of bulk thermoplastics are applied, with about 5% being for special engineering functions. Among these, uses of polyethylene and polybutylene terephthalates are expanding at about 22% per year, polyphenylene oxides by about 11% annually, with polysulpones, polyaryl, other ketones, polyphenylene sulfides and polyamide imides by about 13%. These applications of course, will be extended in ingenious and unforeseen ways in the world of tomorrow. Our point is, however, that new and equally compelling functions of polymers are steadily appearing. These have especially to do with the Age of Information, in which

0097-6156/81/0175-0165$9.25/0

the versatility of the polymeric state, ranging electromagnetically from the superb electrical insulation of polyethylene and polystyrene to the semiconducting properties of polymer carbon, rubbers and widely used silver and gold filled epoxy and polyimide adhesives, makes them vital elements in the vast and growing global network of communication and information processors.

Indeed these qualities have been developed since the mid-Century so that in a particular segment of the information industry such as telephones and communications, our volume of synthetic polymers used annually exceeds that of any other class of materials, although the actual tonnage of metallic and inorganic matter still leads. For the world of tomorrow, we find microelectronics, thin film circuitry and systems and, especially now photonics, with lasers and light guides, to be dominant components. All of these strongly use polymers, for their special physical-chemical as well as familiar mechanical and electro-optical qualities.

We shall illustrate some of these developing roles, as symbolic of major trends in polymer technology. Especially these uses depend also on polymer science. Thus we shall expect to see an even more intimate link than up to now between basic understanding and technical utilization - an interaction of high importance in all modern industrial society.(1) There is, however, an additional element in the vigorous pursuit of polymer science and technology in cybernetics. For it is in these uses of thin membranes, with exposure to electromagnetic fields and charged particles, that the intersections and analogy with living matter, with the tissues of plants and animals, become manifest. And here indeed is a great frontier in the world of tomorrow. Every scientific element of vital processes calls for more discovery, but none more forcefully than how signals are sent in living things, and how the nerves regulate, and ultimately the brain computes.

Polymer behavior is prominent in every one of these actions, yet we still do not know the elementary process in a single one of them. The elegant chemistry of acetylcholine at synapses, and of adenosine in its energy regulations, are still undefined in respect to an actual signal system. And beyond all this, the marvelous mystery of tissue assembly, of cellular communication and intelligence,(2,3) involves yet other polymer behavior whose understanding may indeed be aided by the films and functions of polymers in electro-technology.

Probably we should emphasize the position of polymers in the world of tomorrow also with reference to other strong social and economic factors derived from the growing field of communication and information processing. Prime among these is energy, where the transport of people for communicating as well as moving of records, papers, films etc. are substantial factors in energy usage. For instance, the energy used for a typical telephone conversation has diminished in the past six years by 1/3, from about 320 BTU's to 220. The point is that if, as we shall show, polymers are essential elements of communications and information systems, then polymers will participate in the drive to reduce energy usage. As noted, a telephone call linking people over some typical distance of tens of miles takes about only 224 BTU's or less than the energy in 1/2 tablespoon of gasoline. This usage supports a complete network, including the building costs, network operation, transmission equipment, computers and motor vehicle maintanance resources. If only 10% of personal driving were replaced by telephone services, a national saving of at least 625,000 barrels of crude oil daily or 10% of imports would be possible, with a concordant improvement of balance of payments of more than 6-1/2 billion dollars per year.

So the functions of polymers should be sought ever more widely in the times ahead, despite their extensive part in modern life already. Let us now peer briefly into properties and potentials which provide these new parts in forwarding the output of human minds and meanings.

A primary function in this information system of the world of tomorrow is transforming the communication modes of human beings into some mode capable of high velocity and quantity. This is achievable by methods such as the analog and digital coding of speech and pictures, electromagnetically, including optically. Techniques of conversion for speech and hearing have been the classic schemes. Here, microphones and vibrating membrane receivers and loud speakers are the crucial elements in the historic evolution of telephone, and later of broadcast radio. Polymer-related matter has played a part since the beginning, especially in Edison's carbon particle microphone, the primary speech transducer in telephones up to the present period. In the condenser microphones, types of electret were used more than 50 years ago. In these, waxes were charged and functioned crudely despite their small capacitance. In 1948 it was shown that many films of acrylics and cellulose esters, polystyrene and vinyls could be charged,(4) but in subsequent study and a continuing examination of possibilities, no successful usage was achieved. Then, in work of Sessler and West at Bell Laboratories,(5) a new course was taken, on the basis of modern polymer science and technology.

They produced high performance electrets from thin polymer films metallized so as to yield high capacitance. Both electrical and mechanical properties of these transducers have been remarkable examples of how applications of science of solids, including knowledge of electron traps, conduction processes in insulators and the viscoelastic phenomena of semicrystalline polymers, can be combined.(6) Incidentally, similar ideas have been applied to optimization of the properties of particle microphones, through assemblies of perfectly microspherical polymer carbon systems. These have shown what limits of performance could be expected in that sort of sound-electricity conversion.(7)

These efforts now assure, in the world of tomorrow, a generic capability for transforming voice and other mechanical signals, including even the vibratory movements of pen or stylus on paper, into appropriate electromagnetic signals, through polymer electret films. Modern electronic telephones produced commerically in Japan, and presently (with novel designs) in this country, are based on these low power transducers. These enable for the first time all of the advantages of solid state integrated circuitry also to be included in the telephone instrument. Recorders, tape systems, a host of new high fidelity sound processors use these elements. A large variety of device designs is now available which can exploit the remarkable stability of charged polytetrafluorethylene. Its sensitivity in such membranes for microphones declines by less than one dB after being at 95% relative humidity and 50°C for a year, an accelerated aging which would hardly ever be encountered in practice. Thus is offered a wide range of options for transforming mechanical motion of almost any sort, from the smallest to relatively large amplitude, into electronic signals.

But also in the course of this work, fundamental qualities of 'polymer structures have been found that open wide paths for other future engineering. Basically, it was long recognized that application of electric fields to dielectric material at elevated temperatures, with the dielectric then subsequently cooled, would yield injected space charges and dipole orientation, as seen in Fig. 1.(8) Modern solid state studies indicate, however, that charge injection by corona discharge, high voltages exceeding breakdown in the sample, or by liquid contacts and electron beams are preferable schemes. The direct electron beam radiation, analogous in its charge implantation to what we have developed practically for semiconductors and a whole variety of other solids, is an effective method for electret formation, Fig. 2. Films of about 25 micrometer thickness of polyfluoroethylene propylene (Teflon FEP) and polytetrafluoethylene (Teflon) were found to capture charges effectively, associated with the electronegativity of the bonds and positions of the dipoles. The penetration of the charge into the polymer was followed by measurements with a split Faraday cup. Here it is found that at the end of a given radiation induction, charges q_1 and q_2 on the front and rear electrodes of the film yield mean special depth "d" of the total charge in the form where D is sample thickness. It is seen that at modest electron energies, there is about 10% or so depth of charge injection in the 25 micrometer FEP specimens, Fig. 3. During radiation with beams of about 40 KEV energy and 10^{-8} amperes/cm^2 current density, the conductivity of the volume penetrated rises to about 10^{-12} ohm cm. But after about 10^4 sec., this conductivity reaches a level of about 10^{-17} ohm cm., and would require years for restoration of

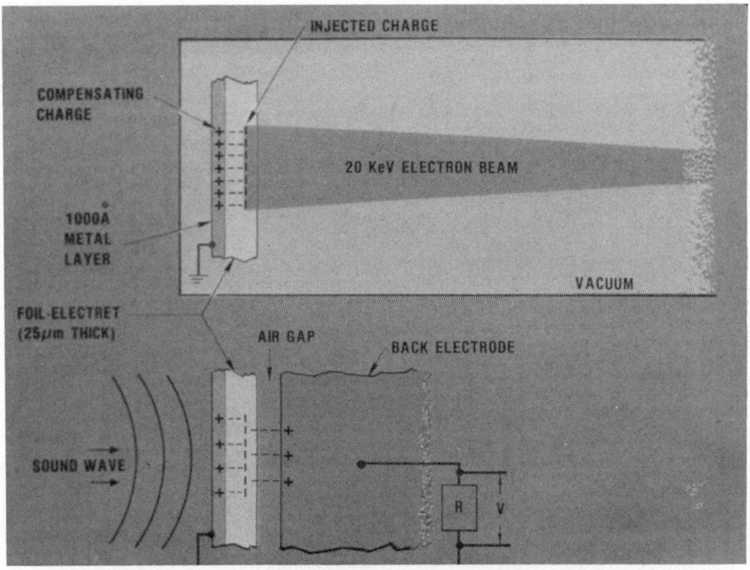

Figure 1. Implantation of electrons in polytetrafluoroethylene films to form electric transducer of high stability and sensitivity.

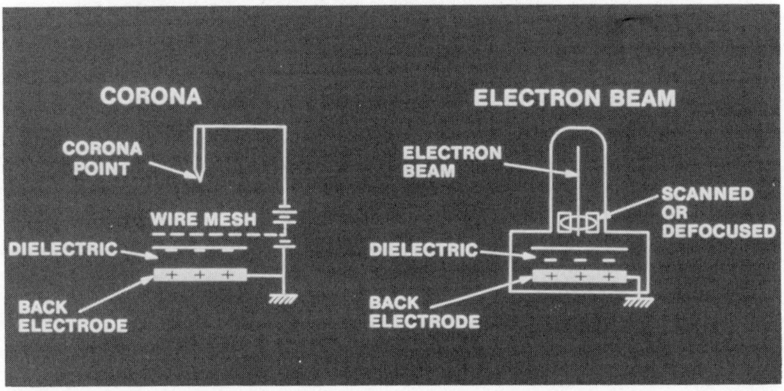

Figure 2. Details of methods of charging polymer films that function as electromechanical transducers and also information storage elements.

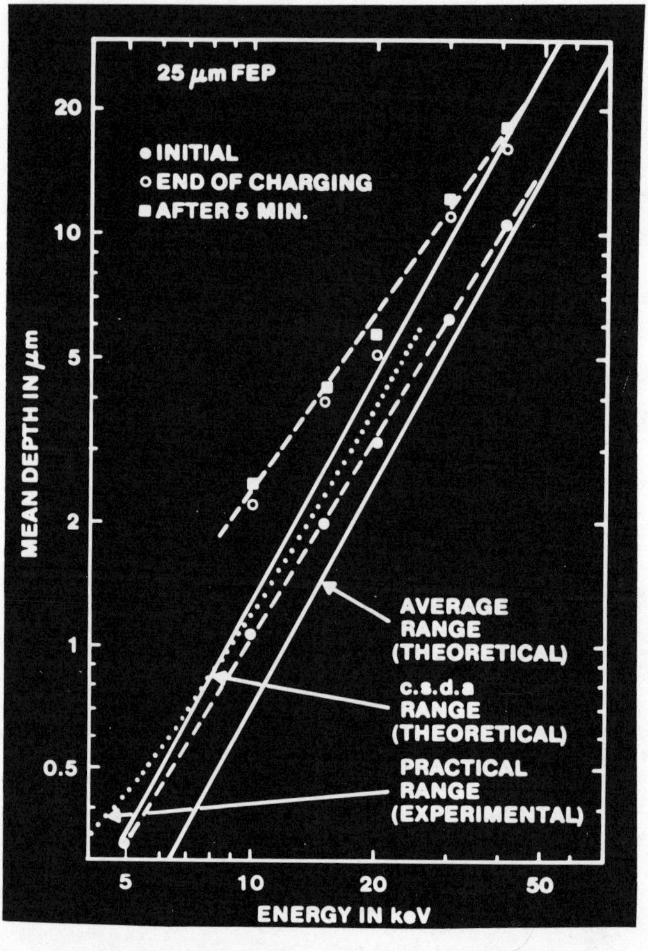

Figure 3. Depth in micrometers of charge injection in a 25-μm fluoroethylene polymer film as a function of energy of injected electrons.

the 10^{-21} ohm cm. of the original unirradiated material. Nevertheless the uncharged portion seems to have retained its original low conductivity. So an interesting trap localization and modification of internal fields seem to be controllable. Levels of breakdown strength of the polymer rather than the density of trap sites seem to limit capabilities for charge storage. Thus polytetrafluorethylene can maintain $.5 \times 10^{-6}C/cm^2$ (ohms per square centimeter), polyethylene terephthalate $1.4 \times 10^{-6}C/cm^2$, each corresponding to a few megavolts per centimeter of electric field. The trap densities, however, exceed $10^{-16}/cm^3$.

Thermally distributed current from these charged films gives further details of the behavior of the charges. Thus, it is indicated that the polyester and the polytetrafluorethylene environments had trapping levels with activation energies for liberating the charges between 0.4 and 2.2 electron volts and with mobilities of about $10^{-10}/cm^2v$.(9)

The complex issue of traps and charge distributions in polymer films suggests much future investigation. Static electricity also remains a puzzling phenomenon in nature. However, the significance of these effects in technical mechanisms is just now being revealed. They appear widely in electrographic and xerographic recording and duplicating processes (10) such as applied in our laboratory so many years ago by Mr. Chester Carlson. It has been assumed that these traps involve various structural units, such as primarily of course the atoms within the polymer, but also groups of associated chain segments and also eventually at interfaces with varying levels of crystallinity. As figures show, a step-like charge decay suggests a series of traps, Fig. 6.(11) Obviously the chemical (especially ionic) behavior of these systems may also be affected by such traps and charging.

As example of continued investigation in this report is measurement of induced conductivity of Teflon FEP film after radiation of 35 micrometer films by 50 to 100 kilovolts x-rays (Mo target) and currents between 2.5 mA and 20 mA. Seventy-five kilovolt x-rays, with dosages between 25 rads per second and 220 rads per second, produced conductivity that was studied under 90 volts of applied external field, equivalent to 3.75×10^4 volts per cm.

After radiation, a strong upsurge in conductivity then decreased and leveled off after about an hour. This was analyzed in terms of trap filling, which showed a linear dependence on the exposure rate, X, at the radiation induced current Ir. For this polymer, mobility of the holes greatly exceeded that of electrons.(12) Additional studies done by electron pulse ionization near the surface yielded consistent behavior of the hole currents.(13) Low hole mobility, $\mu = 2 \times 10^{-9}$ cm.^{2}/Vsec., is in reasonable accord with earlier field independent values. Excellent devices for telephone speech transducing are now manufactured from these polymers, Fig. 5.

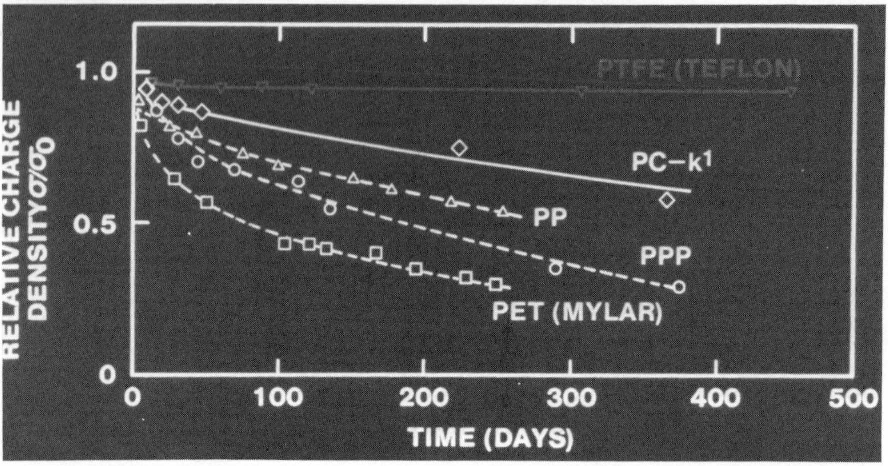

Figure 4. Relative charge density persisting in various polymer compositions as a function of time after initial charging. The polytetrafluoroethylene system traps retain charge indefinitely whereas those in the polyethylene terephthalate decay relatively quickly.

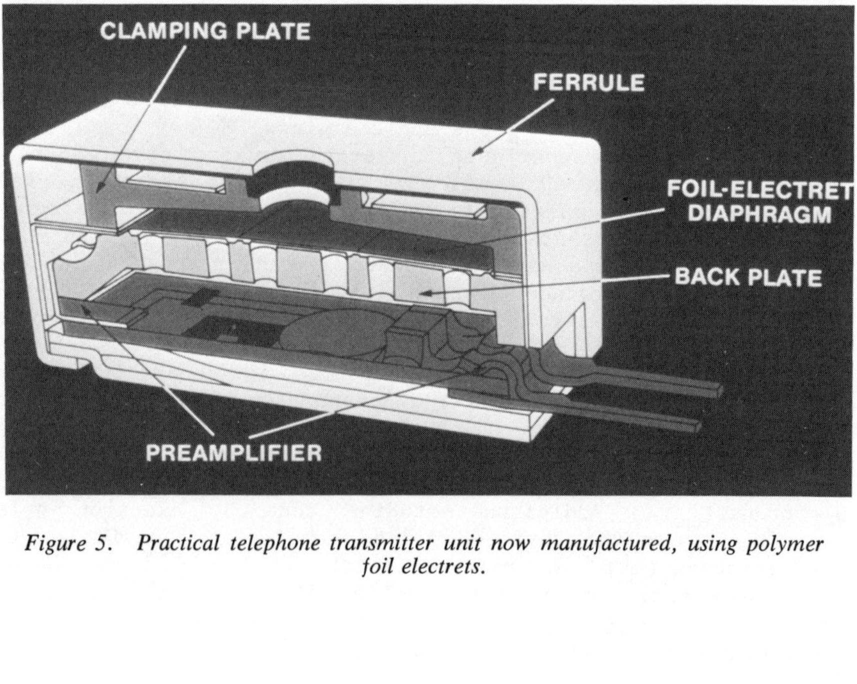

Figure 5. Practical telephone transmitter unit now manufactured, using polymer foil electrets.

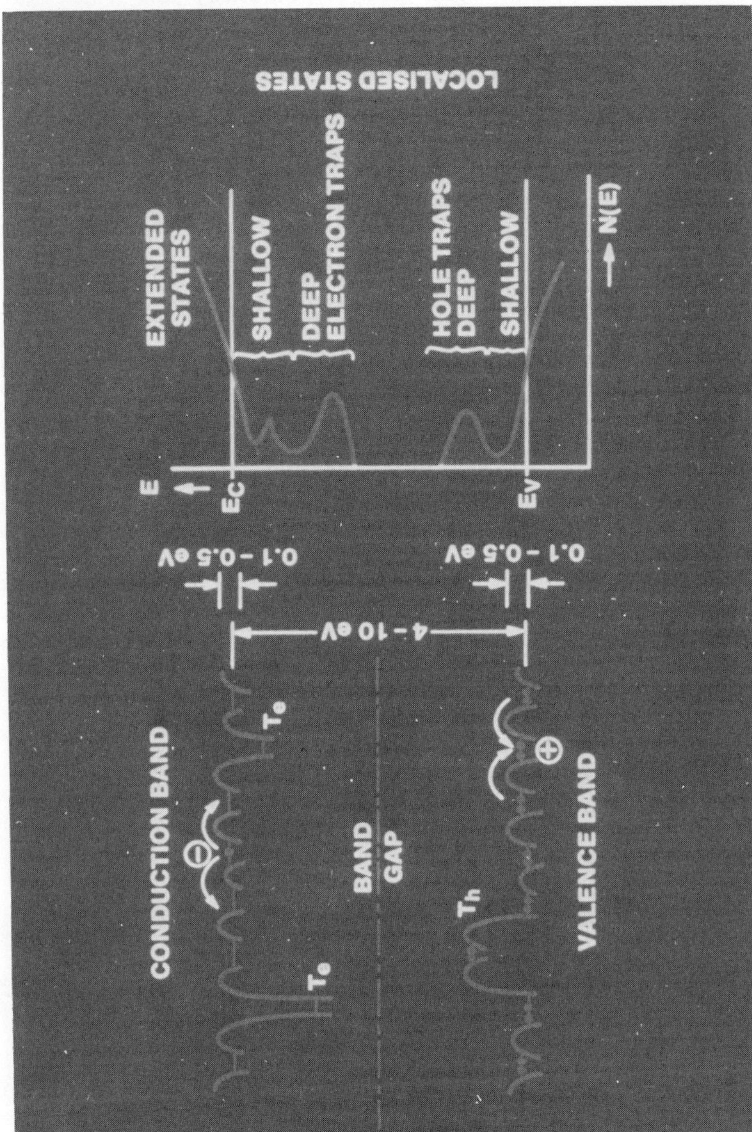

Figure 6. Diagram of the possible charge distribution in insulating polymer films into which electrons have been injected. The states suggested are symbolic of the wide range of electromagnetic effects to be expected in polar structures of macromolecular solids.

These traps, (Fig. 6) and similar effects in the motion of holes and other charges through polymers, would eventually be correlated also with such structural probes as positron lifetimes in macromolecular solids. Extensive recent studies of positron lifetime are based on positronium decay. In this, the lifetime of o-positronium (bound positron-electron pair with total spin one) is reduced from about 140 nanoseconds to a few nanoseconds by "pick-off annihilation" in which some unpaired electron spins in the medium cause conversion quenching of orthopositronium to para-positronium. The speed of the τ_2 effect is supposed, among other things, to represent by pick-off annihilation the presence of defects in the crystalline lattice. In any case, what amounts to empty space between molecules can then be occupied by orthopositronium.(14,15,16) It is now found in linear polyethylene, by T. T. Wang and his co-workers of Bell Laboratories(17) that there is marked shift in positron lifetimes over the temperature range of 80°K to 300°K. For orthopositronium pick-off annihilation rate $\lambda_2 = \frac{1}{\tau_2}$, for a variety of specimens, there is a parallel linear temperature dependence, except in the region 160°K to 230°K. Indeed, outside of this region the temperature dependence is very like the volume change, except with opposite slope. Further the change outside the transition region is insensitive to crystallinity of the specimen, and apparently therefore the λ_2/T curves relate to structural changes in the disordered regions only. It is believed that molecular motion in those regions, with a correlation of frequency ν_c of $10^7 - 10^8$ hertz or more, facilitate the diffusion of electron density into the free volume occupied by the orthopositronium. Self-annihilation rate would then be reduced.(18)

The disordered polyethylene shows a broader distribution of orthopositronium annihilation irregularities than a more crystalline low molecular weight sample. Above 230°C, which is the T_g claimed for polyethylene,(19,20) positron annihilation effects are uniform for all specimens of varying perfection.

Thus, overall, in the future we should look for specific correlations of the charge implantation, electret formation, charge transfer, and positron annihilation as methods of identifying solid state qualities. These in turn would lead to additional functions of polymers for the transformation and storage of information. Someday one could hope that there would be a coupling of this informational content with molecular mechanics of oriented or otherwise configured polymer systems. Accordingly, a structure could be programmed, by appropriate electromagnetic or optical information insertion, then to alter its physical state under highly controlled and timed conditons.

Trapped Charges as Information Codes

Will there be such polymeric machines, based on rather more intimate interactions than even the ion-based, largely long term exercises that analogs of muscle function have already produced? Indeed the influence of external fields on natural macromolecular processes remains also a fascinating arena. For instance, it has been reported recently that the polytetrafluorethylene electrets described above were included in bandages applied to the healing of full-thickness skin incisions in guinea pigs. These charged films appeared to influence the formation of collagen fibers. Compared to the presence of the same materials uncharged, examples showed 128% improvement in tensile strength of the incision after five days and 55% increase after nine days. Similar influences on bone knitting have also been reported by Japanese workers.(21)

Combinations of Polymers; Phases

Indeed, in the world of tomorrow we can expect new aspects of polymer solids to extend the conventional and successful structure ideas of this century. These, of course, were the recognition as molecular identities of the chains of repeating chemical monomers. The circumstances of those entities have resulted in interesting concepts of solubilities, viscosity, and other mechanics, and especially thermodynamic limitations on mutual solubility or compatability of polymer mixtures. But we have known for decades that even homogeneous regular chain polymers such as Carothers' polyesters and polyamides formed solids with manifold imperfections and irregularities, such as order-disorder crystal configurations.(22,23)

Now as we discussed with respect to charge distribution in polymer films, we know that there are dislocations and disorders in marvelous variety. The technology of these imperfection systems will increasingly take advantage of this complexity of states, just as it has for decades in polymer and rubber formulations and compounding, plasticizer miscibility, etc. But now we have a special stimulus for probing polymer mixtures, blends, composites and heterogeneous states, in which even the simplest mixtures are mostly found. For we know that somehow in nature these limited compatibilities are coupled probably with charge distributions from ions or other electrolyte phenomena. This action must lead to such macroassemblies as tubulin, of microtubules, of wondrous membranes fiber nets, and a host of other supermolecular structures, in natural tissue. These segregations of matter presumably represent some surface or interfacial definitions. These may be rudimentary qualities of information content that we also believe must be present in living structures.

Thus, let us look toward identifying what we now can understand about mixtures and surfaces and interfaces of compatible and incompatible polymer systems. Perhaps then we shall see a little of a beginning pattern of how the sublime beauties of D'Arcy Thompson's "shape and form" must ultimately relate to macromolecules. We have seen that charges do distribute, do provide technical qualities of high value, do exist for long times in a polymer environ and thus could perhaps provide these forces at a distance that Coulombic ranges engender, and over which chemical bonding and dispersion forces fall short.

Polymer Interfaces in Composites

Once more, too, as we seek to peer into the future, a lively technology seems to be paralleling the deep challenges of scientific understanding of polymer composites and their components. So let us sample briefly some of the major trends in these applications as a further preface to noting how simpler polymer systems combine or reject combination. In each case we shall be mindful that we are looking for the film, fiber or surface qualities which involve action over some distance beyond the usual dispersion and dipole forces (although these remain presumably the principal interactions). A perceptive survey by T. Alfrey, Jr. and W. J. Shrenk (24) covers the extraordinary qualities of multipolymer systems. These range from the commercial virtues of Noryl polystyrene/polyphenyleneoxide blends miscible in all proportions, to interpenetrating polymer networks having restrained phase separation into submicroscopic regions. Oriented forms of these systems can yield microfibrils, spirals and other morphologies suggestive of natural plant and animal tissues. So once more, in crude comparisons at least, the future needs to probe the fine details of mixed polymer behavior.

Similarly, where inorganic and organic matter are combined as in composites, filled rubbers and laminates, the mechanics of phase, surface and film interaction are often startling.

Epoxy composites containing around 65% by weight of polymer carbons fibers are having a profound impact on the design of structures including aircraft, rockets, spaceships and automobile bodies. In a recent assessment of such engineered structural material, J. J. Harwood and his associates (25) at the Ford Scientific Research Laboratory have shown how assembly of these polymer carbon/epoxy systems can yield tensile strength of more than 200,000 pounds per square inch or 15,000 MPa with stiffnesses of millions of psi or nearly 200 GPa. The essence of these qualities seems consistent with our comments about polymer structure, for the carbon fibers are generated by selective decomposition of carbon chain systems which result in extensive network formation and electronic excitation as in Fig. 7, Fig. 8. These lead to rigid, stable, solids, which show little of the planar anisotropy or crystallinity of graphite.(26) The actual

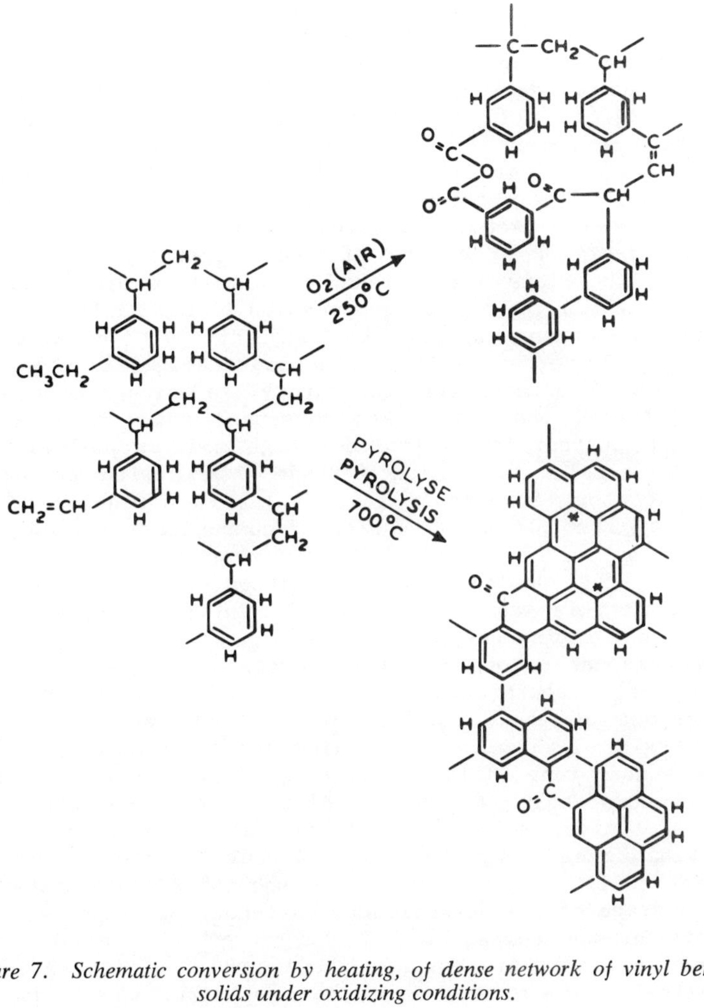

Figure 7. Schematic conversion by heating, of dense network of vinyl benzene solids under oxidizing conditions.

strength of such fibers is enhanced, however, by orientation of the layer planes parallel to the fiber axis.(27)

Polymer carbon composites are fully established as major parts of the new ways in which designers and materials engineers merge to generate novel properties and functions. But as the understanding of polymer carbon systems spreads, the features we have noted about the configuration and bonding of the progenitor polymer will yield enhanced performance. Fifty million psi modulus is now characteristic of commercial products of carbon fibers and uni-directional composites reinforced with it go to more than half of this rigidity. Not only military and rocket components, but most of the structure of the business jet aircraft called Lear Fan are based on the extraordinary properties of polymer carbon composites. Fourteen hundred pounds of such fibers are used in the total structure of this airplane, which can carry 1700 pounds of fuel. Bonded carbon fiber sections of the plane are designed to withstand 8,000 pounds of torsion and 25,000 pounds of compression. The 20 to 30 percent reduction in weight, typical of many aircraft uses, is even surpassed by 70 percent weight reduction achieved in experiments noted, in the Ford Scientific Laboratory, in which a total carbon fiber reinforced body of a six passenger automobile has been created and tested. This includes high modulus polymer carbon fibers of Hercules and Union Carbide types, with strength at breaking of about 2350 Mpa's (340 ksi) with a modulus of about 350,000 Mpa (50×10^6 psi). The cured composites contained about 62 percent by volume of the polymer carbon fiber. (A detailed account of the design and fabrication of this remarkable conceptual vehicle was reported at the International Conference on Composites, Paris, July, 1980, by E. J. Horton.)

With respect to other polymer carbon composites, studies by D. Gloge, Bell Laboratories, indicates that 10% polymer carbon fiber content of the cross section, in our ongoing design of cable carrying lightguide glass fibers for communications, can triple the duct run. This is due to the higher strength, and in fact will completely relieve the glass fibers from stresses on the cable. Also, such carbon fiber reinforcement is 4 times lighter than steel. It further, of course, retains the total insulating qualities of the whole new photonics cable system.

This stage of polymer science and technology thus betokens a period of growth comparable to the surge in earlier decades occasioned by synthetic organic chemistry, with its variety of new chemical monomers and novel bonding structures. In other words, we may see the fields of mixtures, blends, composites and internal structure perfection controls producing novelties of performance in addition to those bestowed by the vast range of chemical variance.(28,29)

Polymer Blends: Linear and Microgel

Accordingly, it is interesting to observe the beginnings of these combinations of new science and already familiar technology in such structures as block copolymers and blends of both drastically varied and chemically similar macromolecular structures. An example of the latter is the various blends of microgel molecules, three dimensional globular types, with the linear systems of a diversity of synthetic rubbers and natural rubber. The globular microgel network structure indeed modifies both the rheology and ultimate mechanics of the rubber.(30)

Improvement in the processing and vulcanized qualities of a range of systems have been reported over the past decades. Modification of natural rubber, due to work in the British Rubber Producers Research Association, yields some of the most striking applications of microgel. A detailed study at the MV Lomonosov Institute of Fine Chemical Technology, in Moscow, on the effect of microgels on mechanical properties of cis-polyisoprene and butadiene-styrene rubbers extensively illustrates the properties of blends from latex combination of microgel and conventional or linear systems.(31)

Regarding simpler mixtures of related but chemically different polymers, the dimensions of future activity are reflected in recent monographs.(32,33)

Polymer Blends: Linear but Varying in Subunits

A recent investigation in our laboratories involves the delicate balances in energy and entropy which determine various levels of compatibility of even closely related polymers. Thus, A. J. Lovinger has examined the tensile properties and morphology of blends of polyethylene and polypropylene, Fig. 9.(34) These blends are known commercially for high impact strengths. Various earlier workers have found a range of qualities yielding a maximum in strength and modulus in a blend of about 90% polypropylene by weight.(35) Others had observed a maximum in strength and modulus at 75% polypropylene.(36)

Current findings are that over a wide range the presence of one regular microcrystalline polymer partly influenced the morphology of the total blend and affected the nucleation and spherulite growth of the other component. So in the present context, the conclusions are that surfaces and interfaces are readily induced in such mixtures such that, in accord with our earlier findings, relatively long range, and even macroscopic effects are produced by relatively intimate mixtures of close chemical similarity.

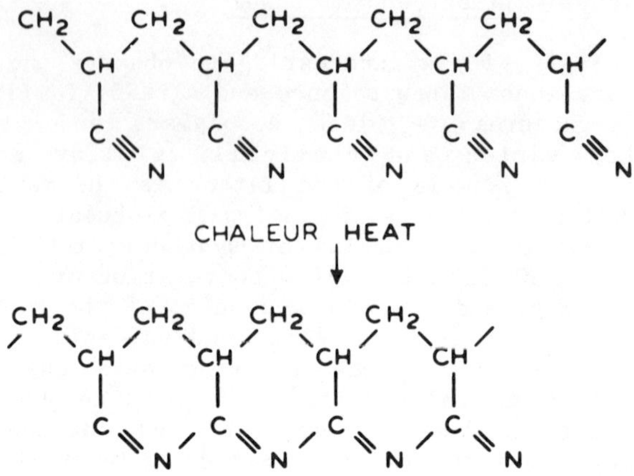

Figure 8. Schematic conversion of polyacrylonitrile by heating, into eventual polymer carbon network.

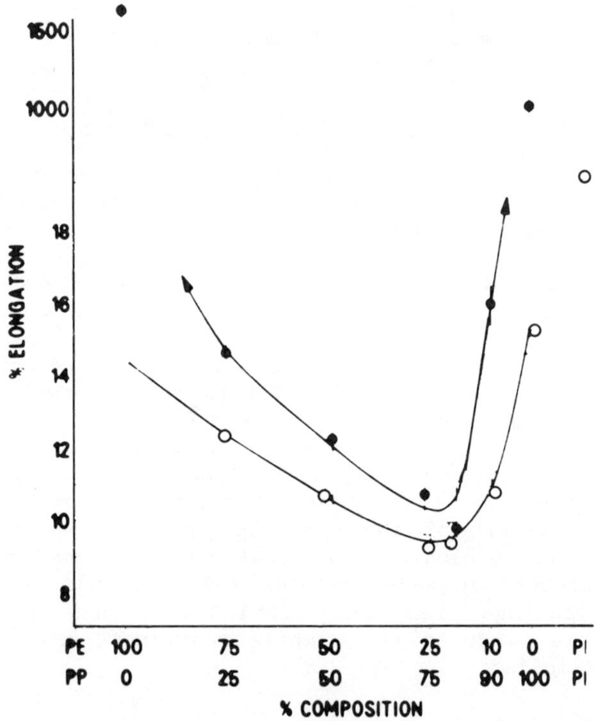

Figure 9. Percent elongation at rupture as a function of the composition of mixed polymers of polypropylene and polyethylene indicating influence of solid state phases on strength and extensibility.

Homogeneity as Reflected in NMR of Blends

In further search of qualities of mixtures which could bear on the ways that changing polymer interfaces regulate and take on certain gross forms, Douglass and McBrierty have applied many studies of nuclear magnetic resonance to appropriate hydrogens and other magnetically polar components of the systems.(37) They have examined in our laboratories a series of alkanes C_6H_{14} through $C_{94}H_{190}$ as a reference frame for relations of the magnetic spins to the host lattice vis-a-vis their coupling, and thus to the exchange of energy between neighbors in the spin system. In this latter, the spin energy moves from a hot or exciting, to cold, relaxing region. The scale of volume involved (that is, the relaxation time required in a specimen having irregular motions and hence inhomogeneity) can reflect on the distances and dimensions of domains when the diffusion coefficient for the spin energy is also known. In the alkanes, the methyl group in terminal molecular planes apparently can reorient rapidly and thus provide an energy sink for the rest of the excited proton system.

It is found that the relaxation parameter $T_1\rho$ as a function of temperature does not follow an increase with chain length, as the square of the number of methylene carbons. Nor is it linear with N, the number of methylene carbons, which should be true if relaxation to the lattice were rate controlling. Rather, it shows a temperature-induced increase of the minimum value of $T_1\rho$ with about the 1.6 of N. So, both spin diffusion and spin lattice coupling are reflected. For a spin diffusion coefficient D of approximately 2×10^{-12} cm.^{2}/sec., the mean square distance for diffusion of spin energy in a time t is the $\sqrt{r^2} = 200\sqrt{T_1}$Å, or about 15Å on a $T_1\rho$ time scale.

When the applied magnetic field H_1 considerably exceeds the local field of the proton dipoles, the small scale structure of inhomogeneity determines shape of the $T_1\rho$ decay. Models can be made of formalized spherical domains and the like. They affirm that a composite decay such as noted above, resulting from the combined spin coupling and spin diffusion effects, does demonstrate inhomogeneities. In even so classic a system as plasticized polyvinyl chloride, recent heat capacity measurements in our laboratories by Bair (Bair, H., Private Communications) indicate at least two glass transitions. Douglass finds that polyvinyl chloride at 75°C containing 1.5% plasticizer (and higher) gives composite signals, one of which appears to be unplasticized PVC. Presumably the plasticizers are not uniformly mixed, and free induction decay studies indicate very small regions of low mobility.

In still other NMR studies of mixtures, Douglass and McBrierty have considered the homogeneity of polyvinylidene fluoride - poly methylmethacrylate blends.(38) Here the energy exchange between proton and fluorine magnetic spins is efficient only if the nuclei are adjacent. The results indicate, when taken around 40°C, that either the fluorine nuclei in a 40 polyvinylidene fluoride/60 poly methylmethacrylate blend are close to a large fraction of methyl groups, or at least that the molecular motion of the polyvinylidene fluoride molecules is altered by the presence of the poly methylmethacrylate molecules. Under both circumstances, the evidence is that the blend seems to be homogeneous in the amorphous regions.

A contrasting conditions occurs as found by Douglass and McBrierty when polypropylene and polyethylene are blended. This is to be expected from our earlier comments on crystalline and mechanical properties. Here, films of high molecular polyethylene and polypropylene were prepared from solution by coprecipitation and fusion, according to the technique of Coombs, Cannon and Keller.(39) The general model derived from these studies shows coexisting crystalline aggregates of polyethylene and polypropylene, with various stages of interpenetration and molecular sharing. Thus it is evident again that polymer blends are highly sensitive to local submolecular structure.

Particularly striking behavior of polymer separations into characteristic domains such as spheres are found when films of binary or more polymer species are cast from solution. In the case of polystyrene and poly (vinyl methyl ether), Davis and Kwei have studied effects of subsequent heating of the heterogeneous film, cast from trichlorethylene. When such films were heated above the critical temperature, they became clear and homogeneous in appearance as long as the elevated temperature was below another temperature curve of between 140° and 160°C, over a compositional range 20 to 80% polystryene in the mixture. The lower temperature curve which had to be exceeded for homogenization was in the 60° to 80°C range.

When these two polymers were cast from toluene, the film was initially homogeneous and showed only one glass transition temperature T_g.(40) Thus, we see in the solid state significant redistribution of interfaces and domains can be achieved, in some intermediate temperature range which was found to depend heavily on moledular weight. The domain sizes ranged from 4 to 140 micrometers in a 50-50 polystyrene (molecular weight 37,000) film blend. After the redistribution, single glass transition temperatures equal those for the corresponding toluene-cast film were determined.

In related work, Kwei, Frisch, Radigan and Vogel have examined ternary polymer mixtures containing the incompatible pair polymethylmethacrylate and polyethylmethacrylate. They found that a third component, polyvinylidene fluoride, would lead to completely homogeneous mixtures. In the single phases, T_g, at least over the range 40 to 70% by weight of polyvinylidene fluoride, equals the volume fraction average of the T_g of the component polymers. (This work followed determination of binary parameters of the polymethylmethacrylate/polyvinylidene fluoride and polyethylmethacrylate/polyvinylidene fluoride methacrylate/ polyvinylidene fluoride.) (41,42)

The primary incompatibility of the two polymethacrylates (43) is itself striking evidence of the sensitivity of the fluorine interaction parameters to subtle chemical and compositional effects.(44) In the intermediate ranges of polyvinylidene fluoride concentration, when the ternary mixture becomes apparently homogeneous and T_g equals the average of $\Sigma_i \phi_i T_{gi}$ over the range of 40 to 70% polyvinylidene fluoride, it appears that the free volumes are additive. Thus, the various segments of differing structure would seem in this case to be extensively mixed. Again, future studies are promising, in revealing how extensively polymer blends can be controlled, and what the kinds and properties of the inhomogeneities are.

Polymer Interfaces and Phases

It also appears, as we shall discuss presently, that some macromolecules, such as polyvinylidene fluoride noted above, have exceptional interaction properties, in which the segments may behave rather differently than the chemical monomer units would imply. For these and many related reasons, the studies of E. Helfand at Bell Laboratories on theoretical concepts of interfaces in polyphase systems of macromolecules give us a keen sense of the scope of future discoveries that are possible in this field.

Once more, of course, we have in mind the pervasive issue of how nature distributes complex polymer compounds and functions in the formation of subcellular and cellular and living systems. Thus a theory has been derived for two immiscible polymers.(45,46) The nature of unsymmetric polymer/polymer interface is assumed to arise from balancing of the local and nonlocal free energies. The nonlocal portion does not tend to sharp boundaries, in opposition to the local portions. The nonlocal factors of free energy come from the conformational entropy, which requires the units to be associated with each other through the chain bonds, and also the corresponding interaction that those conditions impose on the appropriate segments. Reasonably good consistency with known parameters of 15 polymer pairs are obtained. Specifically, interfacial tension calculated is comparable with that observed. The series has been extended to concentrated polymer solution/solvent interfaces by the same workers.(47) The essential parameters of the interface have been identified in this detailed theory. In accord with earlier work, the density profile at the interface for both the lattice model and the Gaussian random walk statistical form (which agree very well) show that a great variety of configurations and interactions could exist, and thus might offer options for novel technical treatment.

Another welcome outcome of the theoretical studies is a stimulus for more experimental information and the identification of parameters suitable for sharpening measurements. A promising aspect of such matters in our laboratories is the expanding use and interpretation of Brillouin scattering by G. D. Patterson and his associates.(48,49) Both the Rayleigh and Brillouin scattering by solids are sensitive to density fluctuation, inpurities, inhomogeneities of the sort that we are emphasizing. Inhomogeneities in films of polyvinylchloride and of cellulose acetate are shown as peaks near the main longitudinal Brillouin spikes. For polyethylene terephthalate film, the scattering spectrum was complex and there seemed to be at least two well defined longitudinal peaks. This suggests heterogeneity of dimensions of about 1,000A and is presumably consistent with partial crystallinity of the polyester. It seems likely that the transverse phonon peaks, which appear relatively strongly compared to longitudinal ones, arise from properties of the crystalline regions.(50) The relaxation phenomena and scattering indicate no large effect on the moduls μ at hypersonic frequencies at T_g. However, at melting, there is a large shift in μ. A study of the mixtures 40/60 and 75/25 indicate single pairs of polarized Brillouin peaks marking a homogeneous amorphous phase.

Block Co Polymers

Now again, a state of inhomogeneity in polymers, so especially interesting in films and interfaces, occur when discontinuities are built into the main valence chains and networks. Block polymers are the classic embodiments of this. Many periodic distances separating domains in such alternating or rhymthic copolymers have been reported. These indicate existence of phases in laminar domains and, in other cases, of spherical domains.(51) Cases are shown experimentally for styrene/isoprene copolymers and also for styrene/butadiene.(52,53,54)

Through Helfand's theory,(55) interface conditions are defined. Transitions from nearly pure component A to component B are expected in the interface regions. Most interestingly, the theory suggests an entropy decrease associated with preferred selection of conformations necessary to keep the density uniform in the interphase area. Thus, we may speculate that localized but supermolecular configurations exists in forms which would allow information to be distributed in such a film, such as through selective diffusion, particular light absorption, dielectric condition or other physical, optical or electromagnetic pecularity. The difference between simple polymers will, of course, be concentrated in the interphase area.

Likewise, conformation entropy decreases as the microdomains enlarge, because the system rejects increasingly the many conformations leading to density inhomogeneity. It favors the special ones which fill the lamellae centers in achieving density uniformity. Minimization of the free energy (Fig. 10) taking these things into account gives a domain size in styrene/isoprene cases of around 40 manometers, quite close to the rather roughly observed figures, Fig. 11.

Polymer Films as Pattern Generators

Another role for polymer film and surfaces in the world to come is already firmly founded in the notion of modern thin film and integrated electronic circuitry. The era of solid state electronics determines nowadays our use of automata and other elements of highest productivity in international economy, as well being increasing factors in science, eduction, and national security. These capabilities are now primarily embodied in micro circuits, whose integrated form is made directly on single crystal surfaces of silicon or similar semiconductor.

A forthcoming era of photonics circuitry embodying laser generators and various photo detectors and storage systems, will provide similar ultramicroscopic circuits for communications, logic and memory.

These elegant instances of the best we know of how to attain electromagnetically the action of digital machines and of analog signals and processes are produced by computer aided design (CAD). They then come out as truly artistic elements in a combination of engineering and operational support of knowledge and communication around the world.

Now the role of polymer films is as the lithographic resists on the semiconductor chip surface. Accordingly, the various layer patterns and diffusion processes in the semiconductor and other electronically active films that comprise eventually the integrated circuit can be regulated as to shape and depth. The present ideal is to have continued down-sizing of the active components, from the present 150,000 on a silicon chip of about a square centimeter size, to a million or so within this decade. (Our aspirations to be comparable to organic efficiencies are still some distance off, since even the 150,000 corresponds roughly in cubic form to an order of magnitude less numbers of active components than there are neurons in a cubic centimeter of the brain. And of course, the neuron arrays work vastly beyond any functions we can achieve on these chips.) However, the fabulous economy of these systems and their unsurpassed efficiency (56,57) depend especially on the more than 500,000 interconnections which must be assured among the 150,000 component-bearing chips. These mean fine lines of conducting matter, perfectly formed, so that metals and other ingredients can be either added or dissolved away, according to the eventual circuit shape. Photography has performed this with the use of positive resists for decades. These are when the polymer becomes soluble because of some photo exposure in contrast, to negative resists when the polymers is cross-linked and its removal is decreased by insolubility, (Fig. 12). The classic positive resists usually have a weakly acidic polymer such as phenol or

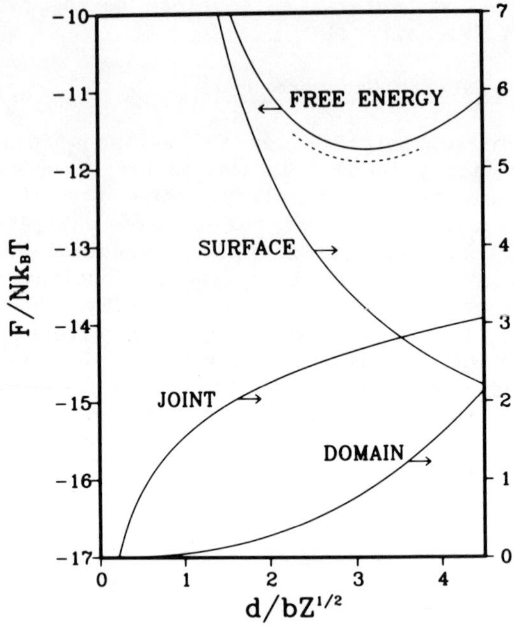

Figure 10. Illustration of influence of domain formation in block copolymers according to the models of Helfand and coworkers. The free energy is shown as a function of the size and separation of the domains of varying composition. (The spontaneous separations may be analogous to the way superstructure is formed in natural polymers of plants and animals.)

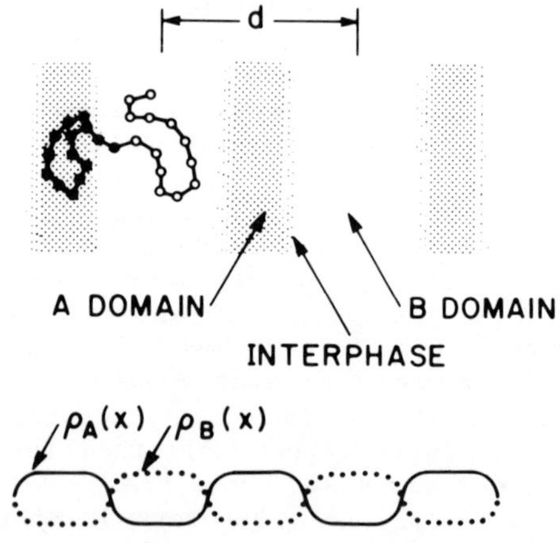

Figure 11. Schematic of the block copolymers domains with density and structure indicated by the Helfand concept and approximated by certain experiments.

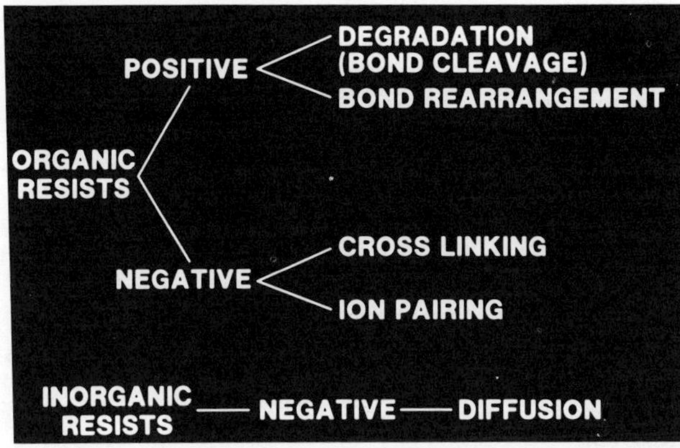

Figure 12. Description of the functions of polymer films forming organic resists, acting as the information pattern in the design and fabrication of electronic integrated circuits.

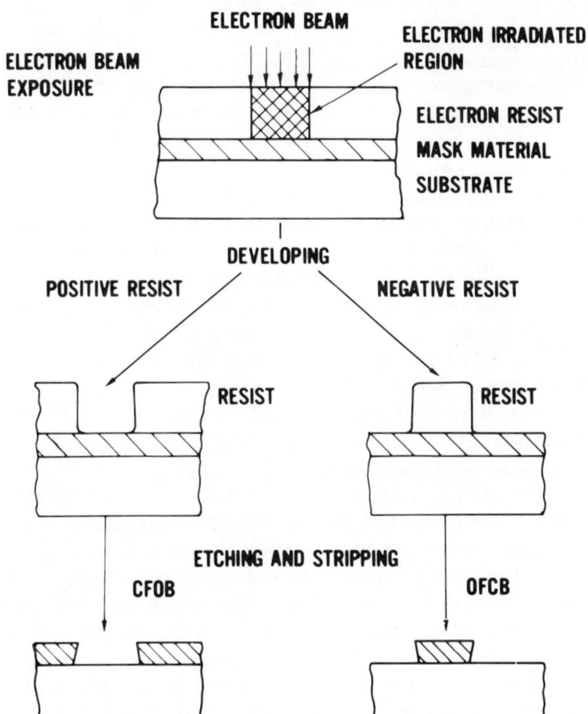

Figure 13. Outline of the pattern generation for integrated circuits provided by electron beam writing on chemically suitable polymer films, which are thus sensitized for development.

formaldehyde condensation systems, or sometimes a strongly acidic acrylic acid containing element.

Evidently, a new area of polymer film activity is arising from this function of information pattern production.(58) Our invention in recent years of relatively high speed electron beam generators (Fig. 13) for primary patterns (mask making) and also direct writing where the circuit design is traced directly on the semiconductor chip has had a revolutionary influence on compactness and quality of integrated circuitry.(59) Correspondingly, it has depended on steady improvement of polymer resists. For this function such film should be after dissolution, or cross-linkage to prevent solubility, (negative resist) capable of submicron resolution with a sensitivity to 10 to 30 kilovolt electrons of about $10^{-6}C/cm^2$.

The removal processes or development may include plasma and sputter etching, ion milling and other supplements to wet etching. Evidently, these factors occasion new polymer science and technology, ranging from initial syntheses to the understanding of phase relations, charge distribution and absorption, and overall chemical reactivity. With electron beams poly-methylmethacrylate has excellent resolution, but dissolves relatively slowly, by a curious nonswelling mechanism long ago remarked.(60) Sensibility of the polyolefin sulfone to electron radiation was earlier described to chain breaking at the CS bond.(61) Polybutene^{-1} sulfone is particularly appropriate and is now extensively used commercially (Fig. 14).

For negative resist, where cross-linkage is required to reduce swelling and prevent solubility, various balances between solubility of the unexposed portions and cross-linkage of the negative itself have been sought in a variety of polymer structures (Fig. 15). A particularly useful one developed by Thompson and coworkers in our laboratories is a copolymer of glycidylmethacrylate and chlorostyrene (Fig. 16). It is capable of resolving one micrometer lines and spaces, but swelling still obscures submicron resolution.

In the world of tomorrow it does seem that widespread commerce, technology and science will proceed from these new information-based applications of polymer films. Obviously, varieties of printing plates are presently using these methods for the conventional patterns of human language. The recording of digital codes, now widely done by polymer films containing magnetic particles, may turn toward direct polymer inscription. The processes involved extend to behavior of polymer films as protection for metals, woods, and other essential elements of civilization's structure. The detailed studies of chain scission, polymerization and of cross-linkage are all fundamental to processes of curing, degradation, stabilization and synthesis which have been studied in other contexts for generations, but which are sure to acquire new attention as this era of information science and technology evolves.

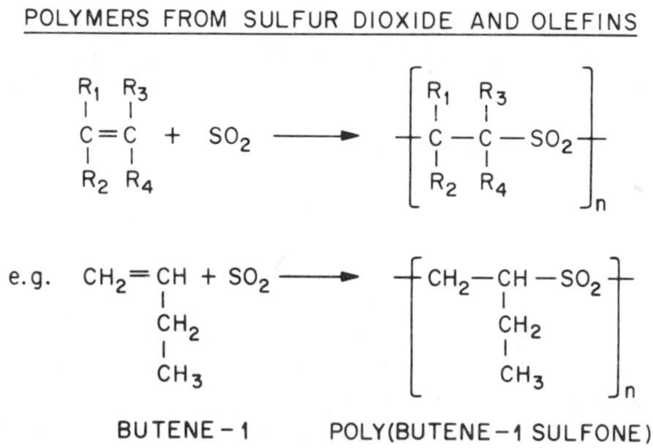

Figure 14. Example of polysulfones exhibiting special response to electron beam bond cleavage, yielding positive resist.

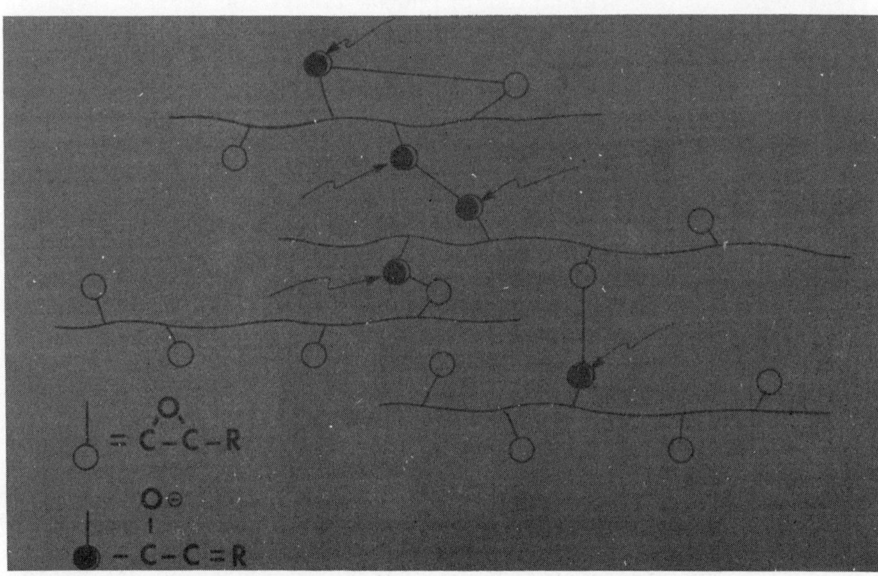

Figure 15. Schematic of systems providing negative resists through precisely controlled cross-linking, such as in epoxy systems.

Economics alone make this probable, for the present industrial applications of integrated circuitry and solid state electronics derived from the original transistor and solid state science now contribute $100 billions per year to the gross national product directly. Further, many elements of our national security depend on the use of these communication and information processing systems in every phase of command, control and weaponry.

Polymers in Photonics

Indeed, a fascinating example of the new wave of science and engineering using the precise formation of polymer films for a function crucial to new resources for handling information is now in full activity. It is that, as a consequence of discovery of the laser by Townes and Schawlow, an era of photonics is coming in to perform the functions of signals, logic and memory provided by electronics and magnetics up to now. The transmission of photons efficiently is an essential element in this new arena. In our laboratories and others, hair-thin fibers of appropriately chemically pure and doped silicon now dramatically exceed the efficiency of copper or other electron conductors in transmitting light waves. The production of these fibers is another saga of modern science, itself derived from our knowledge of silicon and its reactions developed in the origination of silicon semiconductors three decades ago and since.(62)

While the transparency of these fibers has been improved in the last decade as much as in 3,000 years before (Fig. 17), the fragility of glass remains a crucial limitation. It can, however, be drastically reduced by coatings, which fill the surface defects or distribute the stresses so that fractures are reduced. These coatings must be done concurrently with the drawing of the fiber from the preform before there is any contact of it with a solid surface. The application of the coating which follows 2200°C high purity environment of the fiber drawing furnace is a liquid coating system applied with a pressureless reservoir by applicator. It was developed by L. L. Blyer, Jr. and his associates at Bell Laboratories. A compliant applicator, for monitoring of concentricity by laser adjustment of the applicator, hence relieves contact with the coating die (Fig. 18). Solution systems are avoided because of the time required for solvent vaporization.

For highly fluid coatings that have to solidify rapidly, obviously fast cross-linkings by thermal or radiation activation are effective. A characteristic susceptibility of fibers to fracture with increasing length, because of the statistics of flaw distribution and the many ways in which these flaws can be induced, such as by chance particles, inhomogeneous qualities of the coating, etc., demand unprecedented precision in polymer film formation. However, there are additional factors, such as

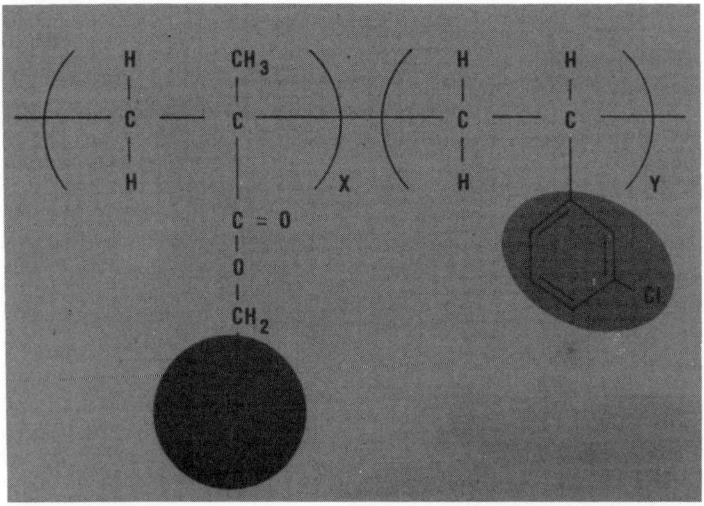

Figure 16. GMC, an example of copolymer selected by Bowden and Thompson for optimal sensitivity in pattern generation, through negative resist formation.

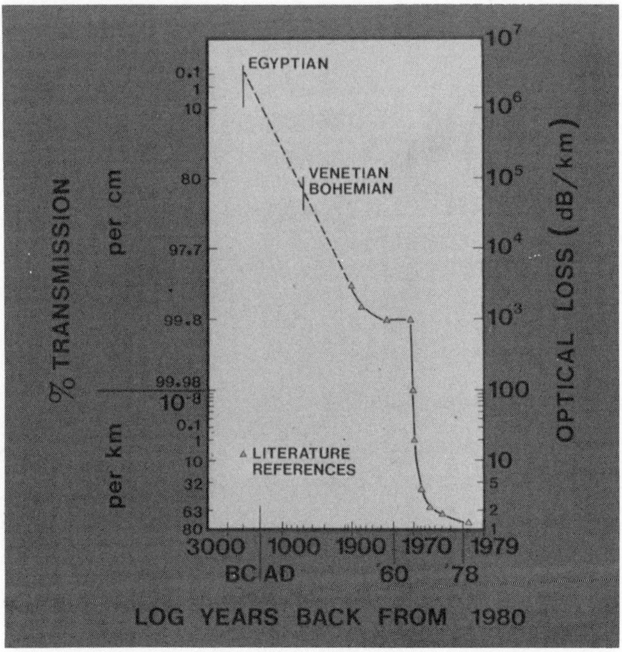

Figure 17. Chronology of improvement in light transmission of glasses over periods of history leading to introduction of photonics technology in the information age. Polymer protection of glass surfaces is an essential present feature.

sensitivity of glass to formation of color centers (F-center) as a result of any kind of radiation, as attractive as that may be in forming the polymer layer (Fig. 19).

Mechanical Factors in Polymer-Coated Light Guides

Beyond that, it is also found that fiberlight guides engender apreciable attenuation from what is called micro bending. This is when micron-sized displacements of the fiber axis occur in small periods of length, like a millimeter. These arise from casual perturbations of the fiber and its cable by nonuniform lateral stresses like winding, handling, suspending, etc. These important optical losses can be vastly reduced or eliminated by cushioning the fiber, by means of the thin polymer protective layer, from these stresses. This introduces, however, a need for low modulus coating which may, of course, be covered by a hard secondary coating, if necessary. The relaxation modulus should be below 10^7 Newtons per square meter to give appropriate protection (Fig. 20). Thus, we see the immediate application of sophisticated polymer rheology in relating structure to these extraordinary demands for a fluid system at 10^3 - 10^4 centipoise when applied, but which can be quickly converted to the solid cushioning and, at the same time, flaw-healing film noted. Obviously, for such silica fiber protection, silicones for thermal cure are useful. The ones normally used are vinyl end blocked polydimethylsiloxane or polymethylphenylsiloxane cross-linked by multifunctional silanes. Cross-linkage at high temperatures is sufficiently rapid so that total fiber drawing rates of greater than one meter per second are practical. But the abrasion resistance of these silicones is so poor that an extruded nylon jacket is put over them. Ethylene/vinyl acetate polymers have also been applied from melt. Poly (ether-urethane acrylate) polymers have tensile relaxation modulus over 30 minutes of about 7×10^7 dynes/cm^2 and yield 0.3db/kilometer excess micro bending loss, whereas even a modified epoxy acrylate coating with a tensile relaxation modulus in 30 minutes of about 4.5×10^8 dynes/cm^2 yields 11.6db/kilometer excess micro bending loss, (resulting from a 132 grams tenison on a 10" diameter drum).

As can be seen (Fig. 21), the appropriate protection of these fibers yields extraordinary strength, exceeding previously achieved qualities of fiberglass. Application of the findings and principles to the making of large-scale composites, as well as to the use of polymers in stabilizing glass contains and structural glass offer new frontiers for future and extensive progress.

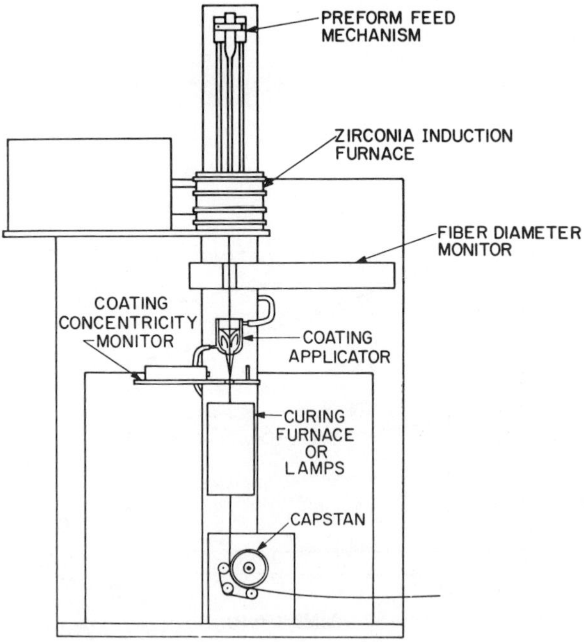

Figure 18. Method of coating ultratransparent light guide fibers of silica with polymer films for strength and stability.

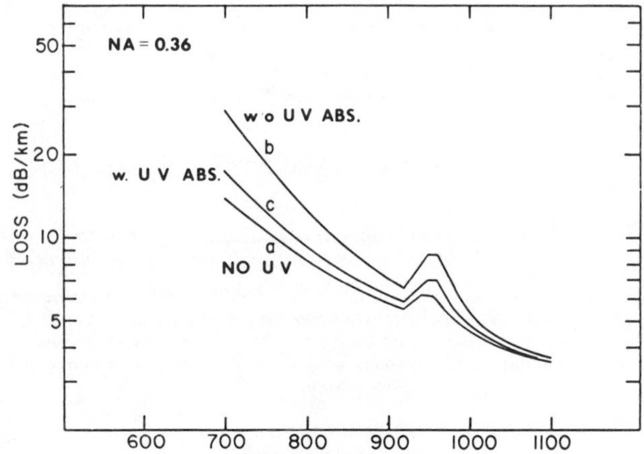

Figure 19. Influence of UV radiation as polymerization agent on the subsequent total photon attenuation in silica fibers, as a function of wavelength in nanometers. When no absorption of the UV is arranged lowest losses are assured. (The loss expressed in decibels per kilometer is a modern figure of merit for light guide systems.)

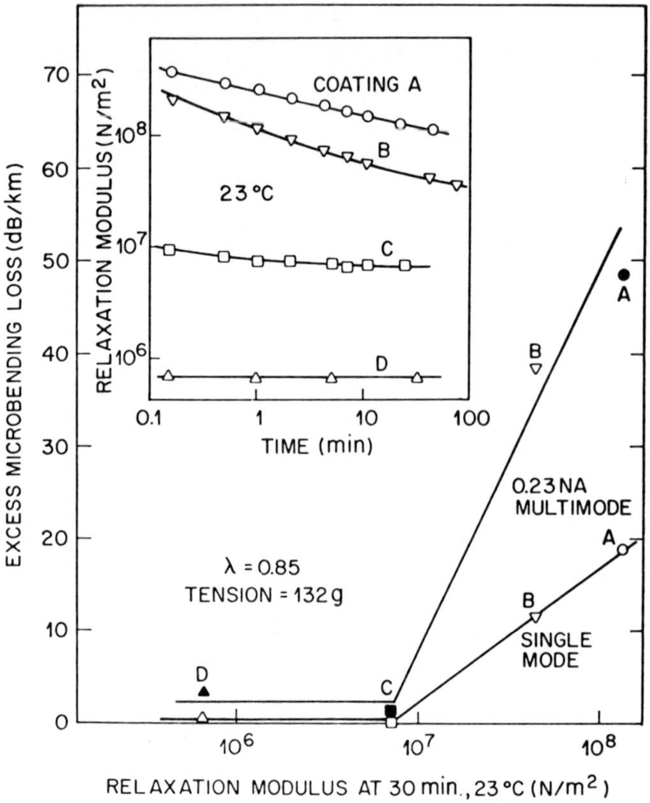

Figure 20. Limitation of excess microbending loss of photon transmission through light guides, as a function of relaxation modules of the stabilizing polymer coating.

As evidence of light guide fiber stabilization and control through processing refinement, plot of frequency of failure against the tensile strength (in hundreds of thousands pounds per square inch) shows extraordinary uniformity for the control system compared to wide distribution of failures from surface defects and other variations in a conventional fiber glass system.

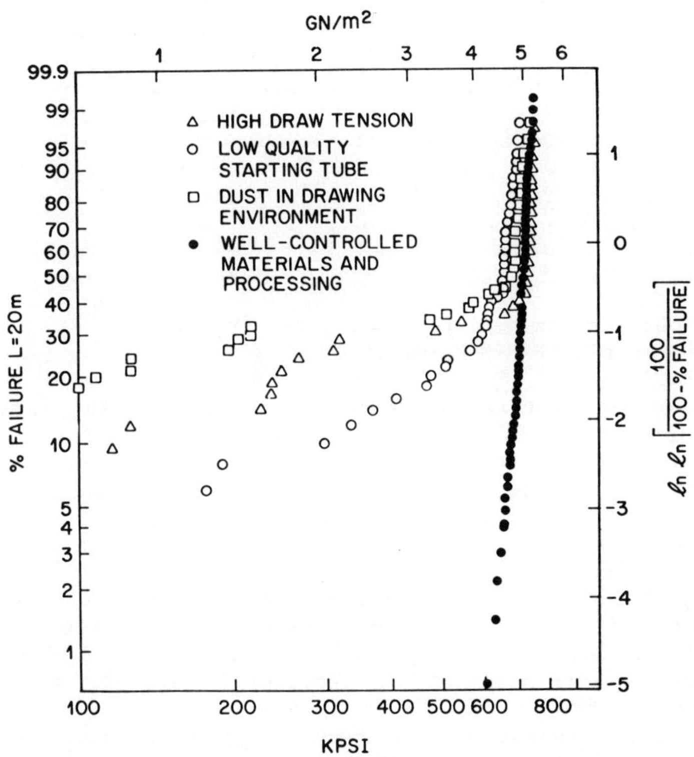

Figure 21. Demonstration of high strength and sharp distribution of tenacity of silica fibers protected and controlled compared to those from conventional processing.

Macroscopic Polarization in Polymer Solids

Our general consideration of domains and phases in the polymer systems have usually involved rather small regions of a given sample or surface. This quality includes even the familiar crystallite and sperulite aspects of macromolecule solids, although conventional orientation in fibers or biaxial sheets then does project these local structures into dimensions that approximate the total sample. There are some interesting cases emerging, which will doubtless be extended in the future, where polymers are further serving as information and electronically-sensing elements, because of qualities which have been induced in the total film by cooperation of all the molecules. A good example is polyvinylidene fluoride, where the β form of its crystal and the configuration of the molecule in that, was found to give a highly dipolar cell.(63) Correspondingly suitably poled films, or those cold drawn at moderate temperature and then exposed to a high electric field at high temperatures, do give a piezoelectric constant (induced charge per unit stress) of d_{33}, about $16 \times 10^{-12} \frac{C}{N}$. (Ultrasonic electronic transducers and other sensing devices can well be constructed from these films and have already been widely explored.)(64)

Consideration of the structure of polyvinylidene fluoride (65) assuming a barrier of 3 kilo cal per mole for rotational minima of conformation of the chain by A. E. Tonelli (66) led to detailed conformation and its implications for dipole structure (Fig. 22). Indeed, the material can approximate a ferro electric. It is thus of interest in our expectations of the environments that polymers can provide for the creation of new phenomena. The total array of dipoles in polyvinylidene fluoride will switch in about 3 microseconds at 20°C with 200 megavolts per meter field. The system becomes much slower at lower temperatures and fields. But we do have a case of macroscopic polarization intrinsic to the polymer molecules, which thus supplements the extensive trapping and other charge of distribution phenomena that we have discussed in connection with electrets.

New Polymers Will Serve New Purposes

Now we have sampled new aspects of the properties and potentials of polymers for an era of information and communications. We should also ever be mindful that new chemical structures and unexpected qualities will continue to appear. In terms of the surface features, we have remarked on charges in films and in electromagnetic properties for signal storage and processing. We shall expect also to find the role of composites and adhesives, where surface interactions are crucial, to be spreading widely in mechanical design. In subtle ways, such as how hydroxyethylmethacrylates serve as soft contact lenses, the surface and bulk properties will function in complex biological and prosthetic replacements.

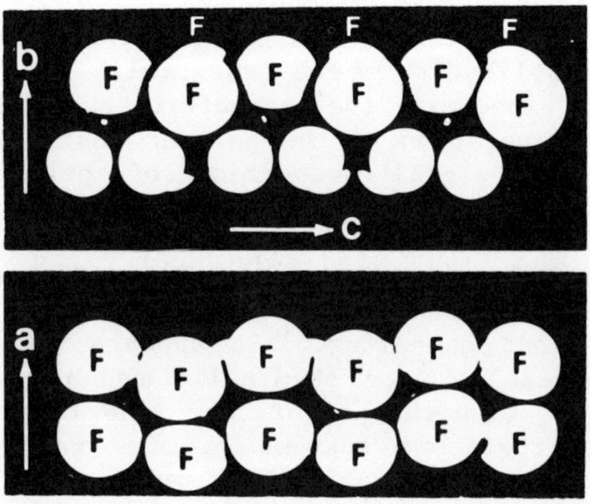

Figure 22. Schematic of poly vinylidene fluoride showing polarity in the unit cell when suitably oriented, so that the system exhibits macroscopic polarity and piezoelectric properties.

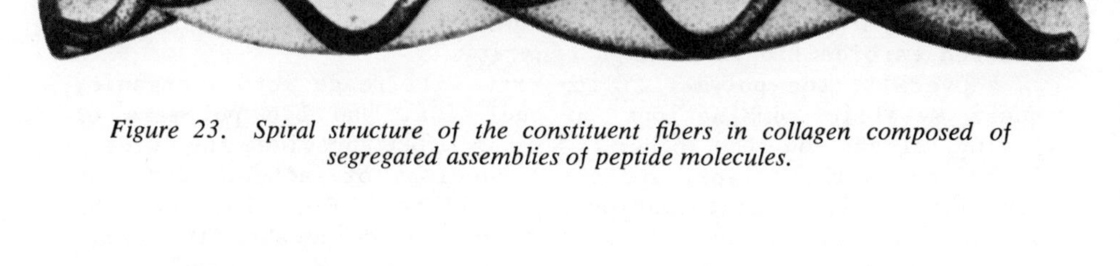

Figure 23. Spiral structure of the constituent fibers in collagen composed of segregated assemblies of peptide molecules.

Still other applications of surface and thin film properties of polymers are as diffusion and slow release agents for drug administration by capsules, for specific release agents injected locally into tumors, and for various herbicides and pesticides. Thus, 2-4-dichlorophenoxy acetic acid has been esterified with poly (2-hydroxy ethyl methacrylate) for field testing of control of aquatic weed growth in ditches and canals. Such polymeric delivery systems, especially for drug therapy, would appear to have extensive application where essentially the information for dosages is contained in the structure chemical composition, dimensions and aggregation of the polymer host. As indicated in the discussion of signal properties of polymers charged by implantation, these and other qualities of polymer films can presumably be modified by yet other implantations of charges, irradiation and selective heat conversion of small surfaces or volumes, such as are being produced in metals and semiconductors by laser fusion and annealing. It appears that these intense thermal impacts can produce chemically and physically modified surfaces. Indeed, with the use of very fine laser beams, certain microporous, heterogeneous systems can be created in this way, whereas larger scale irradiation can modify film properties and structure as well as composition.

Obviously, all of these principles applied to film formation and modification will be of special significance as when the films are used as coatings. Here too, we should expect a strong extension of an old and classic role of polymers.

In the energy domain, new and efficient uses in gas lines, electric cable ducts and the like, will promote surface stabilization and endurance as well as complex stress capability of various extruded or cast systems. Such reactants as acetylene terminated polymers have yielded cross-linked cured, networks of exceptional density and durability. A diimide dianhydride combined with (3) ethynylaniline yields an acetylene terminated tetraimide. On further polymerization at 250°C, the cross-linked structure derived can be used continuously at about 230°C. When this is combined with polymer carbon fibers or filaments, an exceedingly refractory and tough binder is produced.

Pyrolysis analogous to polymer carbon formation has also been applied to methylchlorodisilane. This is converted to beta silicon carbide fibers of high tenacity.

Overall, the polymer of tomorrow will reach into inorganic, quasi metallic combinations on one side, and bio polymers of living tissue on the other. These will provide the widest interface in the science and the technology of matter. Both the wonderful spiral conformation of collagen, Fig. 23, and the subtle information content of its peptide components in muscle action are qualities to be sought in polymers made by people.

LITERATURE CITED

1. See Advanced Technology, AAAS, Wahington, D.C. (1980) pp. vii-xi., Baker, W. O., Science, 1981, 211, 359.

2. Nature, 1980, 285, 287; Evan, W. H., Nature, 1980, 283, 521.

3. Newmark, P., Nature, 1980, 284, 659.

4. Dickenson, T. A., Electrical Manufacturing, August, 1948, 101.

5. Sessler, G. M. and West, J. E., J. Acoust. Soc. of America, 1973, 53, 1589.

6. Sessler, G. M. and West, J. E., Proc. Fifth Intl. Congr. Acoust. Liege, 1965, paper J42; J. Acoust. Soc. of America, 1962, 34, 1787; 1966, 40, 1433.

7. Baker, W. O. and Grisdale, R. O., U.S. Patent 2,697,136, 1954; Baker, W. O. and Winslow, F. H., Twelfth International Congress of Pure and Applied Chemistry, September, 1951; Winslow, F. H., Baker, W. O., Pape, N. R. and Matreyek, W., J. Poly. Sci., 1955, 16, 101.

8. Gross, B., Ann. Acad. Brasil, Cieng, 1945, 17, 219.

9. Perlman, M. M. and Unger, S., Electrets, Charges Storage and Transport in Dielectrics, 1973, Perlman, M. M., ed., 105, Electrochemical Society, Wintle, H. J., Jour. Appl. Phys., 1971, 42, 4724.

10. An expert review covering a range of insulating and semiconducting polymers and their charging behavior is by Mort, J., Science, 1980, 208, 819. Earlier extensive investigations of these effects were done by H. A. Pohl and co-workers, Pohl, H. A. Progress in Solid State Chemistry, H. Ruri, Ed MacMillan, New York 1964, 316.

11. Mort, J. and Lakatos, A. I., Jour. of Non-Crystal. Solids, 1970, 4, 117.

12. Gross, B., West, J. E., Feffeira, G. F. L. and Faria, R., J. Appl. Phys., 1980.

13. Gross, B., Sessler, G. M., Seggern, H. and West, J. E., Appl. Phys. Letters, 1979, 34, 555.

14. Stevens, J. R. and Mao, A., Jour. Appl. Phys., 1970, 41, 4273.

15. Groseclose, G. and Loper, G. D., Phys. Rev., 1965, 137, 939.

16. Wilson, R. K., Johnson, P. O. and Stump, R., Phys. Rev., 1963, 129, 2091.

17. Chuang, S. Y., Tao, S. J. and Wang, T. T., J. Poly. Sci., 1976,

18. Tao, S. J., Appl. Phys., 1974, 3, 1.

19 Davis, G. T. and Eby, R., J. Appl. Phys., 1973, 44, 4274.

20. Change, S. S., Poly. Preparation, 1972, 322.

21. Konikoff, J. J. and West, J. E., "Annual Report of the Conference on Electrical Insulation and Dielectric Phenomenon," National Research Council, 1978.

22. Baker, W. O., Fuller, C. S. and Pape, N. R., J. Am. Chem. Soc., 1940, 62, 3275.

23. Baker, W. O. and Fuller, C. S., J. Am. Chem. Soc., 1942, 64, 2399.

24. Alfrey, T. and Shrenk, W. J., Science, 1980, 208, 813.

25. Beardmore, P., Harwood, J. J., Kinsman, K. R. and Robertson, R. E., Science, 1980, 208, 833.

26. Winslow, F. H., Baker, W. O. and Yager W. A., Proceedings of the Conferences on Carbon, 1956, 93, University of Buffalo, New York.

27. Watt, W., Proc. Roy. Soc. of London, 1970, A319, 5.

28. The continuing vitality of the latter resources is skillfully shown by Anderson, B. C., Bartron, L. R. and Collette, J. W., Science, 1980, 208, 807.

29. Abelson, P. and Dorfman, M., "Advanced Technology," Eds., AAAS, 3, Washington, D.C.

30. Baker, W. O., Ind. Eng. Chem., 1949, 41, 511.

31. Kuleznev, V. M., Elkina, I. A., Vankova, L. N. and Dogadkin, B. A., Kolloidzhurnal, 1970, 32, 381.

32. Paul, D. R. and Newman, S., "Polymer Blends," 1978, Academic Press, New York.

33. Manson, J. A. and Sperling, L. H., Polymer Blends and Composites, 1976, Freedom Press, New York.

34. Lovinger, A. J. and Williams, M. L., J. Appl. Poly. Sci., 1980,

35. Noel, O. F. and Carley, J. F., Poly. Eng. Sci., 1975, 15, 117.

36. Deanin, R. D. and Sansone, J. F., Polymer Preprints, 1978, 19(1), 211.

37. Douglass, D. C., ACS Symposium Volume "Characterization of Macromolecules by ESR/NMR," 1980, Washington, D. C.

38. Douglass, D. C. and McBrierty, V. J., Macromolecules, 1978, 11, 766.

39. Coombs, A. and Keller, A., J. Poly. Sci., Polymer Phys. Ed., 1979,

40. Davis, D. D. and Kwei, T. K., J. Poly. Sci., 1980,

41. Nishi, T. and Wang, T. T., Macromolecules, 1975, 8, 909.

42. Kwei, T. K., Patterson, G. D., and Wang, T. T., Macromolecules, 1976,

43. Hughes, L. J. and Britt, G. E., J. Appl. Poly. Sci., 1961, 5, 337.

44. Flory, P. J., "Principles of Polymer Chemistry," 1953, Cornell University Press, Ithaca, New York.

45. Helfand, E. and Tagami, Y., J. Poly. Sci., 1971, B-9, 741.

46. Helfand, E. and Tagami, Y., J. Chem. Phys., 1971, 56, 3592; 1972, 57, 1812.

47. Helfand, E. and Sapse, Ann Marie, J. Chem. Phys., 1975, 62, 1327.

48. Patterson, G. D., Macromolecules, 1974, 7, 220.

49. Patterson, G. D., J. Poly. Sci. Phys. Ed., 1975, 13, --.

50. Patterson, G. D., Nishi, T. T. and Wang, T. T., Macromolecules, 1975, 8 --.

51. "Block and Graft Co-Polymers," Burke, J. J. and Weiss, V., Editors, 1973, Syracuse University Press, Syracuse, New York.

52. Meyer, R. R., Polymer, 15, 1974, 137.

53. Douy, A., Meyer, R., Rossi, J., and Gallot, B., Molecular Crystals and Liquid Crystals, 1969, 7, 103.

54. Duoy, A. and Gallot, B., Makromol Chemistry, 1972, 156, 81.

55. Helfand, E., and Wasserman, Z., Macromolecules, 1977 ____________ ____________ and Polymer Preprints, American Chemical Society, March 1977.

56. Linvill, J. G. and Hogan, C. L., Science, 1977, 195, 1107; "Advanced Technology".

57. Abelson, P. H. and Dorfman, Mary, Editors, 1980, 110 117, 132, A.A.A.S., Washington, D. C.

58. Bowden, M. J. and Thompson, L. F., Solid State Technology, 1979, May, 72.

59. Herriott, D. P., Collier, R. J., Alles, D. S. and Stafford, J. W., EEG Trans. on Electron Devices, 1975, Ed. 22, 87.

60. Ueberreiter, K., "Diffusion in Polymers," Editors: Crank, J. and Park, G. S., Academic Press, New York, 1968.

61. Brown, J. R. and O'Donnell, J. H., Macromolecules, 1972, 5, 109.

62. Miller, S. E. and Chynoweth, A. G., Editors, "Optical Fiber Telecommunications," 1979, Academic Press, New York.

63. Hayakawa, R. and Wada, Y., "Advances in Polymer Science," 3, 1973, 1, Springer, Berlin.

64. Crane, G. R., J. of Appl. Phys., 1975, ______ ______.

65. Kawai, H., Jap. J. of Appl. Phys., 1969, A, 975.

66. Tonelli, A. E., Macromolecules, 1976, 547.

RECEIVED August 4, 1981.

14

Achievements and Predicted Developments in Enzyme Engineering

EPHRAIM KATCHALSKI-KATZIR

The Weizmann Institute of Science, Rehovot, Israel 76100

Interest in immobilized enzyme derivatives stems primarily from our growing awareness of their potential as industrial catalysts and as a new type of model system for the investigation of isolated aspects of complex biological phenomena (1,2,3,4).

Enzyme immobilization on or within a solid matrix by conjugation with native or synthetic water-insoluble polymeric supports leads to immobilized enzyme preparations which can serve in the laboratory as reusable and removable highly specific catalysts, which often possess storage capacity and operational stability superior to that of the corresponding native enzymes. Continuous large-scale processes can be carried out in immobilized-enzyme reactors. Immobilized enzymes in conjunction with a detector have led to the development of highly specific electrode systems, and enzyme-columns and enzyme-membranes have been employed successfully in the appropriate analytical and monitoring devices. Immobilized enzymes are also being explored for clinical application in the form of extra-corporeal shunts or microcapsules. Moreover, the clarification of some of the principles underlying the kinetic behavior of immobilized enzyme systems, i.e., effects of the microenvironment imposed by the chemical nature of the support polymer material and the effects of diffusional restrictions on the translocation of substrate and product, make possible in principle the modulation of the properties of a bound enzyme by its conjugation to a support of predetermined chemical and physical characteristics.

Immobilization Techniques

Enzymes can be immobilized by matrix entrapment, by microencapsulation, by physical or ionic adsorption, by covalent binding to organic or inorganic polymer-carriers, or by whole cell immobilization (5). Particularly impressive is the great number of chemical reactions developed for the covalent binding of enzymes to inorganic carriers such as glass, to natural polymers such as cellulose or Sepharose, and to synthetic polymers such as nylon, polyacrylamide, and other vinyl polymers and

0097-6156/81/0175-0203$5.00/0

copolymers possessing reactive chemical groups. The covalent attachments lead as a rule to nonspecific binding. It is obvious, however, that the preparation of well-characterized carrier-enzyme conjugates will require the development of highly specific reactions by which an enzyme can be attached to a desired carrier.

Stability of Immobilized Enzymes

Immobilized enzymes have been found in a considerable number of cases to be markedly more stable to denaturation by heat, denaturing agents, or organic solvents, than the corresponding native enzymes (6). The reasons for this effect are still not fully understood, and it is thus recommended that in the future a thorough analysis be carried out of the conformation, ease of denaturation, and conformational fluctuations of immobilized enzymes.

Effect of Microenvironment and Kinetic Behavior

Considerable progress has been made within the last decade in elucidating the effects of the microenvironment (such as electric charge, dielectric constant and lipophilic or hydrophilic nature) and of external and internal diffusion on the kinetics of immobilized enzymes (7). Taking these factors into consideration, quantitative expressions have been derived for the kinetic behavior of relatively simple enzyme systems. In all of these derivations the immobilized enzymes were treated as simple heterogeneous catalysts.

Further progress in understanding the mechanism of action and kinetics of immobilized enzymes will require the study, both theoretical and experimental, of more complex enzyme systems. Of particular interest will be the study of the highly specific interactions between an enzyme embedded in a native membrane and the adjacent high or low molecular weight compounds.

Enzyme Membranes

Enzyme membranes can be prepared by adsorbing the enzyme on the surface of a suitable native or synthetic membrane, or, in the case of membranes with large pores, by impregnating the whole membrane with enzyme. The resulting enzyme membrane can be stabilized by covalently cross-linking the adsorbed protein with a suitable bifunctional reagent (8).

In spite of the theoretical interest in enzyme membranes, they have so far not been used in industry and their use in the clinic and in the laboratory is rather limited. However, as techniques for the preparation and stabilization of immobilized complex enzyme systems develop, one can expect to see an increase in the number of cases in which permeable and impermeable enzyme membranes will be used advantageously.

Immobilized Coenzymes

Many enzymes require the participation of dissociable co-enzymes such as NAD^+, $NADP^+$ or ATP for their catalytic activities. The use of coenzymes to activate immobilized enzymes on a large scale is hampered by their relatively low stability and high cost. Attempts are therefore being made to stabilize the coenzymes and to find suitable means for their continuous regeneration. The principal approach has been to covalently attach a co-enzyme to a polymeric water-soluble matrix, thus making the co-enzyme, like the enzyme, potentially reusable (9,10).

Immobilized Enzyme Cascades

Increasing evidence has accumulated to show that in nature many of the enzymes carrying out a sequence of consecutive reactions of the type:

$$S \xrightarrow{(E_1)} P_1 \xrightarrow{(E_2)} P_2 \xrightarrow{(E_3)} P_3 \cdots\cdots \xrightarrow{(E_n)} P_n$$

(where S denoted the substrate for enzyme E_1, and P_1 to P_n stand for the products formed as a result of enzymatically catalyzed reactions performed by enzyme E_1 to E_n, respectively) are associated in fairly tight aggregates, or jointly embedded in a cell membrane, or act in a gel-like surrounding. The study of the characteristics of artifically produced enzyme cascades seems therefore of both theoretical and practical interest (11,12,13).

Large Scale Enzyme Catalyzed Synthesis

Immobilized enzymes and immobilized cells can be used in principle as heterogeneous catalysts not only for the enhancement of hydrolytic reactions but also in specific synthetic reactions. Many of the enzymes involved in the synthesis of complex biological compounds *in vivo* and *in vitro* require cofactors which have to be provided in suitable form when immobilized enzymes or cells are used.

Adenosine triphosphate (ATP) is one of the most important cofactors involved in many of the synthetic reactions going on within the cell. Its recent large scale *in vitro* enzymatic synthesis from adenosine and acetylphosphate is of particular interest. Three enzymes immobilized in polyacrylamide gel were used: adenosine kinase, adenylate kinase and acetate kinase (14).

Industrial Applications

Immobilized enzymes and immobilized cells are being employed as specific heterogeneous catalysts by several of the chemical industries. An evaluation of the mode of action and efficiency of the various types of enzyme reactors, namely, the stirred tank reactor, the packed bed reactor and the fluidized bed reactor, has been recently carried out.

Aminoacylase electrostatically bound to DEAE-Sephadex has been adopted by the Tanabe Seiyaku Co. in Japan for the production of native optically active L-amino acids from the corresponding racemic acetyl DL-amino acids prepared synthetically (1). Penicillin acylase (penicillin amidase) immobilized by adsorption or covalent binding to organic or inorganic carriers is used for the production of 6-aminopenicillanic acid (6APA) from penicillin G (1). Immobilized glucose isomerase is being used extensively in the U.S.A., Japan and Europe in the production of high fructose syrups by the partial isomerization of glucose (1).

An industrial process worked out by the Japanese has enabled the continuous production of L-aspartic acid from ammonium fumarate by using columns packed with immobilized *E. coli* cells (15).

Use in the Analytical Area

The techniques developed in enzyme immobilization have facilitated the development of enzyme electrodes and of novel enzyme -based, automated, analytical methods (16,17,18). Enzyme electrodes have resulted from the combination of an enzyme membrane and an ion-selective electrode; they were used successfully to assay directly appropriate substrates. Enzyme columns or enzyme tubes, prepared in a conventional manner, were used as a specific auxiliary component in the indirect assay of substrates in many of the novel automated analytical procedures.

Immobilization of Microbial Cells

The immobilization of microbial cells under conditions where an activity or set of enzymic activities remains intact, but the normal metabolic processes cease, represents a novel technique for enzyme immobilization. Moreover, immobilized cells might enable the standard fermentation methods to be replaced by immobilized -cell -based continuous processes.

Within the last decade procedures have been worked out for microbial cell immobilization by cross-linking with bi- or multifunctional reagents, or by entrapment in collagen, gelatin, agar, cellulose triacetate, alginate, K-carrageenan, polyacrylamide and polystyrene (1,15,19). The effect of the matrix on the metabolically inactive cells is still not understood. One might expect, however, that further experience and information in this area will lead to the formulation of some practical generalizations which will serve as guidelines for choosing the most suitable matrix for entrapment, and the most appropriate conditions for immobilization.

Concluding Remarks

Future progress in enzyme engineering in general and in the use of immobilized enzymes and immobilized cells in particular might be expected in organic synthesis and degradation, in analysis, in bioelectrochemistry and especially in bioenergetics, i.e. the transformation of cellulose and of agricultural waste products into gas or liquid fuels. The anticipated developments are likely to emerge through cooperation between biochemists, bacteriologists, geneticists, molecular biologists, bioengineers and polymer chemists.

Literature Cited

1. Chibata, I. "Immobilized Enzymes": John Wiley & Sons: New York, 1978.
2. Methods in Enzymology, Vol. 44; Mosbach, K., Ed.; Academic Press: New York, 1976.
3. "Applied Biochemistry and Bioengineering", Vol. 2; Wingard, L.B., Jr.; Katchalski-Katzir, E., Ed.; Academic Press; New York, 1979.
4. Katchalski-Katzir, E. "Enzyme Engineering"; Weetall, H.H.; Royer, G.P., Ed.; Plenum: New York, 1980; p. 3.
5. Goldstein, L.; Manecke, G. "Applied Biochemistry and Bioengineering", Vol. 1; Wingard, L.B., Jr.; Katchalski-Katzir,E; Goldstein, L., Ed.; Academic Press: New York, 1976; p. 23.
6. Klibanov, A.M. Anal. Biochem. 1979, 93, 1.
7. Goldstein, L. Meth. in Enzymology, 1976, 44, 397.
8. Katchalski, E.; Silman, I.; Goldman, R. Adv. Enzymol. 1971, 34, 445.
9. Mosbach, K.; Larsson, P.O.; Lowe, C.R. Meth. in Enzymology, 1976; 44, 859.
10. Lowe, C.R. Trends Biochem. Sci. 1978, 3, 134.
11. Goldman, R.; Katchalski, E. J. Theor. Biol. 1971, 32, 243.
12. Mattiasson, B.; Mosbach, K. Biochim. Biophys. Acta 1971, 235, 253.
13. Haggerty, C.; Jablonski, E.; Stav, L.; Deluca, M. Anal. Biochem. 1978, 88, 162.
14. Baughn, R.L.; Adalsteinsson, O.; Whitesides, G.M. J. Am. Chem. Soc. 1978, 100 304.
15. Tosa, T.; Sato, T.; Mori, T.; Matuo, Y.; Chibata, I. Biotechnol. Bioeng. 1973, 15, 69.
16. Guilbault, G.G. "Handbook of Enzymatic Methods of Analysis"; Marcel Dekker: New York, 1976.
17. Keyes, M.H.; Semersky, F.E.; Gray, D.N. Enzyme Microb. Techn. 1979, 1, 91.
18. Danielson, B.; Mosbach, K. Meth. in Enzymology, 1976, 44, 667.
19. Yamamoto, K.; Tosa, T.; Yamashita, K.: Chibata, I. Europ. J. Appl. Microbiol. 1976, 3, 169.

RECEIVED April 17, 1981.

15

Polymeric Monolayers and Liposomes as Models for Biomembranes and Cells

BERND HUPFER and HELMUT RINGSDORF

University of Mainz, Institute of Organic Chemistry, D-6500 Mainz, West Germany

Biological membranes consist of a bilayer of phospholipids in which membrane proteins are either embedded (integral proteins) or simply adsorbed (boundary proteins) (1) (Figure 1.). These systems fulfill a variety of functions of basic importance. One of the most significant is the compartimentation via the formation of cells and cell subunits based on the self organization of membranes (hydrophobic effect (2)).

Models for biological membranes have either been realized as planar lipid monolayers at the gas-water interface (3) or as bimolecular lipid membranes (BLM) (4) and spherical liposomes (vesicles), respectively (5,6) (Figure 2.). All these models that are only composed of lipid molecules exhibit a diminished stability compared to natural cell membranes. Obviously the protein part besides being functionally important plays a role in terms of stability of biomembranes. This is the case not only for the integral but especially for the boundary proteins (7).

To build up a stable cell model according to this concept would mean to isolate membrane proteins and lipids and try to put them together as mother nature does. This idea to use membrane proteins for membrane stabilization does not yet seem to be realizable and therefore simpler possiblities for constructing stable membrane and cell models are desirable.

One approach could be the attempt to include the lipids into the stabilization process. Lipid molecules bearing polymerizable groups can actually be arranged as planar monolayers or as spherical vesicles and polymerized by high energy irradiation within these membrane like structures under retention of the orientation of the molecules (8,9,36).

Besides the possible use of such stabilized liposomes as drug carriers (10) another application could be their use as models for the cell-cell interaction and as polymeric antitumor agents on a cellular level (11) i.e. trying to mimic the body defense against tumor cells (12). We hope that this article which in many terms is more a pre- than a review will help to stimulate the discussions on polymeric antitumor agents not only on the al-

0097-6156/81/0175-0209$5.75/0

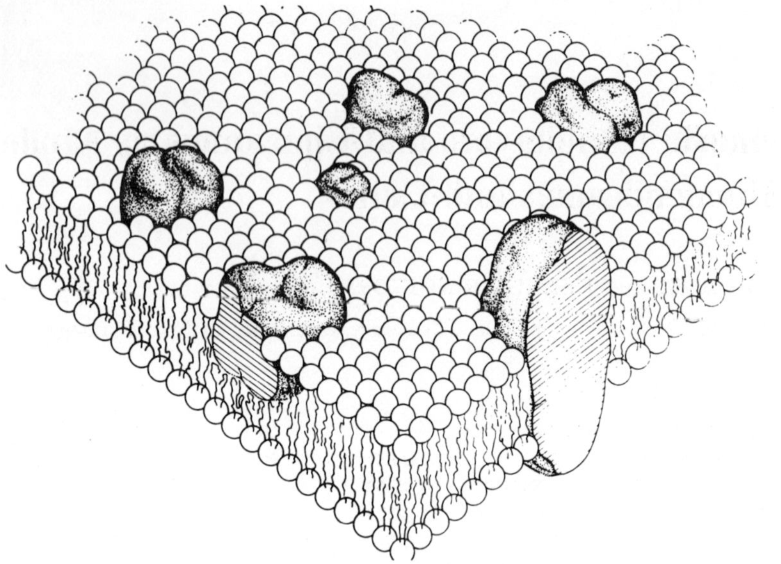

Figure 1. Fluid mosaic model of a cell membrane (1).

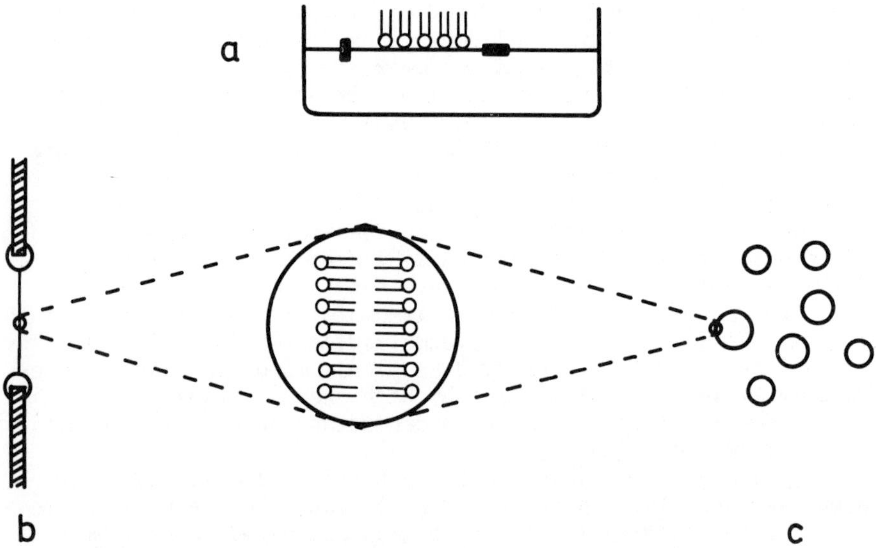

Figure 2. Model membrane systems. (a) Monolayer at the gas–water interface. (b) Planar bimolecular lipid membrane (BLM). (c) Liposomes.

ready well established molecular (13,14,15) but also on the cellular (11,12) level.

The Concept of Polymeric Model Membranes

Conventional vesicles are either prepared by lyophilization of isolated natural membrane components (e.g. egg yolk lecithin (16)) or of synthetic phospholipids (17). In addition, synthetic amphiphiles containing two alkyl chains but not bearing naturally occurring head groups can form membrane like bilayer assemblies (18,19). Vesicles from single chain amphiphiles (20,21,22) have been reported to be unstable and can only exist under special conditions. Recently, however, more stable vesicles were prepared from single chain lipids bearing two polar head groups and a rigid segment in the center of the molecules (23,24). Yet, all these models exhibit a limited lifetime compared to natural membranes mainly due to fusion of these vesicles on prolonged standing (19). Attempts have been made to stabilize these model systems via incorporation of e.g. cholesterol (25). Another approach to stabilize model membranes has been described by Khorana et al. (26), who synthesized phospholipids bearing photoreactive groups and linked them by UV-irradiation. A polymerization, however, of the lipid molecules of a vesicle as a stabilization concept has only recently been achieved (8,9,27,28,29). This concept has now been generalized to a great extent (30).

There are different possibilities for the localization of the polymerizable group in the lipid molecules. A polyreaction in the hydrophobic part of the monomers (Figure 3., a - c) influences the phase transition of the hydrocarbon chains (where they turn from the gel state to the liquid crystalline state), whereas a polymerization in the hydrophilic part of the molecules (Figure 3., d) changes the properties of the head groups. In our opinion the properties of biological membranes can be approached either by making the right choice of polymerizable groups or by preparing mixed systems of natural and polymerizable lipids. Especially the rigid, fully conjugated polymers from the diacetylene monomers (Table I.) do not exhibit phase transitions any more (27).

Monomers for polymerization according to Figure 3. are collected in Table I. (9,30).

Monolayer Experiments

Spreading Behavior of the Monomers. Spreading of monolayers of the monomers (Table I.) was carried out at the gas-water interface of a commercially available Langmuir type film balance according to earlier investigations (31,32). The principle of the characterization of lipids via surface pressure-area isotherms is illustrated in Figure 4. (3). If the monomers are spread on the surface, at large surface areas and low surface pressures they lie flat on the water and form a gaseous type phase. If this layer of

Table I.
Polymerizable and liposome forming lipids (9,30).
Type a - d cf. Figure 3. ($R = CH_2{=}C(CH_3)-CO$)

			F/°C
a	$R-NH-(CH_2)_{10}-CO-O-(CH_2)_2$ $R-NH-(CH_2)_{10}-CO-O-(CH_2)_2$ (both chains bound to $\overset{\oplus}{N}(CH_3)_2$) $Br^{\ominus}$	(1)	99
b	$H_2O_3PO-(CH_2)_9-C{\equiv}C-C{\equiv}C-(CH_2)_9-OPO_3H_2$	(2)	93
c	$H_3C-(CH_2)_{12}-CH{=}CH-CH{=}CH-CO-O$ $H_3C-(CH_2)_{12}-CH{=}CH-CH{=}CH-CO-O$ (both bound to X)		
	X : $-(CH_2)_2-\overset{\oplus}{N}(CH_3)_2-(CH_2)_2-$ $Br^{\ominus}$	(3)	94
	X : $-CH_2-\underset{\vert}{C}H-CH_2-O-\underset{O^{\ominus}}{PO}-O-(CH_2)_2-\overset{\oplus}{N}(CH_3)_3$	(4)	220
	$H_3C-(CH_2)_{12}-C{\equiv}C-C{\equiv}C-(CH_2)_8-CO-O$ $H_3C-(CH_2)_{12}-C{\equiv}C-C{\equiv}C-(CH_2)_8-CO-O$ (both bound to X)		
	X : $-(CH_2)_2-\overset{\oplus}{N}(CH_3)_2-(CH_2)_2-$ $Br^{\ominus}$	(5)	93,5-94,5
	X : $-CH_2-\underset{\vert}{C}H-CH_2-O-\underset{O^{\ominus}}{PO}-O-(CH_2)_2-\overset{\oplus}{N}(CH_3)_3$	(6)	215
	X : $-CH_2-\underset{\vert}{C}H-CH_2-O-\underset{OH}{PO}-O-(CH_2)_2-NH_2$	(7)	181
	X : $-(CH_2)_2-\overset{\oplus}{N}H\left[(CH_2)_2-SO_3^{\ominus}\right]-(CH_2)_2-$	(8)	109
	X : $-\left[(CH_2)_2-O-\right]_3(CH_2)_2-$	(9)	49,0-49,5
d	$H_3C-(CH_2)_{17}-O-CH_2$ $H_3C-(CH_2)_{17}-O-CH$ CH_2-O-Y Y : R	(10)	36
	Y : $-CO-NH-(CH_2)_5-NH-R$	(11)	54

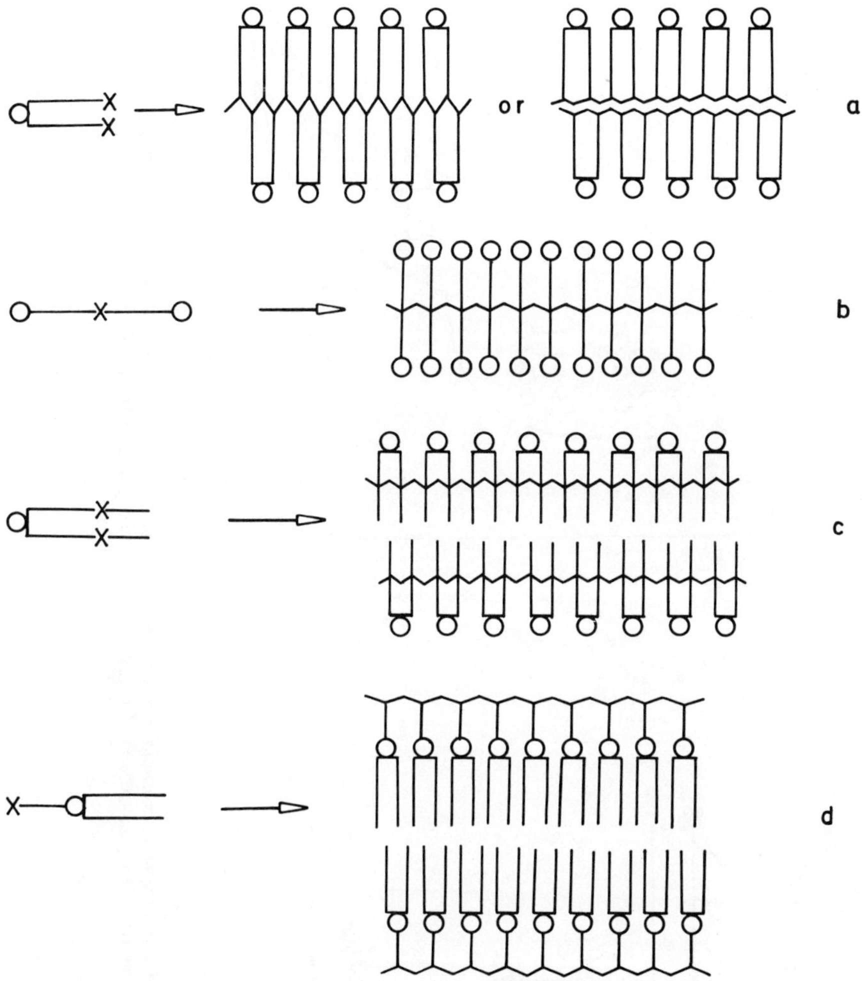

Angewandte Chemie

Figure 3. Possible preparation of polymeric model membranes (X = polymerizable group). (a)–(c) Polymerization preserving head group properties. (d) Polymerization preserving chain mobility (30). Corresponding monomers: see Table I.

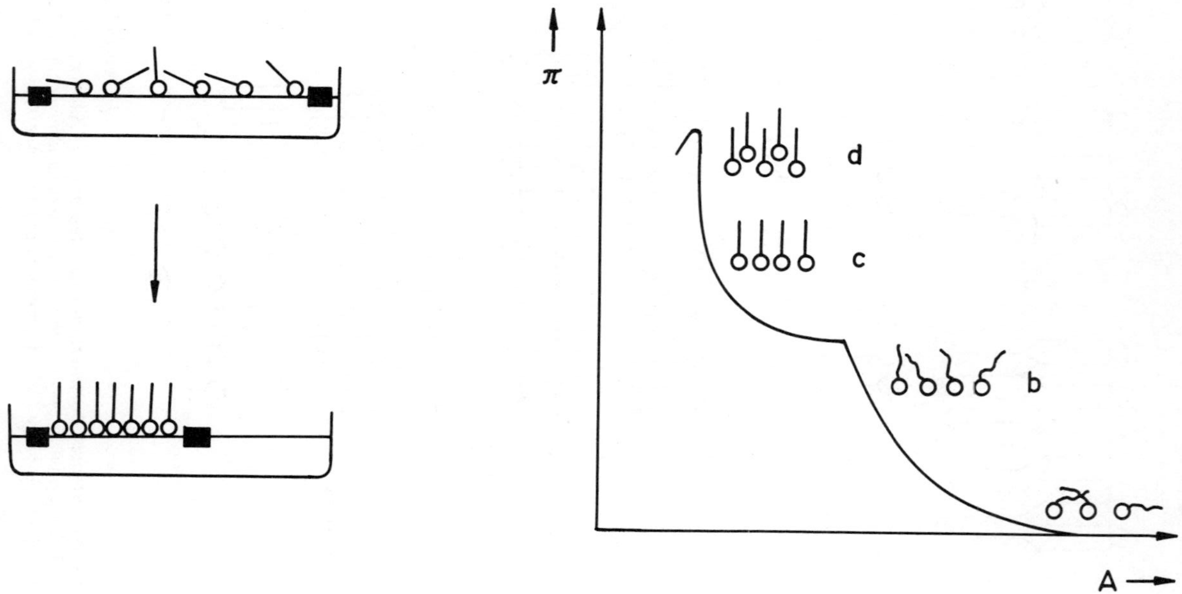

Figure 4. Principle of monolayer characterization via surface pressure (π)-area (A) isotherms; (a) *gaseous phase,* (b) *liquid expanded phase,* (c) *condensed phase (head packing),* (d) *condensed phase (chain packing).*

of molecules is compressed, the chains start their interaction, which is documented by a relatively slow increase of surface pressure. The phase formed is called expanded or liquid like, because the alkyl chains are in a liquid crystalline like state. Further compression leads to an additional phase change and a sharp increase of surface pressure. The state formed is called solid like or condensed and is characterized by a dense packing of the alkyl chains (chain packing) or by the head groups (head packing) and can be seen as a two dimensional crystal. Further decrease of surface area results in the collapse of the monolayer under formation of undefined bi- and multilayers.

Influence of subphase temperature, pH, and molecular structure of the lipids on their phase behavior can easily be studied by means of this method. The effect of chain length and structure of polymerizable and natural lecithins is illustrated in Figure 5. At 30°C distearoyllecithin is still fully in the condensed state (33), whereas butadiene lecithin (4), which carries the same number of C-atoms per alkyl chain, is already completely in the expanded state (34). Although diacetylene lecithin (6) bears 26 C-atoms per chain, it forms both an expanded and a condensed phase at 30°C. The reason for these marked differences is the disturbance of the packing of the hydrophobic side chains by the double and triple bonds of the polymerizable lipids. At 2°C, however, all three lecithins are in the condensed state. Chapman (27) reports about the surface pressure area isotherms of two homologs of (6) containing 23 and 25 C-atoms per chain. These compounds exhibit expanded phases even at subphase temperatures as low as 7°C.

On the other hand, changes of subphase pH have no considerable influence on the shape of the π -A curves of lecithins, since they do not contain movable protons. This is different in the case of cephalins (e.g. (7)), where the isotherms at high pH's are considerably broadened (34).

A more drastic influence of subphase $H^{\oplus}$ concentration can be observed for the sulfolipid analog (8) (Figure 7.). At pH values from 2 up to 5,5 (8) forms two condensed phases, whereas at pH 12, where the nitrogen is deprotonated, there is only one phase present (35). This behavior has a considerable influence on the monolayer polymerization properties of (8), which is discussed later in this chapter.

Monolayer Polymerization. Polymerization of the highly oriented monomer films can simply be achieved by UV irradiation under nitrogen (Figure 8.). The polymerization of the diacetylene monomers (2, 5-9) is a topochemical reaction (37,38) that only takes place, if the monomers are perfectly ordered, i.e. in the crystalline state or in oriented mono- (32) or multilayers (39) and leads to the formation of conjugated blue and red colored polymer backbones (Eqn. I.).

In monolayers the polyreaction is only possible, if the monomers are in the condensed phase. (That this, however, is some-

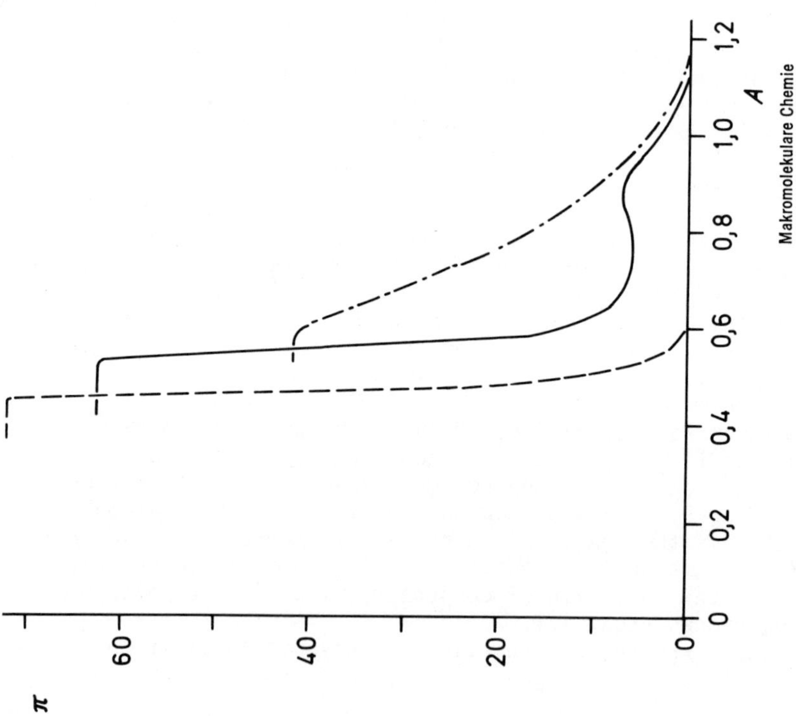

Figure 5. Surface pressure area isotherms of polymerizable and natural lecithins at 30°C (34). Key: (– – –) distearoyllecithin, (——) diacetylene lecithin (6), (—·—·—) butadiene lecithin (4). π: Surface pressure in $mN \cdot m^{-1}$; A: surface area in nm^2/molecule.

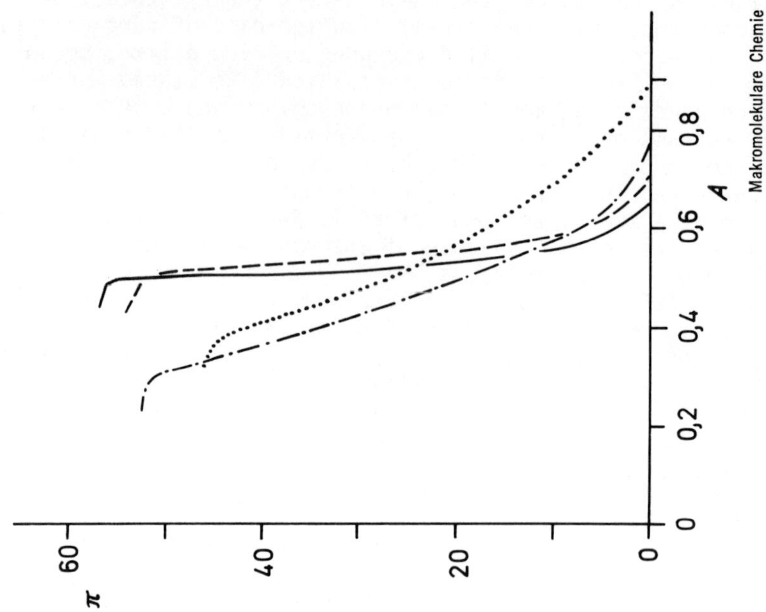

Figure 6. Surface pressure area isotherms of polymeric lecithins at 20°C (34). Key: (——) poly-(6), (—·—·—) poly-(4), monomers, (– – –) (6), (· · ·) (4) for comparison.

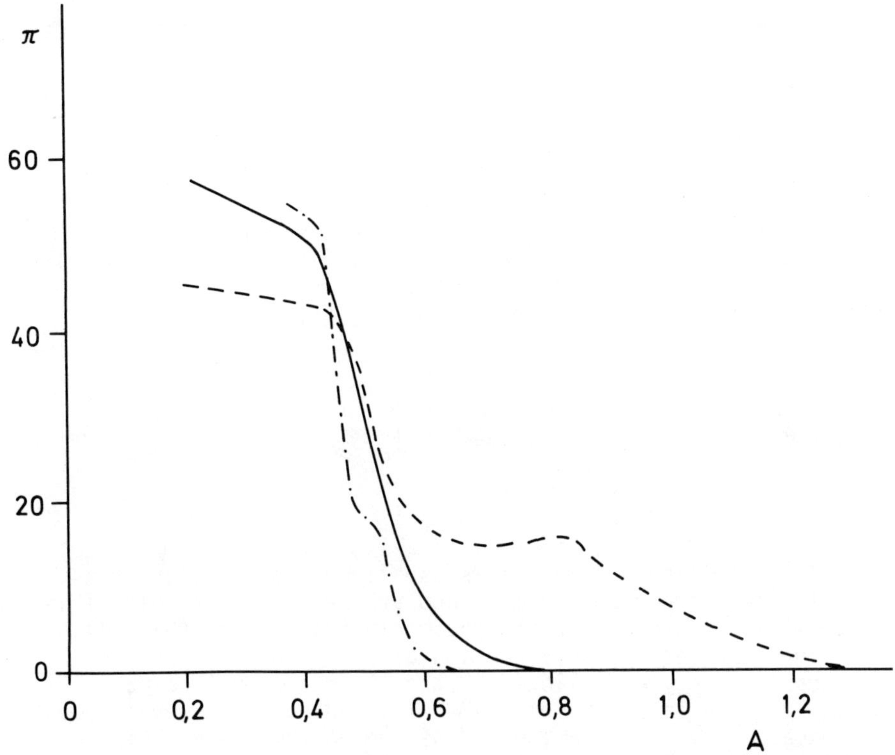

Makromolekulare Chemie

Figure 7. Surface pressure area isotherms of sulfolipid (8) at different pHs (35). Key: (— — — —) pH 5.5, 41°C; (——) pH 12, 10°C; (– – –) pH 12, 21°C. π: surface pressure in mN · m⁻¹; A: surface area in nm²/molecule.

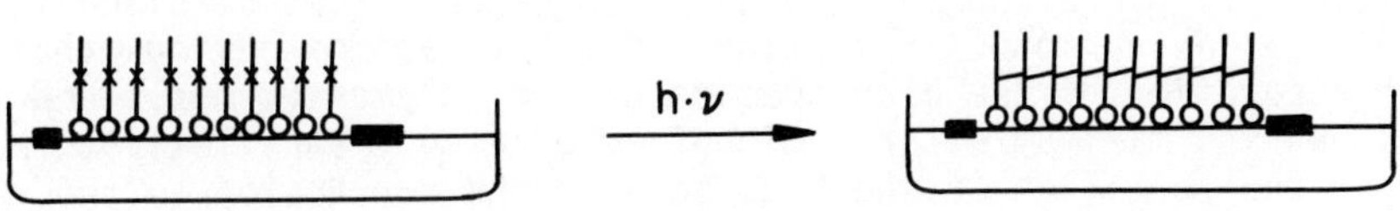

Figure 8. Formation of polymeric monolayers from polymerizable lipids (X = polymerizable group).

times only partly true will be shown below.)

(I)

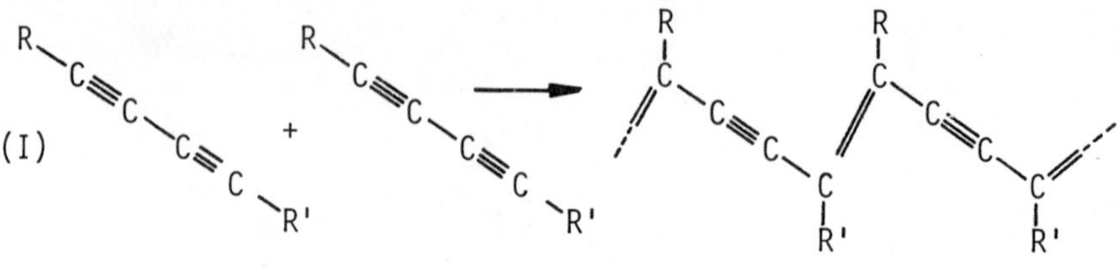

On the contrary, butadiene and methacryloyl monomers (1,3,4, 10,11) can also be polymerized in the liquid expanded phase. The butadiene lipids have previously been shown to form 1,4-trans-poly(butadiene)s (40) in the monolayer (Eqn. II.).

(II)

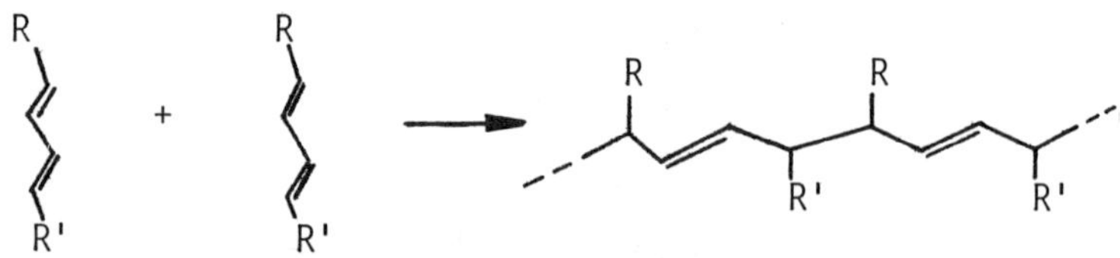

The resulting polymer films usually occupy smaller areas than the corresponding monomers and are less compressible. This is shown in Figure 6. for the phospholipids (4) and (6) (34).

In all cases the polyreaction at a constant surface pressure can qualitatively be followed by the change of surface area (32). An unaltered area after some time of irradiation, however, does not necessarily indicate the end of the polyreaction. This can be proved by following the monolayer polymerization spectroscopically. A special small Langmuir trough that fits into the sample chamber of a commercial UV/VIS spectrometer has been developed for this purpose (41). Since polydiacetylenes have strong absorption coefficients in the visible spectrum, one monolayer already creates O.D.'s of about 0,08 A, whereas the monomers only absorb very weakly in the UV region (254 nm). Therefore, the degree of absorption in the visible spectrum is a criterion for the rate of conversion. Figure 9. summarizes the absorbance curves of diacetylene lipid (5) at various irradiation times. It becomes obvious that the absorption of the blue species of the polymer (λ max.: 660 nm) has its maximum after 5 min of irradiation. Parallelly to and after the formation of the blue polymer the conversion to the red species (λ max.: 540, 500 nm) takes place. The absorption spectrum of the final red polymer film is identical to those of polymer multilayers (39). The color change from blue to red during polymerization is also observed in the solid state and in multilayers and is thought to be due to a conformation change of the polymer backbone (42,43). All the investigated diacetylene monomer monolayers show the described color change except the phospholipids

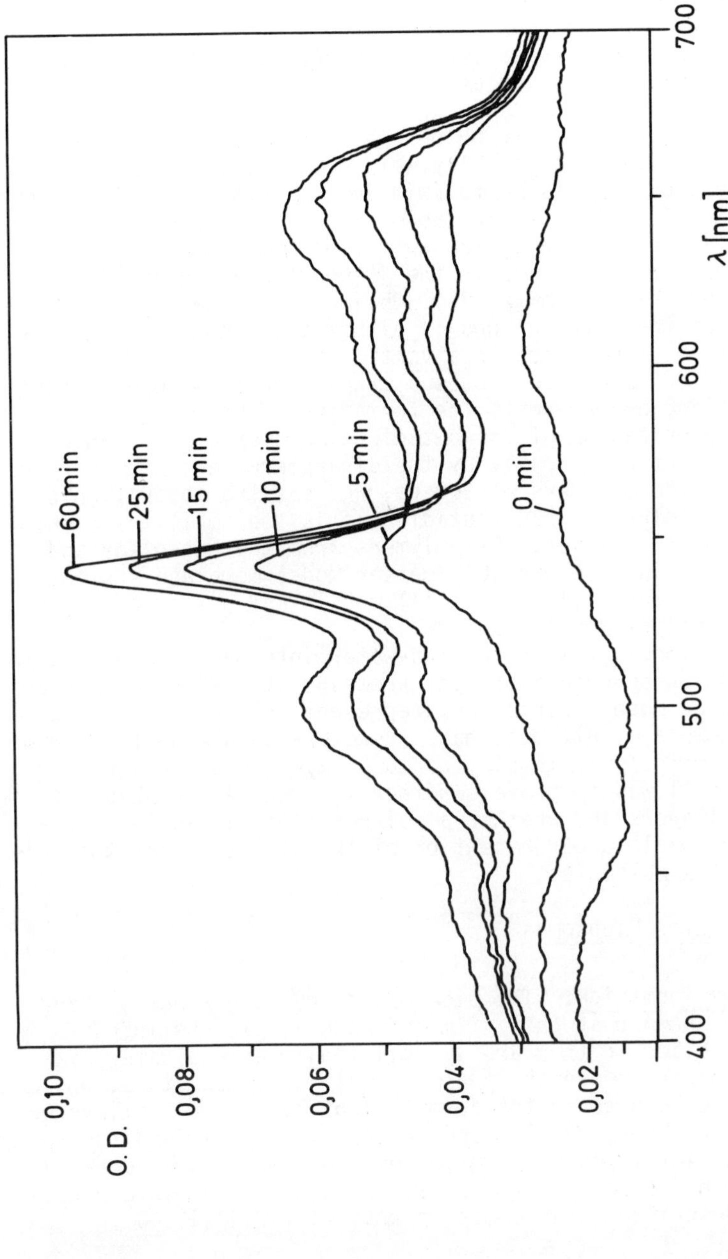

Figure 9. VIS Multiplot of absorbance of a monolayer of diacetylene lipid (5) vs. polymerization time. Constant surface pressure 10 mN · m^{-1}; 20°C; N_2 atmosphere.

(6) and (7), which turn red immediately (λ max.: 472 nm). In contrast to the lipids not based on glycerol in these molecules the diyne moieties have different distances from the hydrophilic head group and are therefore shifted towards each other. O'Brien (29) postulates that the reaction is principally intermolecular in these cases, with diacetylenes on one molecule able to polymerize only with diacetylenes which are in the same chain of adjacent molecules.

The strong influence of pH on the packing density of (8) (compare Figure 7.) is especially well documented by the polymerization behavior of this sulfolipid in the monolayer. (8) at 41°C at pH 2 to 5,5 is only polymerizable in the "first" condensed phase, i.e. at surface pressures up to 15 mN/m, whereas a polyreaction does not take place in the "second" phase at surface pressures from 20 to 50 mN/m (35). Obviously in the second phase the packing of the diyne groups is either too tight to permit a topochemical polymerization or a vertical shift of the molecules at the gas-water interface causes a transition from head packing to chain packing and prevents the formation of polymer.

The polymerization of the butadiene monomers (3,4) can also be followed spectroscopically by the disappearance of the strong absorption of the monomers at 260 nm, whereas the absorption of the resulting poly-1,4-trans(butadiene)s is too small to be observed in a single monolayer. The polymers from the butadiene and methacryloyl lipids are probably better model membrane systems, because the polymer chains are still mobile and not excessively rigid as the polydiacetylenes.

Although monolayers at the gas-water interface are useful to study adsorption phenomena of e.g. proteins at membranes they are not a very good model, since they represent only one half of a biological membrane. Attempts have therefore been made to extend this concept of polymer monolayers to bilayers and particularly to liposomes. It was to prove, whether the monomers (Table I.) could form bilayers and whether a polyreaction within these bilayers was possible under retention of the structure and the orientation of the molecules.

Polymeric Bilayer Liposomes

Liposome Formation. The pioneering investigations of Bangham (5) have shown that thin films of natural phospholipids form bilayer assemblies if they are lyophilized in excess water by simple handshaking above the phase transition temperature. While this procedure results in the formation of large, multibilayered spherical structures, by ultrasonication of such lipid dispersions small unilamellar liposomes are formed (16), which are schematically shown in Figure 10. Additional methods for liposome preparation are described in a number of reviews (17,44,45,46).

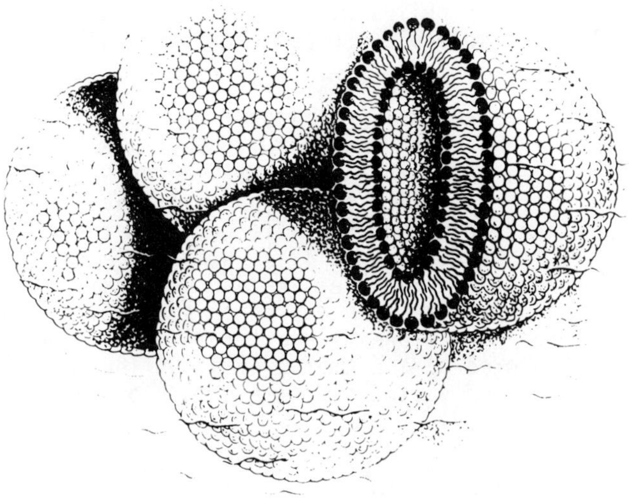

Figure 10. Schematic of small single bilayered liposomes.

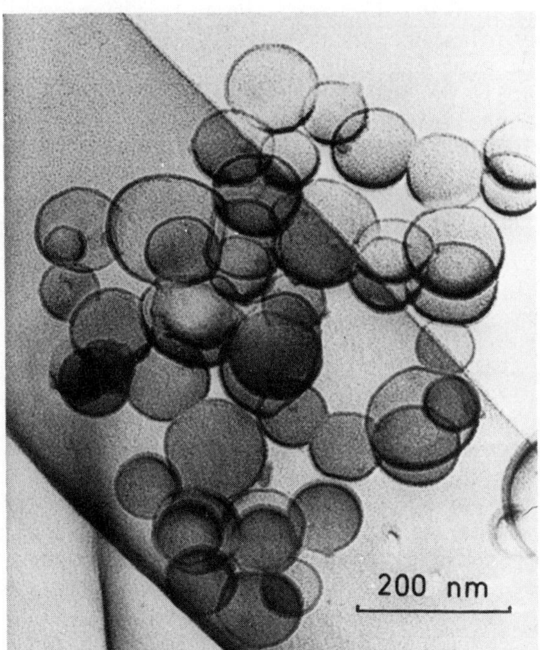

Angewandte Chemie

Figure 11. Electron micrograph of monomeric liposomes from (9), negatively stained with uranylacetate (9).

Recently it could also be shown that a great variety of simple structured amphiphiles with either one (20-25) or two (18, 19) long chains per molecule are able to form vesicles upon sonication with ultrasound.

According to these investigations sonication of aqueous suspensions of all the polymerizable lipids (Table I.) results in the formation of clear colorless solutions of monomer liposomes, which turn turbid after some days of storage (9,30). Electron microscopy proves the formation of bi- and multilayered vesicles of a defined spherical shape and of different diameters (range: 100 nm up to several microns, Figure 11.). Whether cephalin (7) really forms vesicles is currently investigated. The diameter and hence the number of bilayers strongly depend on sonication time and intensity. Under the conditions used after 30 min of sonication only bilayer liposomes of a diameter of about 100 nm are formed. We could also build up vesicles from a number of single chain amphiphiles (12,13,14) (9).

$H_3C-(CH_2)_{12}-C\equiv C-C\equiv C-(CH_2)_8-R$ (12), R = COOH

(13), R = CH_2OH

(14), R = $CH_2-OPO_3H_2$

Polymerization in Bilayers. Upon irradiation with UV light the monomer vesicles are transferred to polymer vesicles (Figure 12.). In the case of the diyne monomers (2,5-9,12,13,14) the polyreaction can again be followed by the color change via blue to red except phospholipids (5,6), which turn red without going through the blue intermediate as observed in monolayers. The VIS spectra of these polymer vesicle dispersions are qualitatively identical to those of the polymer monolayers (Figure 13.).

Polymer formation from monomeric butadiene lipid vesicles is demonstrated by the decrease of the monomer absorption at 260 nm as well as by GPC of the residue of a freeze dried polymer vesicle dispersion. The latter method was also used for proving the formation of polymeric vesicles from the methacryloyl lipids.

Evidence that the liposomes remain intact during the UV initiated polymerization is given by electron microscopy showing that the irradiated solutions still consist of spherical vesicles of almost unaltered shape and size. Also high pressure gel filtration (column: G6000PW; Toyo Soda Manufactoring Co., Ltd.; Shinnayo, Yamaguchi (Japan)) was used for showing that the size distribution of monomer and polymer vesicles remains almost the same.

The great stability of polymerized vesicles can be demonstrated by several experiments (9): dilution of polymer vesicle solutions with 50% of ethanol does not result in a precipitation. Turbidity measured at 300 nm remains the same (28), whereas monomer vesiclesare destroyed under these conditions followed by a considerable decrease of turbidity. A precipitation of stabilized vesicles can, however, be achieved by the addition of salts (KCl), but again it has to be pointed out that the polymeric vesicles

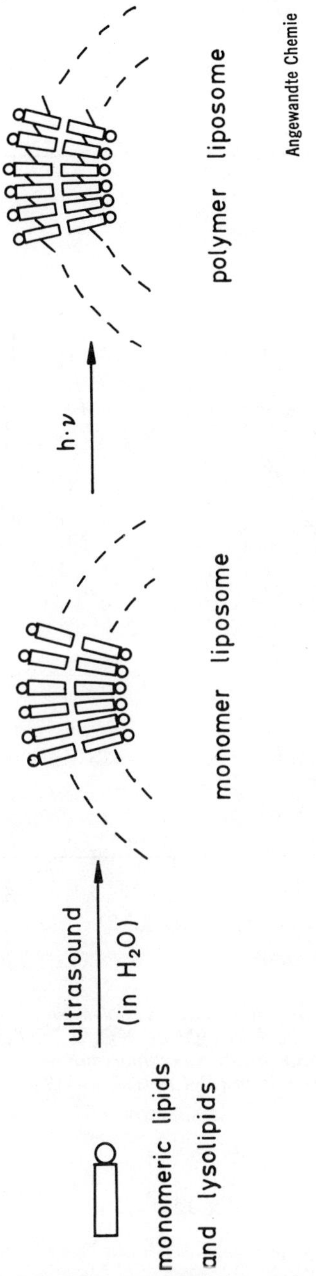

Figure 12. Formation of polymeric vesicles from polymerizable lipids (9).

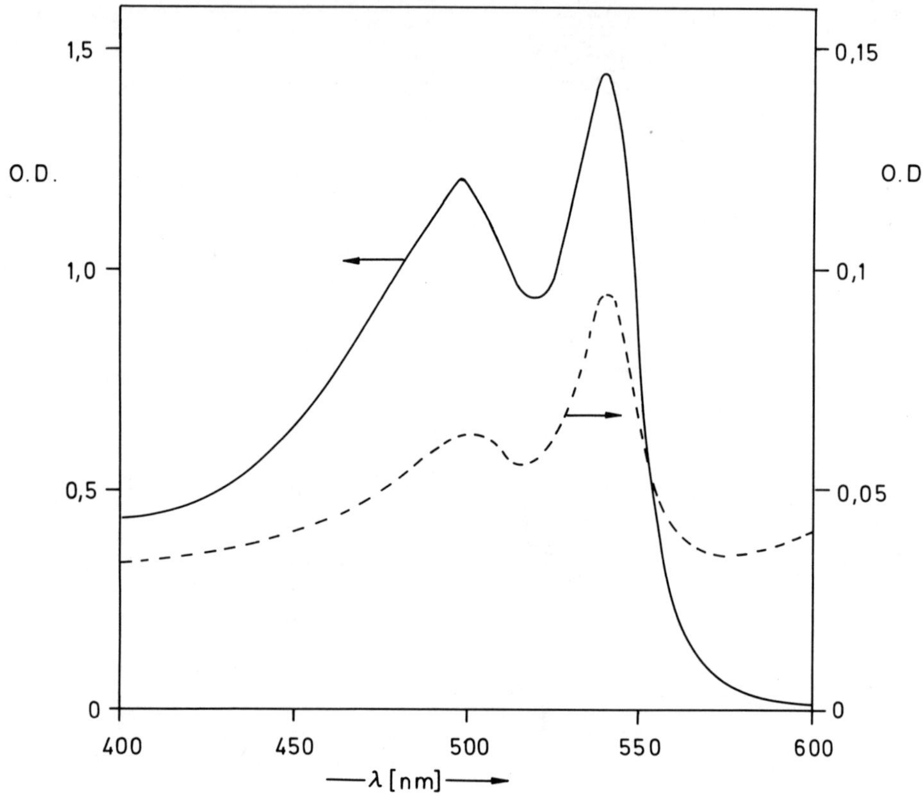

Figure 13. VIS spectrum of polymerized vesicles from (5) after 15 min of UV irradiation with multichromatic UV light at 0°C. c = 0.5 mg/mL; 1 cm cuvet (——). For comparison: spectrum of a polymer monolayer of (5) after 60 min of irradiation (254 nm) (– – –) (9).

are not destroyed by the osmotic shock. Electron microscopy shows that the precipitate consists of spherical vesicles of unaltered size and shape. In addition, the possibility of obtaining scanning electron micrographs of the polymeric liposomes (Figure 14.) (9) is a hint at their extraordinary stability. Monomer vesicles, on the contrary, are destroyed during sample preparation.

Application of Polymerized Vesicles?

Stabilized polymeric liposomes can be used for any problem where stable compartments are necessary. The build up of microenvironments (47,48) is one of the basic principles of living nature. The biomimetic application (49) of this principle has already found increasing interest and first results have been obtained e.g. on the areas of conversion of sunlight into energy (19,50,51), the synthesis of enzyme models (52) as well as the use of liposomes as drug carriers (10,54,53). One of the shortcomings of conventional liposomes, however, is their high rate of leakage when in contact with blood plasma (55), which is mainly due to their destabilization by protein adsorption and their destruction by lipases - a problem that can probably be overcome with polymerized vesicles. An additional problem is the high take up rate of liposomes by the recticuloendothelial system (RES), i.e. liver, spleen, lymph system, (56). For the transport of drugs by liposomes to any other tissue the membrane of vesicles must be provided with a cell specific "homing device" that only permits the transport of liposomes to one kind of tissue. A few results obtained with monomeric liposomes and initial experiments with polymerized vesicles are discussed in the next chapter.

Targeting of Liposomes

Attempts to achieve a tissue specific transport of liposomes have recently been described. Cohen et al. coated liposomes with aggregated immunoglobulin (57). The take up rate of these vesicles into phagocytes could be increased by a factor of 60 compared to uncoated liposomes. 3 to 25 times more liposomes are taken up by the corresponding cells, if they are loaded with the appropriate antibodies (58). Similar attempts to achieve a "homing" have been carried out using lipid fixed antibodies (59).

Nevertheless, a specific targeting has not yet been achieved. For the most part liposomes are accumulated in the RES and are removed too fast out of the circulation of blood.

So far only simple model reactions have been carried out to show the feasability of a specific recognition of polymeric liposomes. Since it has been known for a long time that glycolipids and glycoproteins play an important role in the cell-cell recognition, attempts have been made to incorporate sugars in the membrane of liposomes. Because letins (particular plant proteins) are able to recognize specific sugars, they agglutinate sugar bearing cells (60) and liposomes (61).

Monomers (15) and (16) have been used for the investigation of the recognition of polymeric liposomes by lectins (62).

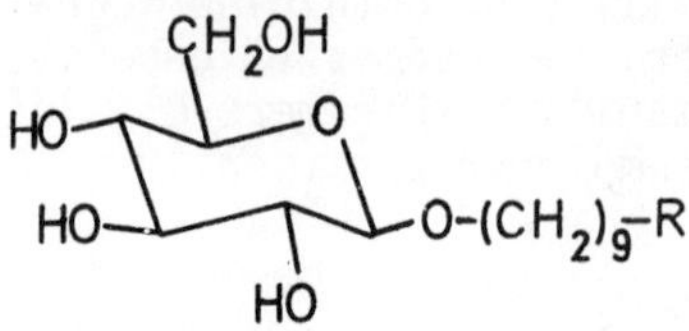

(15)

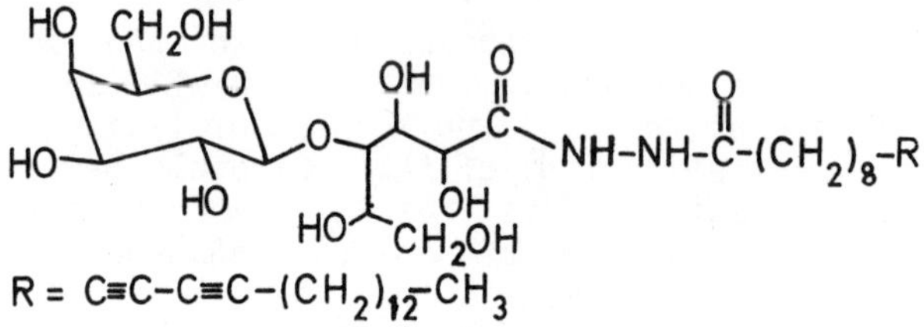

(16)

Glucose bearing compound (15) forms liposomes upon sonication and these can be polymerized by UV irradiation yielding the typical red solution. On addition of the lectin Concanavalin A within a few seconds an agglutination takes place under precipitation of the polymeric vesicles (62). This finding did not occur with polymerized vesicles not containing sugars at their surface. Presently attempts are being made to incorporate lectins in stabilized liposomes and to investigate the recognition of these targeted vesicles by sugar containing liposomes.

Perspectives

In the previous chapters it has been shown that stable cell membrane models can be realized via polymerization of appropriate lipids in planar monolayers at the gas-water interface as well as in spherical vesicles. Moreover, initial experiments demonstrate that polymeric liposomes carrying sugar moieties on their surface can be recognized by lectins, which is a first approach for a successful targeting of stabilized vesicles being one of the preconditions of their use as specific drug carriers in vivo.

In addition, however, one can speculate about another possible application of stabilized vesicles: in living systems tumor cells usually are not attacked by cells of the immune system. On the other hand it is impressive to see what happens if they are not able to elude the cells of the immune system. Antigens of cancer cells can be recognized e.g. by sensitized mice lymphocytes and the result is a destruction of the membrane of the malignant cell as demonstrated via electron micrographs by Old (12).

The process of cell recognition is a molecular one, the final destruction of the cell membrane, however, is a physical process. The exact mechanism is still unclear. At least three possibilities are discussed: a locally high concentration of lysophospholipids

(63), a production of cell toxic i.e. membrane destructing substances by the lymphocyte (64), or a spontaneous membrane destabilization via conformation changes of proteins (65). In addition, there are several hints at the possibility of a mechanical destabilization of cell membranes. The adsorption of liposomes to cells destabilizes the membranes to such an extent that low molecular weight substances can penetrate into the cells (66). A cell fusion, which is started by a great destabilization of the membrane, can be induced by means of several methods e.g. biologically by Sendai virus (67), chemically by calcium ions (68) or lysophospholipids (63), or physically by an electric field (69, 70). Considering the impressive pictures of the death of a tumor cell (12) the question is, whether it is possible to mimic such a cell-cell interaction by synthetic cell models such as liposomes. Conventional vesicles cannot be used for this purpose, because they are too unstable and during an interaction with a malignant cell are destroyed by fusion, endocytosis, or membrane destabilization (71). Also a cell specific recognition has not yet fully been achieved with liposomes.

Polymeric vesicles could be of better use for such an antitumor therapy on a cellular level, since they have at least one of the properties required, namely an extraordinary membrane stability. For a successful application, however, the simple systems prepared so far must be varied to a great extent, because the stability of a model cell membrane is not the only condition to be fulfilled. Besides stability and possibilities for cell recognition as discussed above the presence of cell membrane destructing substances such as lysophospholipids is necessary. These could e.g. be incorporated into the membrane of stabilized liposomes without destruction of the polymeric vesicles. There have already been reports about thekilling of tumor cells by synthetic alkyl lysophospholipids (72).

An additional possibility could be the transport of polymeric tensides, which are themselves not able to destroy the polymeric liposomes, to the membrane of the cancer cell. For the release of the incorporated tenside the membrane of the carrier liposome must contain destabilizable areas comparable to cork-stoppers (11). The destabilizable areas could eventually be opened by photochemical destabilization of the membrane (73), variation of pH (74), increase of temperature (75), or enzymatic processes.

Four possibilities to obtain such a stable cell model that fulfills all the conditions mentioned above are summarized in Figure 15.

1. Mixtures of proteins, natural and polymerizable lipids can be transferred into liposomes and polymerized hereafter. Initial experiments have shown that even very complex proteins such as F_0F_1-ATPase can be incorporated in polymeric liposomes by this method under retention of the activity of the protein (76).

2. Polymerizable lipids can be incorporated into e.g. erythrocyte ghost cells by means of hemolysis and polymerized here-

Angewandte Chemie

Figure 14. Scanning electron micrograph of large multilamellar polymeric liposomes from (5) (9).

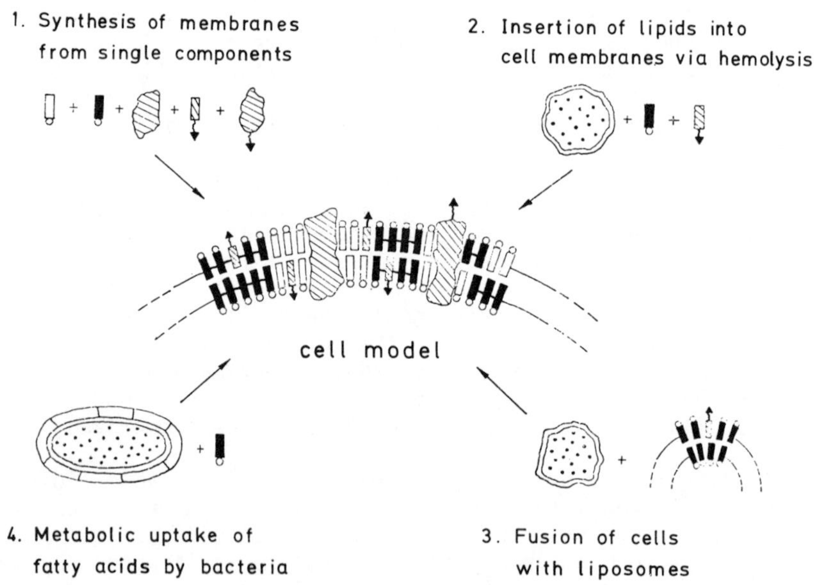

Figure 15. Schematic of the build up of stable cell models via partial polymerization of the membrane. Key: , natural or synthetic lipids; , polymerizable lipids; , proteins; , lipids or proteins bearing cell recoginizing groups.

after. While osmotic hemolysis exhibits several disadvantages (77), electrical hemolysis is a better candidate for this kind of application (78). The method is based on the reversible breakdown of the cell membrane in an electric field.

3. The dielectric breakdown offers a third method for the build up of cell models. Zimmermann and coworkers could show that under appropriate conditions cells can be fused with other cells and liposomes (79). By means of this method lipids from artificial liposomes can be incorporated into the membrane of cells (80).

4. Another possibility is described by Chapman et al. (27). Bacteria cells can incorporate diacetylene carbonic acids into their membrane, if they are grown on these lipids. Brief irradiation of the cells causes visible spectral changes similar to those observed when synthetic lipid liposomes are irradiated.

It can be expected that the discussed possibilities may help to synthesize stabilized cell models and to simulate biological processes.

There is no doubt that the proposed application of polymeric vesicles as antitumor agents on a cellular level is just an interesting speculation. Particularly the problem of mimicking the cell-cell recognition must still be viewed as only very roughly approximated. Since some investigators recently succeeded in fixing monoclonal antibodies to liposomes (81,82), one can feel optimistic that this problem may be solved in the near future. Furthermore, the reader may object that it is senseless "to load the cancer gun with further ammunition as long as we lack information about the target" (83). Without joint efforts, however, cancer will never be defeated. One can just hope that polymer chemistry can contribute to dissolve this big problem.

Literature Cited

1. Singer, S. J.; Nicolson, G. L. Science 1972, 175, 720.
2. Tanford, C. "The Hydrophobic Effect"; Wiley: New York, 1980.
3. Gaines, G. L. "Insoluble Monolayers at Liquid Gas Interfaces"; Interscience: New York, 1966.
4. Tien, H. Ti "Bilayer Lipid Membranes"; Dekker: New York, 1974.
5. Bangham, A. D.; Hill, M. W.; Hill, N. G. Methods Membr. Biol. 1974, 11, 38.
6. Papahadjopoulos, D. Ann. N.Y. Acad. Sci. 1978, 308, 1.
7. Singer, S. J. "Cell Membranes"; Weissmann, G.; Clairborne, R., Ed.; HP Publishing Co.: New York, 1975; p. 44.
8. Day. D.; Hub, H.-H.; Ringsdorf, H. Isr. J. Chem. 1979, 18, 325.
9. Hub, H.-H.; Hupfer, B.; Koch, H.; Ringsdorf, H. Angew. Chem. Int. Ed. Engl. 1980, 19, 938.
10. Gregoriadis, G.; Allison, A. C., Ed. "Liposomes in Biological Systems"; Wiley: New York, 1980.
11. Gros, L.; Ringsdorf, H.; Schupp, H. Angew. Chem. Int. Ed. Engl. 1981, 20, in press.

12. Old, L. J. Sci. Am. 1977, 236 (5), 62.
13. Zaharko, D. S.; Przybylski, M.; Oliverio, V. T. Meth. Cancer Res. 1979, 16, 347.
14. Ringsdorf, H. J. Polym. Sci. Polym. Symp. 1975, 51, 135.
15. Kostelnik, R. J., Ed. "Polymeric Delivery Systems"; Midland Macromolecular Monographs, Vol. 5; Gordon & Breach: New York, 1977.
16. Huang, H. C. Biochemistry 1969, 8, 344.
17. Tyrrell,D. A.; Heath, T. D.; Colley, C. M.; Ryman, B. E. Biochim. Biophys. Acta 1976, 457, 259.
18. Kunitake, T. J. Macromol. Sci.-Chem. 1979, A-13, 587.
19. Fendler, J. H. Acc. Chem. Res. 1980, 13, 7.
20. Gebicki, J. M.; Hicks, M. Chem. Phys. Lipids 1976, 16, 142.
21. Hargreaves, W. R.; Deamer, D. W. Biochemistry 1978, 17, 3759.
22. Jain, M. K.; van Echteld, C. J. A.; Ramirez, F.; de Gier, J.; de Haas, G. H. Nature (London) 1980, 284, 486.
23. Okahata, Y.; Kunitake, T. J. Am. Chem. Soc. 1979, 101, 5231.
24. Baumgartner, E.; Fuhrhop, J.-H. Angew. Chem.Int. Ed. Engl. 1980, 19, 550.
25. Gregoriadis, G. "Liposomes in Biological Systems"; Gregoriadis, G.; Allison, A. C., Ed.; Wiley: New York, 1980; p. 25.
26. Gupta, C. M.; Costello, C. E.; Khorana, H. G. Proc. Natl. Acad. Sci. U.S.A. 1979, 76, 3139.
27. Johnston, D. S.; Sanghera, S.; Pons, M.; Chapman, D. Biochim. Biophys. Acta. 1980, 602, 57.
28. Regen, S. L.; Czech, B.; Singh, A. J. Am. Chem. Soc. 1980, 102, 6638.
29. O'Brien, D.; private communication.
30. Akimoto, A.; Dorn, K.; Gros, L.; Ringsdorf, H.; Schupp, H. Angew. Chem. 1981, 93, 108; Angew. Chem. Int. Ed. Engl. 1981, 20, in press.
31. Naegele, D.; Ringsdorf, H. J. Polym. Sci. Polym. Chem. Ed. 1977, 15, 2821.
32. Day, D.; Ringsdorf, H.; J. Polym. Sci. Polym. Lett. Ed. 1978, 16, 205.
33. Dörfler, H.-D.; Rettig, W. Colloid Polym. Sci. 1980, 258, 839.
34. Hupfer, B.; Ringsdorf, H.; Schupp, H. Makromol. Chem. 1981, 182, 247.
35. Koch, H.; Ringsdorf, H. Makromol. Chem. 1981, 182, 255.
36. Hub, H.-H.; Hupfer, B.; Ringsdorf, H. Am. Chem. Soc., Org. Coat. Plast. Chem. 1980, 42, 2.
37. Wegner, G. Makromol. Chem. 1972, 154, 35.
38. Baughman, R. H. Contemp. Top. Polym. Sci. 1977, 2, 205.
39. Tieke, B.; Lieser, G.; Wegner, G. J. Polym. Sci. Polym. Chem. Ed. 1979, 17, 1631.
40. Ringsdorf, H.; Schupp, H. Am. Chem. Soc., Org. Coat. Plast. Chem. 1980, 42, 379.
41. Day, D. R.; Ringsdorf, H. Makromol. Chem. 1979, 180, 1059.

42. Tieke, B.; Enkelmann, V.; Kapp, H.; Lieser, G.; Wegner, G. Am. Chem. Soc., Org. Coat. Plast. Chem. 1980, 42, 396.
43. Chance, R. R. Macromolecules 1980, 13, 396.
44. Pagano, R. E.; Weinstein, J. N. Ann. Rev. Biophys. Bioeng. 1978, 7, 435.
45. Szoka, F.; Papahadjopoulos, D. Ann. Rev.Biophys. Bioeng. 1980, 9, 467.
46. Ryman, B. E.; Tyrrell, D. A. Frontiers Biol. 1979, 48, 549.
47. Morawetz, H. J. Macromol. Sci. Chem. 1979, A13, 311.
48. Whitten, D. G. Angew. Chem. Int. Ed. Engl. 1979, 18, 440.
49. Breslow, R. Acc. Chem. Res. 1980, 13, 170.
50. Grätzel, M. Ber. Bunsenges. Phys. Chem. 1980, 84, 981.
51. Tien, H. Separ. Sci. Technol. 1980, 15, 1035.
52. Moss, R. A.; Lee, Y. S.; Alwis, K. W. J. Am. Chem. Soc. 1980, 102, 6646.
53. Fendler, J. H.; Romero, A. Life Sci. 1977, 20, 1109.
54. Poste, G.; Papahadjopoulos. D.; Vail, W. J. Meth. Cell Biol. 1976, 14, 33.
55. Scherphof, G.; Roerdink, F.; Hoekstra, D.; Zborowski, J.; Wisse, E. "Liposomes in Biological Systems"; Gregoriadis, G.; Allison, A. C., Ed.; Wiley: New York, 1980; p. 179.
56. Wisse, E.; Gregoriadis, G.; Daems, W. T. "The Recticuloendothelial System in Health and Disease: Functions and Characteristics"; Reichard, S. M.; Escobar, M. R.; Friedman, H., Ed.; Plenum Publishing Co.: New York, 1976; p. 237.
57. Cohen, C. M.; Weissmann, G.; Hoffstein, S.; Awasthi, Y.; Srivastava, S. K. Biochemistry 1976, 15, 452.
58. Gregoriadis, G.; Neerunjun, E. D. Biochem. Biophys. Res. Commun. 1975, 65, 537.
59. Torchilin, V. P.; Khaw, B. A.; Smirnov, V. N.; Haber, E. Biochem. Biophys. Res. Commun. 1979, 89, 1114.
60. Goldstein, I. J., Ed. "Carbohydrate-Protein Interaction"; ACS Symposium Series, Vol. 88, Washington, D.C., 1979.
61. Hampton, R. Y.; Holz, R. W.; Goldstein, I. J. J. Biol. Chem. 1980, 255, 6766,
62. Bader, H.; Ringsdorf, H.; Skura, J. Angew. Chem. 1981, 93, 109; Angew. Chem. Int. Ed. Engl. 1981, 20, in press.
63. Poste. G.; Allison, A. C. Biochim. Biophys. Acta 1973, 300, 421.
64. Sell, S. "Immunologie, Immunopathie und Immunität"; Verlag Chemie: Weinheim (West-Germany), 1977, p. 199.
65. Papahadjopoulos, D.; Hui, S.; Vail, W. J.; Poste, G. Biochim. Biophys. Acta 1976, 448, 245.
66. Szoka, F.; Papahadjopoulos, D.; Jacobson, K. Biochim. Biophys. Acta 1979, 551, 295.
67. Poste, G.; Pasternak, C. A. "Membrane Fusion"; Nicolson, G. L. Ed.; Cell Surface Reviews, Vol. 5; Elsevier: Amsterdam, 1978; p. 305.
68. Papahadjopoulos, D.; ibid.; p. 766.
69. Benz, R.; Zimmermann, U. Biochim. Biophys. Acta 1980, 597, 637.

70. Zimmermann, U.; Vienken, J.; Pilwat, G. Bioelectrochem. Bioenerg. 1980, 7, 332.
71. Poste, G. "Liposomes in Biological Systems"; Gregoriadis, G.; Allison, A. C., Ed.; Wiley: New York, 1980; p. 105.
72. Andreesen, R.; Modolell, M.; Weltzien, H. U.; Eibl, H. J.; Common, H. H.; Löhr, G. W.; Munder, P. G. Cancer Res. 1978, 38, 3894.
73. Kano, K.; Tanaka, Y.; Ogawa, T.; Shimomura, M.; Okahata, J.; Kunitake, T. Chem. Lett. 1980, 421.
74. Yatvin, M. B.; Kreutz, W.; Horwitz, B. A.; Shinitzky, M. Science 1980, 210, 1253.
75. Weinstein, J. N.; Magin, R. L.; Yatvin, M. B.; Zaharko, D. S. Science 1979, 204, 188.
76. Nawroth, T.; Ph. D. Thesis; University of Mainz, 1980.
77. Bodemann, H.; Passow, H. J. Membr. Biol. 1972, 8, 1.
78. Zimmermann, U.; Schultz, J.; Pilwat, G. Biophys. J. 1973, 13, 1005.
79. Scheurich, P.; Zimmermann, U.; Mischel, M.; Lamprecht, J. Z. Naturforsch. 1980, 35c, in press.
80. Zimmermann, U.; Scheurich, P.; Pilwat, G.; Benz, R. Angew. Chem. Int. Ed. Engl. 1981, 20, in press.
81. Huang, A.; Huang, L.; Kennel, S. J. J. Biol. Chem. 1980, 255, 8015.
82. Leserman, L. D.; Barbet, J.; Kourilski, F.; Weinstein, J. N. Nature (London) 1980, 288, 602.
83. Harper, N. J. J. Med. Pharm. Chem. 1959, 1, 467.

RECEIVED May 11, 1981.

16

Auto-Acceleration in Free-Radical Polymerizations Under Precipitating Conditions

ADOLPHE CHAPIRO

Laboratoire de Chimie Macromoléculaire sous Rayonnement, C.N.R.S., 94320 Thiais, France

When a monomer polymerizes under "normal" conditions the reaction follows the rules of stationary state and the rate remains constant over a fairly broad range of conversions. It is well known, however, from the study of numerous systems, that after a certain conversion is reached the reaction medium turns highly viscous and the rate then gradually increases leading to kinetic anomalies generally referred to as the "gel effect". This effect which was reported for the first time in 1939 by NORRISH and BROOKMAN (1) to occur in the bulk polymerization of methyl methacrylate is sometimes erroneously called "Trommsdorff effect" based on a report on German Science during world war II published by the allied forces in 1945. The "gel-effect" is caracterized by auto-accelerated conversion curves and, under extreme conditions, the heat of the reaction can no longer be dissipated which may result in thermal explosions.

Kinetic effects very similar to those occurring in the gel-effect are sometimes observed in polymerizing systems when the polymer is not soluble in the reaction medium and precipitates as it forms.

It is most generally accepted that the auto-accelerated character of these reactions lies in non-stationary conditions brought about by a sterically hindered termination step. The growing chains are assumed to be immobilized in the viscous medium ("gel-effect") or tightly coiled and occluded in dead polymer particles (under precipitating conditions). This leads to two consequences : on the one hand, a gradual increase in the number of growing chains (non steady-state conditions) which is assumed to be the actual cause of auto-acceleration and on the other hand, long-lived radicals build up in the system which are responsible for long-lasting ESR signals and for a "post-polymerization", a process which goes on for a considerable length of time once the sample has been removed from the light source or from the radiation field in the case of photochemical or radiation chemical initiation. It was further noticed in several systems that the molecular weight of the polymer rises at the point where auto-acceleration sets in.

0097-6156/81/0175-0233$5.00/0

It is clear, however, that a simultaneous increase in polymerization rate and molecular weight could either follow from a reduction in the rate of termination or from an increase in the rate of propagation. This last possibility has seldom been considered, except in some of the very early studies such as in the work of BENGOUGH and NORRISH (2) on the bulk polymerization of vinyl chloride where a "catalytic" action was attributed to the precipitated polymer.

A concept along similar lines was recently developed to account for the auto-accelerated character of the polymerization of carboxylic monomers and of acrylonitrile. Accelerated propagation is assumed to occur in oriented monomer-polymer association complexes. This conclusion is reached on the basis of kinetic evidence and the investigation of molecular associations present in these systems.

Polymerization of acrylic acid

Poly(acrylic acid) is not soluble in its monomer and in the course of the bulk polymerization of acrylic acid the polymer separates as a fine powder. The conversion curves exhibit an initial auto-acceleration followed by a long pseudo-stationary process (3). This behaviour is very similar to that observed earlier in the bulk polymerization of acrylonitrile. The non-ideal kinetic relationships determined experimentally in the polymerization of these two monomers are summarized in Table I. It clearly appears that the kinetic features observed in both systems are strikingly similar. In addition, the poly(acrylic acid) formed in bulk over a fairly broad range of temperatures (20 to 76°C) exhibits a high degree of syndiotacticity and can be crystallized readily (3).

A detailed investigation of the polymerization of this monomer in a series of solvents has shown, however, that the auto-accelerated character of the reaction is not related to the precipitation of the polymer. Thus, linear conversion curves and atactic polymers are obtained if the monomer is diluted in such non-associating solvents as toluene, n-hexane, carbon tetrachloride and chloroform, in spite of the precipitation of the polymer, whereas, both auto-acceleration and syndiotacticity persist for fairly high dilutions in water, methanol and dioxane even under conditions where the reaction medium turns homogeneous (4).

Figures 1 and 2 show the corresponding conversion curves in toluene and in methanol solutions respectively. In the latter case log-log coordinates are used to represent the data. The conversion curves are then linear and their slope β, which is the exponent of time in the relation : per cent conversion = Kt^{β}, measures the extent of auto-acceleration. β is referred to as the "auto-acceleration index". For pure acrylic acid β = 1.8 - 2.0 ; in non polar solvents β tends towards unity.

Table I

Comparison of the kinetic features observed in the bulk polymerization of acrylic acid and acrylonitrile

	Acrylonitrile	Acrylic acid
Polymer phase	Precipitates as a fine powder	Precipitates as a fine powder
Conversion curves	Auto-accelerated until 2-3 per cent conversion ; thereafter linear ("Auto-acceleration index" "β" = 1.3 at 20°C).	Auto-accelerated until 1-2 per cent conversion ; thereafter linear ("β" = 1.8 - 2.0 at 20°C).
Order with respect to initiation (at 20°C)	0.8	0.8
Post-polymerization (at 20°C)	Goes on for several month	Goes on for a few days
Over all activation energy	45 KJ/mole between 10 and 60°C. 12.5 KJ/mole above 60°C.	11.3 KJ/mole between 20 and 76°C.

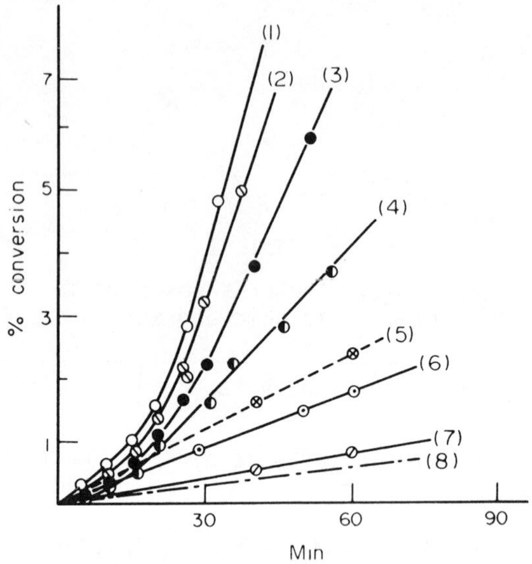

Figure 1. Conversion curves of the polymerization of acrylic acid in toluene solutions (4). Monomer concentrations: (volume per cent) (1) 100%; (2) 95%; (3) 90%; (4) 85%; (5) 80%; (6) 65%; (7) 50%; (8) 27%. The polymer precipitates as a fine powder at all concentrations. Initiation by gamma-rays at 20°C and 12 rad/min.

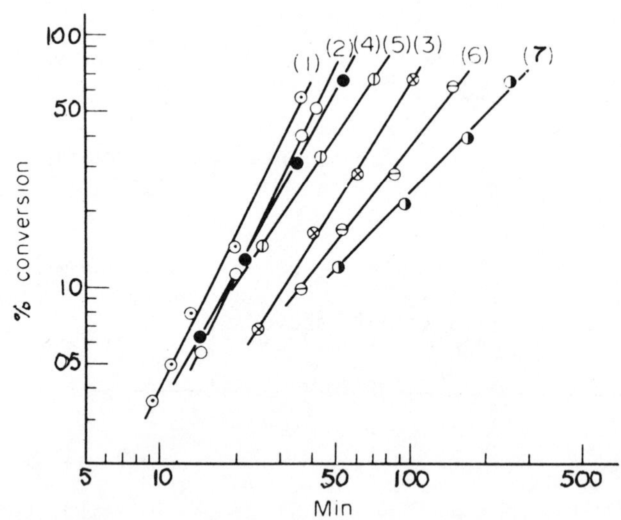

Figure 2. Log–log plots of the conversion curves of acrylic acid in methanol solutions (4). Monomer concentrations: (volume per cent): (1) 100%; (2) 80%; (3) 75%; (4) 60%; (5) 50%; (6) 25%; (7) 15%. The polymer precipitates as a fine powder for monomer concentrations of 100–75%; it forms a swollen gel for 75–50% monomer; the reaction medium is homogeneous for less than 50% monomer.

On the other hand, a good correlation was established between auto-acceleration and the type of molecular association involving the monomer in the system. Pure acrylic acid associates by hydrogen bonds to form "cyclic dimers" and "linear oligomers". The two species are in equilibrium.

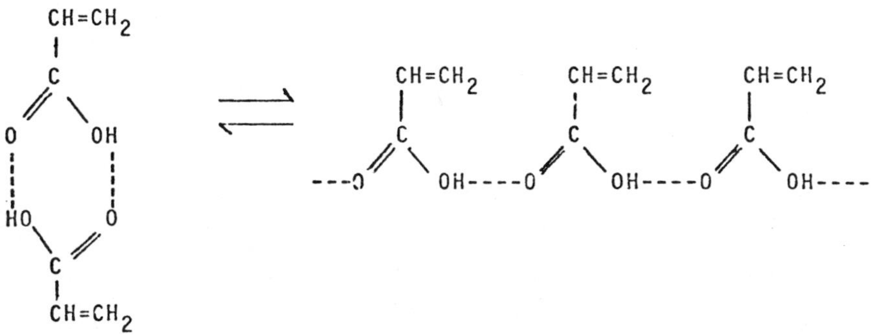

(I) Cyclic dimers

(II) Linear oligomers

The relative abundance of each of these species can be followed for instance by infrared analysis (4, 5). Figure 3 shows the I.R. spectrum of acrylic acid in the 1700 cm^{-1} region. The main peak at 1705 cm^{-1} is due to the vibration of the carbonyl group in the cyclodimeric form, whereas the shoulders at 1730 and 1740 cm^{-1} correspond to "open dimers" and "linear oligomers".

Dilution in non-polar solvents shifts the equilibrium to the dimeric form (Figure 4) while an increase in temperature favors the linear oligomers (Figure 5). Experiments conclusively demonstrated that auto-acceleration vanishes upon dilution when the oligomers are replaced by cyclodimers (4) while an increase in temperature, which restores oligomers again brings about auto-acceleration and syndiotacticity (5). This good correlation between the presence of oligomers and auto-acceleration does not, however, explain why auto-acceleration occurs. Since dimers and oligomers are present in the system right from the beginning the mere presence of oligomers cannot be taken as the cause of auto-acceleration. It was therefore assumed that such oligomers may associate with the polymer arising in the initial stages of the reaction to form a structure III in which the monomer is aligned in a regular manner on the polymeric matrix. In this ordered, "pre-oriented" structure an ultra-fast propagation is assumed to occur.

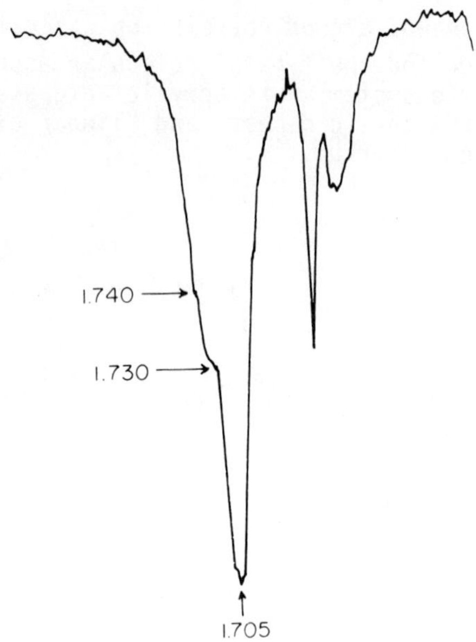

Figure 3. *IR spectrum of acrylic acid in the 1700 cm⁻¹ region.*

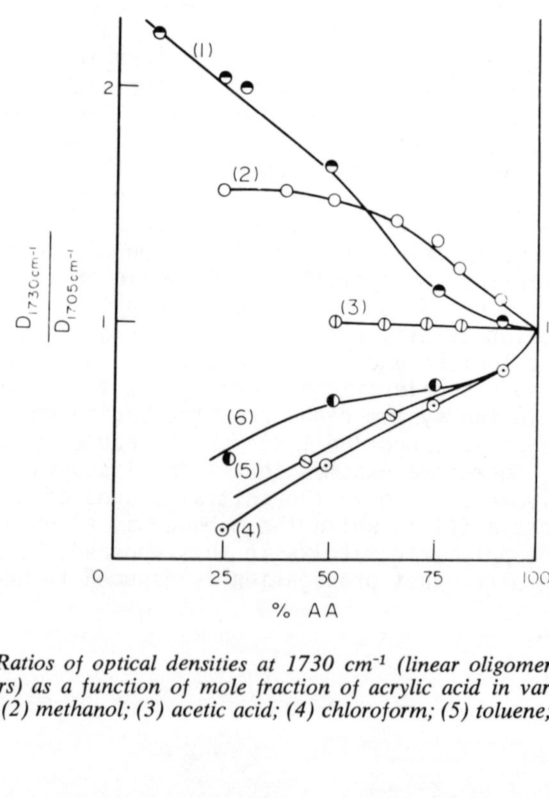

Figure 4. *Ratios of optical densities at 1730 cm^{-1} (linear oligomers)/1705 cm^{-1} (cyclic dimers) as a function of mole fraction of acrylic acid in various solvents: (1) dioxane; (2) methanol; (3) acetic acid; (4) chloroform; (5) toluene; (6) CCl_4 (4).*

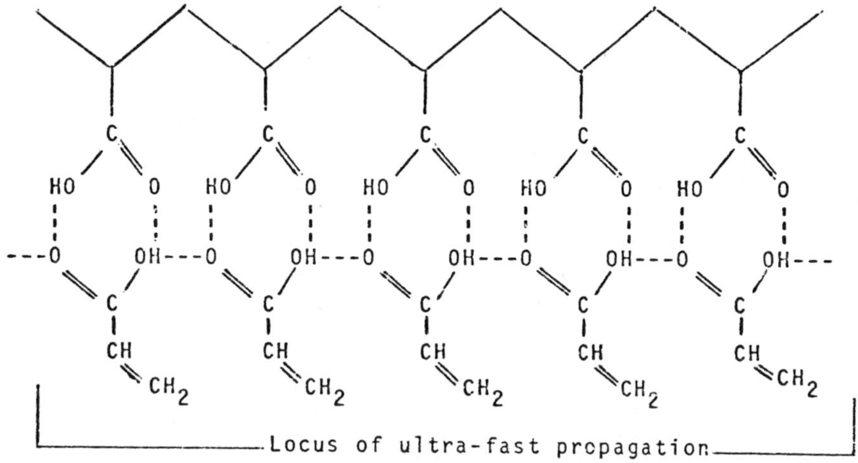

(Structure III)

The gradual accumulation of structures III in the initial stages of the reaction could account for its auto-catalytic character until a "stationary-state" is reached in which the surface of the precipitated polymer available for such association with monomer remains practically constant. A structure such as III could also be responsible for the stereo-control of the propagation step. Indeed, the polymer formed at conversions below 2 per cent exhibits much lower stereoregularity than at higher conversions (see Figure 6). These various peculiar consequences connected with the arizing of structure III will be referred to as "matrix effect".

It should be noted that the absence of matrix effect when the monomer is only associated in the cyclodimeric form suggests that structure I is more stable than the association of the monomer with the polymer and that therefore long ordered structures of type III do not arise. In very dilute acrylic acid solutions in hydrocarbons or chlorinated compouds a matrix effect (auto-acceleration) is again observed. This is believed to arise as a result of a structure III formed by the association of the polymer with "free" monomer (4).

Upon dilution in solvents which may associate via hydrogen bonds (water, methanol, dioxane) the situation is more complex. I.R. and Raman spectroscopy indicate the formation of various monomer-solvent complexes (4, 6). The corresponding absorption bands are in the same range as the characteristic bands for open dimers and oligomers and the latter cannot therefore be determined quantitatively. However, the viscosity of carboxylic acids was found to rise upon addition of water or methanol (4, 7) suggesting that these solvents bind together "oligomers". The persis-

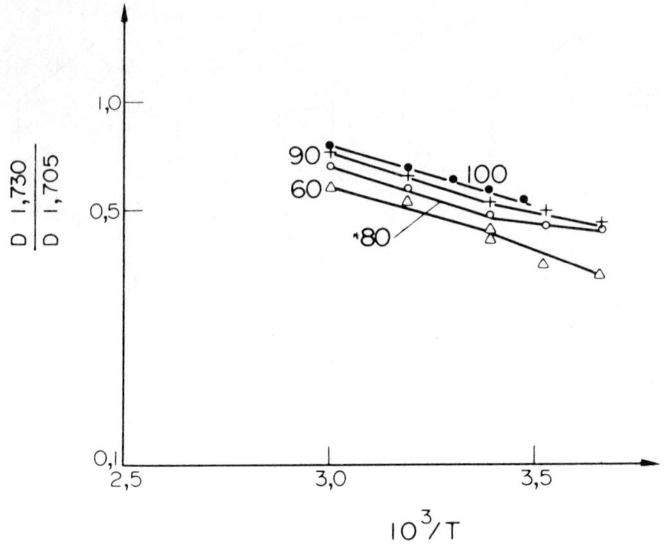

Figure 5. Ratios of optical densities at 1730 cm^{-1}/1705 cm^{-1} plotted in Arrhenius coordinates for pure acrylic acid (100) and acrylic acid solutions in toluene at 90, 80, and 60 volume per cent of monomer (5).

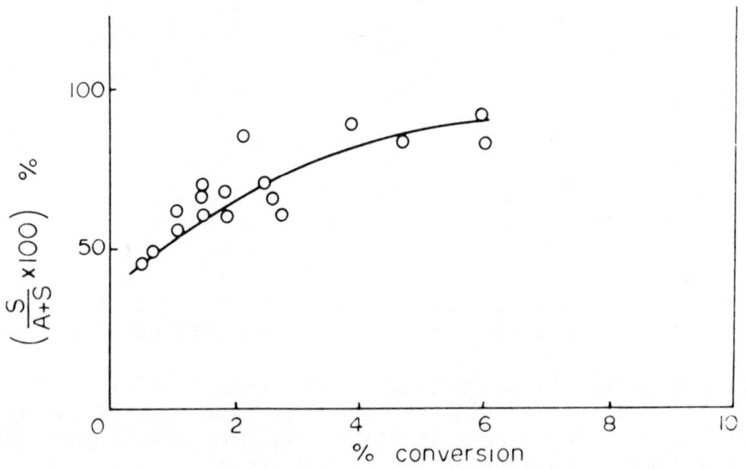

Figure 6. Fraction of syndiotactic polymer formed at very low conversions in the bulk polymerization of acrylic acid (4).

tence of auto-acceleration and syndiotacticity down to fairly high dilution further indicates that the oligomeric structure is maintained. One could also assume that monomer-solvent association complexes are formed which generate more easily an ordered structure such as III. Under very special conditions an "exaltation" of the matrix effect was observed leading to auto-acceleration indexes as high as 10 or even 15. The species responsible for this exaltation was assumed to be the complex (acrylic acid)$_2$-methanol which was characterized in such solutions (6).

Polymerization of methacrylic acid

Methacrylic acid also polymerizes in bulk under precipitating conditions. It forms molecular associations very similar to those of acrylic acid. However, the conversion curves were found to be linear under a variety of experimental conditions : temperatures of 16.5 to 60°C and broad ranges of initiation rates and monomer concentration in numerous solvents (7). It was assumed that structures of type III do arise but owing to steric hindrance and to the rigidity of the poly(methacrylic acid) molecule the monomer cannot align to form a "pre-oriented" complex as in the case of acrylic acid and propagation is not favored.

Investigation of the copolymerization in bulk of acrylic and methacrylic acids showed that auto-acceleration vanishes in the monomer mixture which generates a copolymer containing on the average one methacrylic for each acrylic unit (8). Figure 7 is a plot of various conversions curves in log-log coordinates. This result conclusively demonstrates that sequences of acrylic acid units of a certain length are required for auto-acceleration to set in, i.e. for the alignment of oriented monomers along the polymeric matrix.

Auto-acceleration was observed in the homopolymerization of methacrylic acid solutions over limited concentration ranges in methanol and in water. Perhaps under such conditions swelling of the polymer favors monomer diffusion leading to a larger amount of pre-oriented structures III. Alternatively, a monomer-solvent complex may arise which favors a pre-oriented structure and thus, may be responsible for the onset of a matrix effect (9).

Polymerization of acrylonitrile

The bulk polymerization of acrylonitrile has been studied by numerous workers (for a literature survey on the problem see ref. 10 and 11). The kinetic features of this reaction at room temperature are summarized in Table I. It is one of the typical examples of polymerization under heterogeneous conditions in which the anomalies are generally assumed to arise as a result of non-stationary conditions caused by the "occlusion" of growing chains in the precipitated polymer (10). The presence of occluded radicals was indeed demonstrated by ESR measurements (12) and by

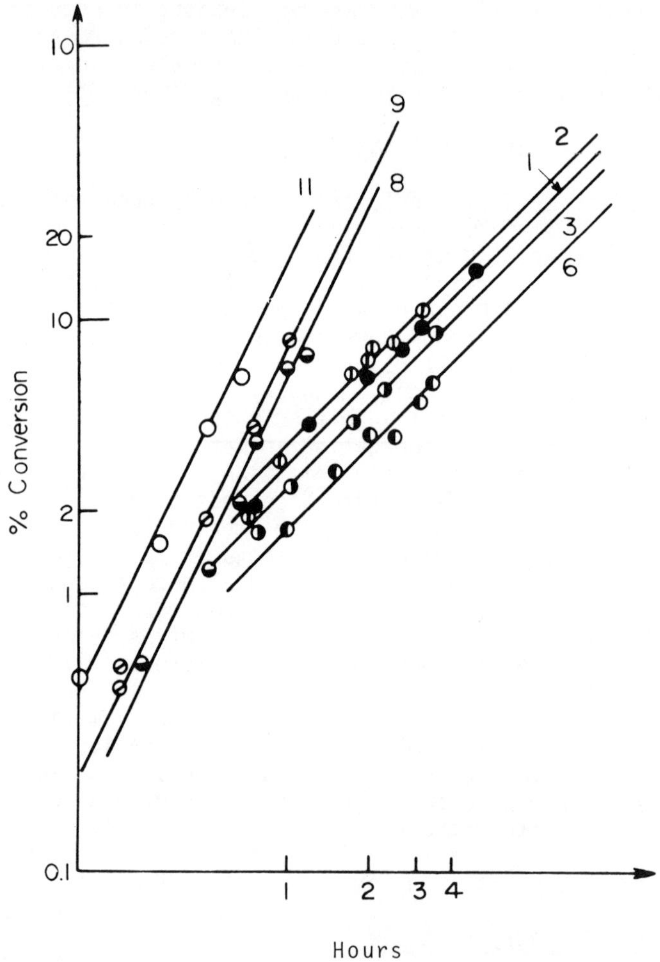

Figure 7. Log–log plots of the conversion curves of the copolymerization of acrylic acid with methacrylic acid (8). Mol % acrylic acid in the mixture: (1) 0%; (2) 23.5%; (3) 45.5%; (6) 71%; (8) 83.2%; (9) 86%; (11) 100%. The copolymer formed in a mixture of 75 mol % acrylic acid contains 50% acrylic acid and 50% methacrylic acid units.

reaction with DPPH (13). These long-lived species are also responsible for the post-polymerization observed in the photo-or radiation-initiated reaction after the radiation has been switched off.

On the other hand, it should be noted, however, that organic nitriles are known to associate to form dipole-dipole complexes (14). The bond dissociation energy of such associated dimers is ca. 20 $Kj.mole^{-1}$ (15) a value of the same order of magnitude as that of hydrogen bonds. The lack of solubility of polyacrylonitrile in its monomer is at least in part due to the formation of -CN---CN- bridges which link polymer chains together. This structure is also responsible for the high melting temperature of the polymer (15).

The data presented in Table I show that the kinetic features of the bulk polymerization of acrylonitrile are very similar to those observed with acrylic acid. It therefore seems pertinent to query whether a matrix effect could not arise in the polymerization of acrylonitrile through a regular orientation of monomer molecules along the polymeric matrix involving dipole interaction of the -C≡N groups (structure IV).

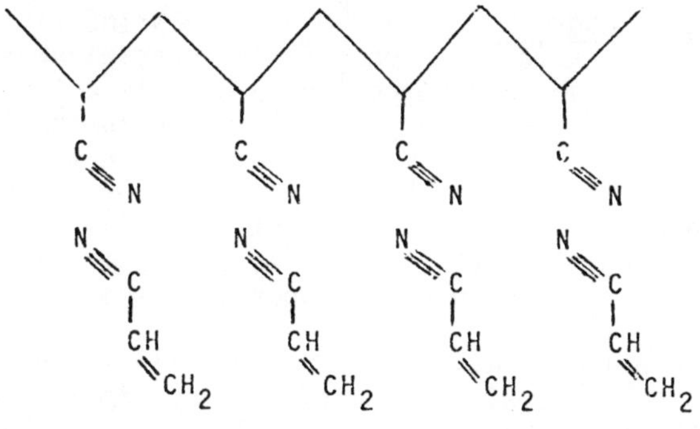

(Structure IV)

Several recent studies carried out in these laboratories did provide evidence for a matrix effect operating in the polymerization of acrylonitrile. These include :

1. The bulk polymerization of acrylonitrile at various temperatures.
2. The polymerization of acrylonitrile adsorbed on polyacrylonitrile.
3. The swelling of polyacrylonitrile in nitriles.
4. The polymerization in polar solvents.

Bulk polymerization of acrylonitrile at various temperatures. The bulk polymerization of acrylonitrile was recently reinvestigated at temperatures ranging from -9 to 85°C, using gamma-rays for initiation (16). In order to evaluate semi-quantitatively the extent of auto-acceleration all conversion curves were plotted in log-log coordinates. The slope "β" of the resulting straight lines, defined as the "auto-acceleration index", was found to be essentially constant at all temperatures from 20 to 85°C and was 1.25 to 1.30. Below 20°C, the value of β dropped to 1.15. The Arrhenius diagrams based on the "initial" and "pseudo-stationary" rates are shown in figure 8. These diagrams exhibit two breaks at ca. 10 and 60°C respectively, the activation energies are 12.5 $Kj.mole^{-1}$ above 60°C, 44-46 $Kj.mole^{-1}$ between 10 and 60°C and are poorly defined below 10°C. It was further found (18) that the "post-polymerization" which is very marked at 20 and 25°C (see also ref. 17) is very slow at 0°C and strongly reduced above 60°C. Figure 9 shows some of these results. Finally, the order of the reaction with respect to rate of initiation (dose-rate of gamma-rays) is 0.8 at 0, 20 and 50°C but drops to 0.6 at 75°C (16).

From these observations it was concluded that the bulk polymerization of acrylonitrile proceeds by different mechanisms depending on the temperature at which the reaction is carried out.

A. The reaction above 60°C. BAMFORD and JENKINS (12) have shown earlier that the radicals trapped in the precipitated polymer disappear rapidly above 60°C suggesting a significant bimolecular interaction of growing chains-ends. This conclusion is supported by the drop to 0.6 of the dose-rate exponent in the gamma-ray initiated reaction at 75°C and by the reduction of the post-polymerization above 60°C (Figure 9). POPOV et al. (19) further found that the swelling of polyacrylonitrile by its monomer suddenly increases above 70°C which corresponds to a transition temperature of the system.

The bulk polymerization of acrylonitrile in this range of temperatures exhibits kinetic features very similar to those observed with acrylic acid (cf. Table I). The very low over-all activation energies (11.3 and 12.5 $Kj.mole^{-1}$) found in both systems suggest a high temperature coefficient for the termination step such as would be expected for a diffusion controlled bimolecular reaction involving two polymeric radicals. It follows that for these systems, in which radicals disappear rapidly and where the post-polymerization is strongly reduced, the concepts of non-steady-state and of occluded polymer chains can hardly explain the observed auto-acceleration. Hence the auto-acceleration of acrylonitrile which persists above 60°C and exhibits the same "auto-acceleration index" as at lower temperatures has to be accounted for by another cause.

In view of the striking similarities between the kinetic parameters found for acrylonitrile in this temperature range and for acrylic acid it is suggested that in both systems the auto-acceleration originates from the same cause, namely the ability

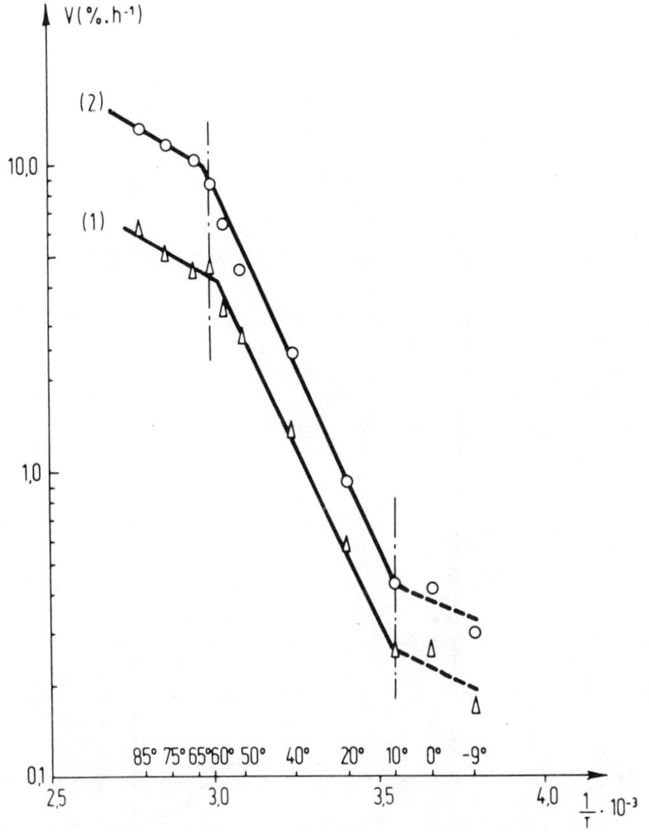

Figure 8. Arrhenius diagram of the gamma-ray initiated polymerization of acrylonitrile in bulk (16) Curve 1: "Initial rates"; curve 2: "Pseudostationary rates."

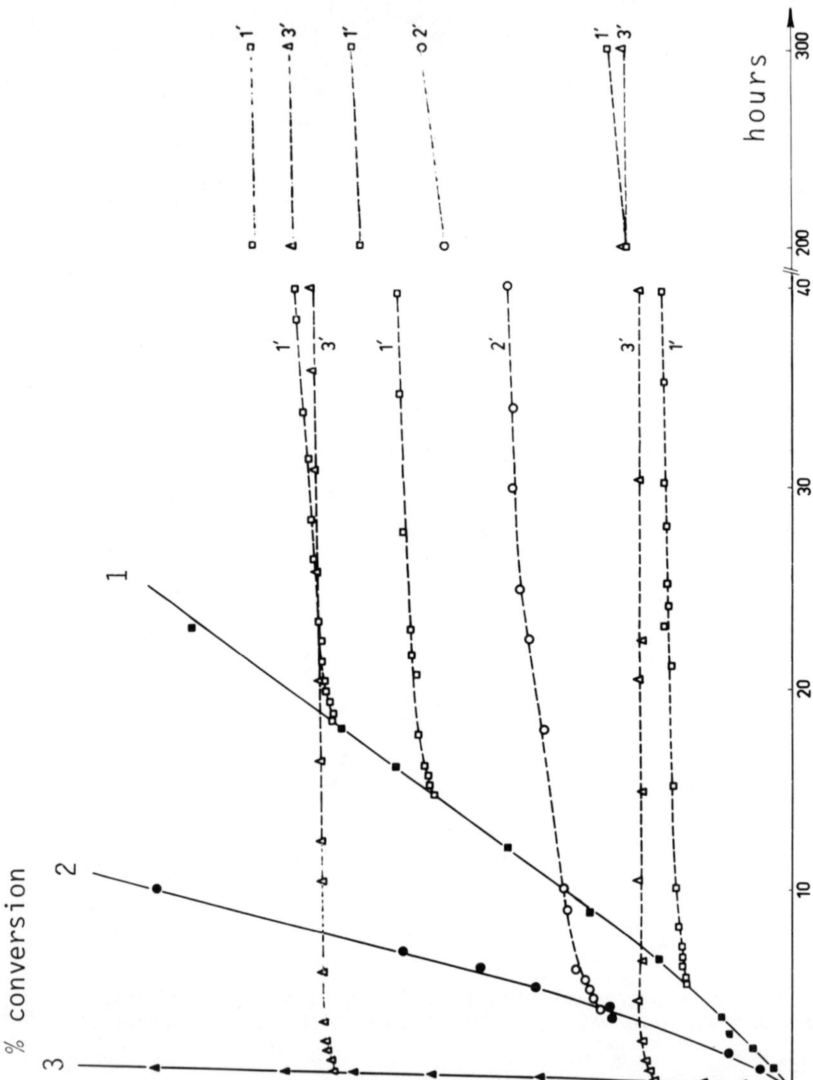

Figure 9. Post-polymerization of acrylonitrile following the gamma-ray initiated reaction (shown by the steep curves) at 0°C (curve 1), 20°C (curve 2) and 75°C (curve 3) (18).

of the polymer formed in the very early stages of the reaction to act as a matrix onto which the monomer associates to form regularly oriented structures such as III and IV in which propagation is assisted by the pre-orientation of the monomer molecules.

B. The reaction below 10°C. In this temperature range the "occlusion effect" dominates. Numerous growing chains become occluded in the dead polymer as demonstrated by long lasting ESR signals. The very small post-polymerization observed in this systems further indicates that monomer diffusion towards the active chain-ends is very slow. This is presumably related to the very low degree of swelling of the polymer. The kinetics are dominated by this "occlusion effect". The over-all activation energy is poorly defined and is presumably determined by a complex ratio of propagation to monomolecular (?) termination which strongly depends on external parameters.

The auto-acceleration observed under such conditions is reduced (β = 1.15) and could partially result from non-steady-conditions but also from a "matrix effect" operating on the surface of unswollen polymer particles. It should be noted in this respect that the post-polymerization which is induced by the growing chains occluded in the precipitated polymer exhibits an initial rate very much lower than the rate observed during irradiation (Curve 1' in Figure 9) which suggests that the contribution of the growth of occluded chains to the over-all rate is small.

C. The reaction between 10 and 60°C. In this range of temperatures the activation energy is very high (44-46 $Kj.mole^{-1}$). Since the polymer gradually swells in its monomer as the temperature rises the additional increase in rate cannot result from a reduction in chain termination. It seems much more likely to assume that the gradual swelling of the polymer favors chain propagation which generates an additional term to the activation energy. This term, which amounts to ca. 30 $Kj.mole^{-1}$, presumably corresponds to the increasing concentration of monomer around the growing chains as the temperature rises and perhaps also to an increased contribution of propagation in oriented monomer aggregates ("matrix effect").

Polymerization of acrylonitrile adsorbed on polyacrylonitrile. An intimate mixture of polyacrylonitrile solvated by its monomer is obtained if one melts acrylonitrile crystals which have been subjected to high energy radiation at low temperatures. The polymer forms under irradiation within the crystal lattice and upon melting, a gel-like phase is obtained in which the individual polymer molecules do not aggregate, presumably because most of the CN groups are then associated in pairs with the -CN groups of the monomer. Such a polyacrylonitrile solvated by its monomer should indeed be an ideal medium for the matrix effect to operate.

Experiments were therefore performed in which crystalline acrylonitrile was preirradiated to a certain dose at -196 and

-95°C. The reaction mixture was then warmed to room temperature and the excess monomer separated by gravity (by keeping the irradiated ampoule up-side down for 16 hours). The gel-like mixture of acrylonitrile and polyacrylonitrile was then irradiated at 20°C and the rate of conversion of monomer determined gravimetrically (20). Figure 10 shows some of the conversion curves obtained under such conditions. It appears that the rate of polymerization of acrylonitrile can be increased tenfold in the presence of highly divided polyacrylonitrile. To ensure that no "catalyst" was created during the irradiation of crystalline acrylonitrile, the monomer separated from the "gel" was also irradiated. Its rate of polymerization at 20°C was found to be the same as that of pure acrylonitrile. Moreover, irradiation of acrylonitrile in the presence of polymer formed at 20°C did not exhibit any significant acceleration over the reaction in bulk (curve 13 in Figure 10).

Swelling of polyacrylonitrile in nitriles. If acrylonitrile associates with its polymer to form oriented structures of type IV one should be able to detect a certain amount of swelling of the polymer in its monomer even at room temperature. Experiments along these lines were carried out using a film of polyacrylonitrile cast from DMF solution and carefully dried. It was indeed found that the film slowly absorbs monomer. The limiting swelling ratio was reached within ca.60 hours and was 16 per cent at 25°C (curve 2 in Figure 11) and 18 per cent at 75°C (curve 3 in Figure 11). This corresponds on the average to one monomer molecule per 6-7 monomer units in the polymer. More pronounced swelling was observed in 50-50 per cent mixtures of acrylonitrile with acetonitrile. The swelling ratio was 28 per cent after 100 hours which corresponds to 1 monomer per 4-5 units (curve 4 in Figure 11).

The very low rate of swelling suggests a complex process which could be the gradual (and slow) exchange of polymer-polymer to polymer-monomer association via dipole-dipole interactions of CN groups. A similar effect was observed in the swelling of PTFE membranes grafted with acrylic and methacrylic acid (21). Such membranes reached their limiting swelling within 10-15 minutes in water, whereas, 100 to 500 hours were required to swell the same membranes in carboxylic acids. The limiting swelling reached in these experiments corresponded to a "molar solvation" of 0.6 to 0.9 carboxylic monomer per carboxylic unit in the polymer (21). Taking advantage of the swelling of polyacrylonitrile in its monomer, experiments were recently performed in which acrylonitrile was grafted in the bulk of PTFE films using gamma-ray initiation. In this process the grafted front moves inwards into the bulk of the film by a progressive swelling of the grafted zones. 50 μm thick PTFE films were successfully grafted in the bulk by irradiation at 20°C in pure acrylonitrile to a grafting ratio of 30 per cent. Such a process can only occur if the grafted zones actually swell in the reacting monomer.

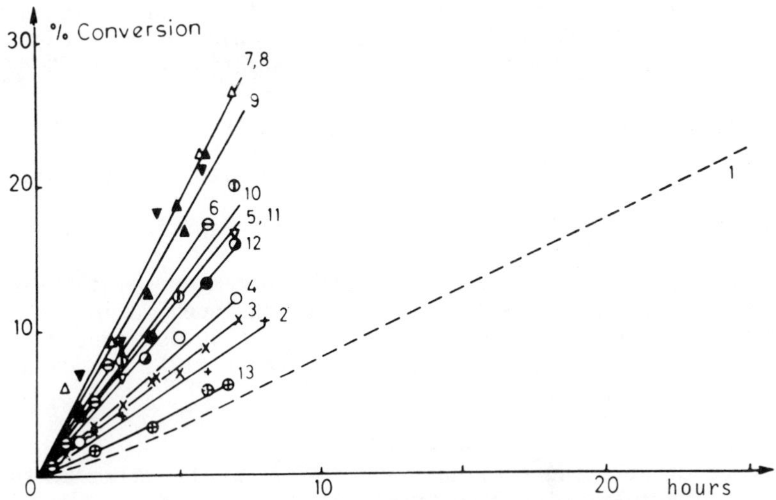

Figure 10. Polymerization of acrylonitrile at 20°C in an intimate mixture with a highly divided polyacrylonitrile obtained by pre-irradiation of the crystalline monomer at −95°C (20). Doses of pre-irradiation of 0.11 Mrad (curve 2) to 3.14 Mrad (curve 12). The broken curve 1 pertains to the polymerization of pure acrylonitrile; curve 13 is obtained in the presence of polyacrylonitrile pre-polymerized at 20°C.

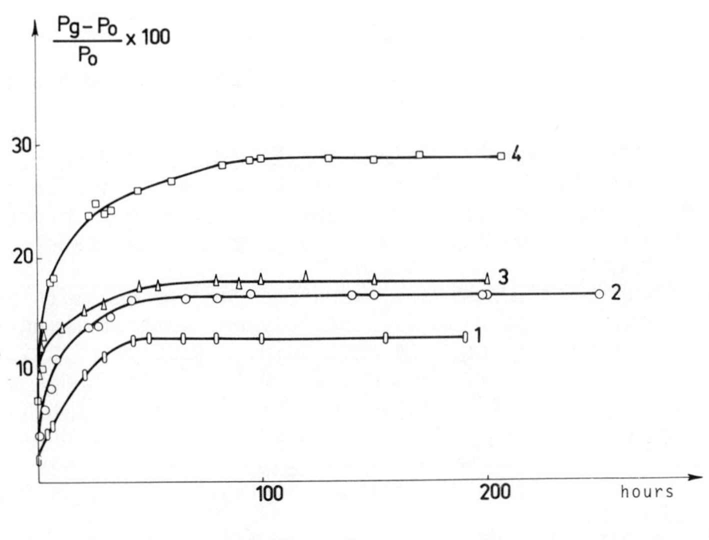

Figure 11. Kinetics of the swelling of a polyacrylonitrile film in (1) a 50–50% mixture of acrylonitrile with toluene at 25°C; (2) pure acrylonitrile at 25°C; (3) pure acrylonitrile at 75°C; (4) a 50–50% mixture of acrylonitrile with acetonitrile at 25°C (18).

The above results demonstrate that polyacrylonitrile is solvated by its monomer and that the process involves molecular association presumably via dipole-dipole interaction of the CN groups.

Polymerization of acrylonitrile in polar solvents. Polar solvents are expected to interfere with the association of nitrile groups in pairs and to replace the nitrile-nitrile association complex by a nitrile-solvent association. Under such conditions structures such as IV should no longer arise and the "matrix effect" should disappear.

Experiments conducted with dimethylformamide solutions at 20°C have shown that the "auto-acceleration index" β indeed decreases upon dilution and becomes 1.0 in a 40 per cent monomer solution where the polymer still precipitates (18). The data are summarized in Table II.

Table II
Influence of diluting acrylonitrile in DMF on the value of the "auto-acceleration index" β (18).

Acrylonitrile volume per cent	100	90	80	70	60	50	40	30	20	15	10
	Precipitating							Homogeneous			
β	1,30	1,30	1,25	1,15	1,10	1,10	1,0	1,0	1,0	1,0	1,0

However, DMF is a solvent for polyacrylonitrile and the polymerization occurs in a homogeneous medium for solutions containing 30 per cent monomer or less. This reduces the value of these experiments as an argument to show the influence of a matrix effect. Indeed the fact that auto-acceleration disappears when DMF is added to acrylonitrile was considered as a proof for the fact that precipitation of the polymer was the cause of auto-acceleration.

Further experiments were therefore carried out with polar solvents which do not dissolve the polymer. Most striking results were obtained with trichloroacetic acid. The polymerization of acrylonitrile in this solvent was found to proceed under precipitating conditions at all concentrations. In spite of this, the conversion curves were perfectly linear in solutions containing 60 volume per cent monomer or less (18). Moreover, these systems exhibit marked post-polymerization showing the presence of long-lived radicals.

These results conclusively demonstrate that precipitation of polyacrylonitrile as a fine powder and occlusion of growing chains resulting in post-polymerization do not bring about auto-acceleration if a highly polar solvent is present in the system.

This solvent prevents the dipole-dipole interaction of the -CN groups and thus, prevents the formation of the pre-oriented association complex which favours propagation and is thus responsible for auto-acceleration.

From the results presented above it can be concluded that the auto-accelerated conversion curves observed in the polymerization of acrylic acid, methacrylic acid and acrylonitrile are not caused by non-steady conditions arising as a result of the occlusion of growing chains in the precipitated polymer. This occlusion which is responsible for the post-polymerization observed in these systems only contributes to a limited extent to the over-all rates.

Auto-acceleration is determined by a "catalytic" action of the polymer formed in the early stages of the reaction. The monomer selectively "solvates" the polymer to form a pre-oriented monomer-polymer complex in which propagation occurs at a much higher rate. At this point it seems difficult to determine to what extent the conclusions reached above can be generalized to other systems. Experiments along these lines are in progress.

Abstract. Auto-accelerated polymerization is known to occur in viscous reaction media ("gel-effect") and also when the polymer precipitates as it forms. It is generally assumed that the cause of auto-acceleration is the arising of non-steady-state kinetics created by a diffusion controlled termination step. Recent work has shown that the polymerization of acrylic acid in bulk and in solution proceeds under steady or auto-accelered conditions irrespective of the precipitation of the polymer. On the other hand, a close correlation is established between auto-acceleration and the type of H-bonded molecular association involving acrylic acid in the system. On the basis of numerous data it is concluded that auto-acceleration is determined by the formation of an oriented monomer-polymer association complex which favors an ultra-fast propagation process. Similar conclusions are derived for the polymerization of methacrylic acid and acrylonitrile based on studies of polymerization kinetics in bulk and in solution and on evidence of molecular associations. In the case of acrylonitrile a dipole-dipole complex involving the nitrile groups is assumed to be responsible for the observed auto-acceleration.

Literature Cited

1. Norrish, R.G.W. ; Brookman, E.F. Proc. Roy. Soc. (London) 1939 , A-171, 147.
2. Bengough, W.I. ; Norrish, R.G.W. Proc. Roy. Soc. (London) 1950, A-200, 301.
3. Chapiro, A. ; Sommerlatte, T. Europ. Polym. J. 1969, 5, 707, 725.
4. Chapiro, A. ; Dulieu, J. Europ. Polym. J. 1977, 13, 563.

5. Chapiro, A. ; Goldfeld-Freilich, D. ; Périchon, J. Europ. Polym. J. 1975, 11, 515.
6. Ansarian, M. Thesis, Paris 1980.
7. Chapiro, A. ; Gouloubandi, R. Europ. Polym. J. 1974, 10, 1159.
8. Chapiro, A. ; Mankowski, Z. ; Renauld, N. Europ. Polym. J. 1977, 13, 401.
9. Gouloubandi, R. ; Chapiro, A. Europ. Polym. J. 1980, 16, .
10. Bamford, C.H. ; Barb, W.G. ; Jenkins, A.D. ; Onyon, P.F. "The kinetics of vinyl polymerization by radical mechanisms"; Butterworth Scientific London, 1958, pp. 103-117.
11. Chapiro, A. "Radiation Chemistry of Polymeric Systems" ; John Wiley & Sons, New York, 1962, pp. 147-149, 204-216.
12. Bamford, C.H. ; Jenkins, A.D. Proc. Roy. Soc. (London) 1953, A-216, 515 ; 1955, A-228, 220.
13. Bamford, C.H. ; Jenkins, A.D. ; Symons, M.C.R. ; Townsend, M.G. J. Polym. Sci. 1959, 34, 181.
14. Saum, A.M. J. Polym. Sci. 1960, 42, 57.
15. Henrici-Olivé, G. ; Olivé, S. Adv. Polym. Sci. 1979, 32, 123.
16. Burillo, G. ; Chapiro, A. ; Mankowski, Z. J. Polym. Sci. Chem. Ed. 1980, 18, 327.
17. Châpiro, A. J. Chim. Phys. 1950, 47, 764. ; ref. 11, pp. 209-214.
18. Mankowski, Z. Europ. Pôlym. J., to be published.
19. Popov, V.A. ; Ostrovoya, N.B. ; Gladyshev, G.P. ; Turekhanov, T.M. Vysokomol. Soed. Ser. B, 1976, 18, 739.
20. Chapiro, A. ; Mankowski, Z. Europ. Polym. J. 1978, 14, 15.
21. Ali-Miraftab, H. ; Chapiro, A. ; Le Doan, T. Europ. Polym. J. 1978, 14, 639.

RECEIVED July 29, 1981.

17

New Graft Copolymerizations onto Cellulosic Fibers

BENGT RÅNBY and CENITA RODEHED

The Royal Institute of Technology, Department of Polymer Technology, Stockholm, Sweden

A graft copolymer is defined as a polymer containing backbone chains - linear or branched, homo- or copolymer - onto which sidechains are grown or connected in a secondary reaction. The concept of graft copolymerization was formulated in the 1940:ies. The first graft copolymers described in the literature appear to be vinyl and allyl ethers of cellulose which were copolymerized with maleic acid esters[1]. The graft copolymer concept was actively promoted by H. Mark et al. at The Polytechnic Institute of Brooklyn, N.Y. New graft copolymers were prepared[2] and the research field thoroughly reviewed[3]. The research on grafting remained active through the 1960:ies and into the early 1970:ies as reviewed in several books[4,5,6]. In spite of the extensive research and a vast number of graft copolymers prepared and described in papers and patents, the successful industrial applications of graft copolymers are rather few. This is in particular true for graft copolymers of cellulosic fibers which are well studied but seldom used commercially.

The main purpose of preparing graft copolymers is to modify existing polymers - native or synthetic. A graft copolymer can retain properties of the substrate polymer and add properties of the grafted sidechains. This is of great interest for native polymers which occur in a few basic types, e.g. polysaccharides, polyuronides, polypeptides (proteins) and polynucleic acids. In addition, native polymers belong to our renewable resources of great potential value which has given new impetus to their wider use in diverse applications. This has meant a recently renewed interest in graft copolymerization onto cellulose and other polysaccharides.

Graft copolymerization can be initiated by physical or chemical processes. This paper describes methods for chemical initiation of grafting onto cellulosic fibers with special emphasis on recent developments, i.e. studies from the last 10-15 years. The initiation reactions may be free radical or ionic processes. The grafting reaction may also be a condensation or addition process.

0097-6156/81/0175-0253$5.00/0

Grafting by free radical polymerization is the most widely used technique.

GENERAL CONSIDERATIONS

Cellulosic fibers contain a partly crystalline cellulose phase as the main constituent. The grafting reactions can usually occur only with the amorphous or disordered cellulose. Only highly swollen cellulose with an expanded lattice may react throughout the ordered regions.

The presence of lignin, resins or other extractives in the fibers may interfere with the initiation or polymerization reactions, e.g. by termination or chain transfer of free radical reactions from phenolic groups. In some cases, lignin has no adverse effect and may even be grafted[7].

Most grafting methods initiate mainly long chains which grow out from a few active sites on the fiber substrate. It is usually difficult to influence the number and the length of the grafted chains, grown by free radical processes.

Initiation of graft copolymerization by radical mechanisms can occur by (a) a redox process on the substrate or (b) a chain transfer process to the substrate. In addition to grafting, formation of homopolymer may occur in both cases.

An example of a <u>redox process</u> is initiation by a metal ion which is oxidized by hydrogen peroxide:

$$Me^{n+} + H_2O_2 \longrightarrow Me^{(n+1)+} + OH^- + HO\cdot \qquad (1)$$

The hydroxyl radicals may abstract hydrogen which gives a radical on the substrate (P·) that can initiate polymerization by adding a vinyl monomer (M):

$$HO\cdot + PH \longrightarrow H_2O + P\cdot \qquad (2)$$

$$P\cdot + M \longrightarrow PM\cdot \qquad (3)$$

The hydroxyl radical may also initiate homopolymerization by adding to a vinyl monomer:

$$HO\cdot + M \longrightarrow HOM\cdot \qquad (4)$$

Certain metal ions with high oxidation potential in aqueous solution may oxidize the substrate (PH) directly at certain sites and form radicals which initiate polymerization:

$$Me^{n+} + PH \longrightarrow M^{(n-1)^+} + H^+ + P\cdot \qquad (5)$$

$$P\cdot + M \longrightarrow PM\cdot \qquad (6)$$

In the ideal case, no homopolymer is formed in this reaction.

A radical chain transfer process for grafting has an initiating step:

$$R\cdot + \sim PH\sim \longrightarrow RH + \sim\dot{P}\sim \quad (7)$$

and a grafting step:

$$\sim\dot{P}\sim + CH_2{=}CHY \longrightarrow \sim\underset{|}{P}\sim \text{ with side chain } CH_2\text{-}\dot{C}HY \quad (8)$$

where R· may be a radical formed by decomposition of an initiator or a growing radical chain end. R· radicals which do not transfer to substrate will form homopolymer.

The efficiency of grafting by chain transfer can be enhanced by introducing active substituents onto the substrate, e.g. thiol groups -SH by reaction with ethylenesulfide onto cellulose chains[8]:

$$\text{Cell-OH} + CH_2\text{-}CH_2 \text{ (bridged by S)} \longrightarrow \text{Cell-OCH}_2CH_2SH \quad (9)$$

$$\text{Cell-OCH}_2CH_2SH + R\cdot \longrightarrow \sim S\cdot + RH \quad (10)$$

$$\sim S\cdot + CH_2{=}CHY \longrightarrow \sim S\text{-}CH_2\text{-}\dot{C}HY \quad (11)$$

Hydrogen abstraction from -SH is faster than from -OH groups. It is generally of interest to increase both the yield of polymer and the grafting efficiency and decrease the formation of homopolymer. This can be achieved by proper selection of the grafting conditions, e.g. monomer concentration, initiating system and its application, reaction temperature and time.

SOME RECENT GRAFTING PROCESSES

A few examples of recently developed grafting reactions applied to cellulosic fibers are described to illustrate typical processes and discuss the mechanisms involved.

The Xanthate Grafting Method

A method of grafting vinyl monomers to substrates of cellulose xanthate was invented by Faessinger and Conté[9]. The initiation is a reaction of ferrated (12) or acidic (13) cellulose xanthate with hydrogen peroxide according to the following scheme (HO-OH and Fe^{2+} give HO· radicals):

$$\text{Cell-CH}_2\text{-O-}\overset{S}{\overset{\|}{C}}\text{-S-Fe} + HO\cdot \longrightarrow \text{Cell-}\dot{C}H\text{-O-}\overset{S}{\overset{\|}{C}}\text{-S-Fe} + H_2O \quad (12)$$

$$\text{Cell-CH}_2\text{-O-}\overset{S}{\overset{\|}{C}}\text{-SH} + HO\cdot \longrightarrow \text{Cell-CH}_2\text{-}\overset{S}{\overset{\|}{C}}\text{-S}\cdot + H_2O \quad (13)$$

Vinyl monomers are added to $-\dot{C}H-$ (xanthate group is split off) and to $-S\cdot$ to form grafted chains. The method can be applied either to a cellulosic substrate onto which a few xanthate groups are introduced or to a cellulose xanthate, dissolved as viscose, from which most of the xanthate groups are removed, e.g. in the fiber spinning process[10]. The maximum number of grafted chains is in this case determined by the number of remaining xanthate groups, e.g. related to the regeneration process of the viscose. With a large number of xanthate groups, short grafted chains can be obtained, which is unusual in grafting. Modified rayon fibers can be prepared by a simultaneous grafting and regeneration of the viscose. The reaction conditions to be controlled are the number of xanthate groups (introduced or remaining), initiator (peroxide + Fe^{2+} ions) and monomer concentrations, reaction temperature and time, and proper byproduct removal. Under optimal conditions conversions of monomer to polymer of close to 100% can be obtained. The homopolymer formation in this process seldom exceeds 5% of the observed add-on by grafting. The main advantage of the xanthate grafting method is the large number of short side-chains that can be introduced. This is usually not possible with other grafting methods.

Redox Initiating Methods

In grafting reactions initiated by redox processes, there is an interaction of the intermediate products with the substrate polymer chains. There may also be a chain transfer reaction involved. It is sometimes difficult to make a clear distinction between a direct redox reaction and a chain transfer process. Three redox systems have been studied extensively in grafting onto cellulose: persulfate ions[11], hydroperoxide/ferrous ions[12,13] and cerium (IV) ions[14].

Persulfate ions in aqueous solution when heated decompose to sulfate radical ions (14) which may react with water (15) or a hydroxyl-containing substrate (16):

$$S_2O_8^{2-} \longrightarrow 2\ SO_4^{\overline{\cdot}} \qquad (14)$$

$$SO_4^{\overline{\cdot}} + H_2O \longrightarrow HSO_4^{-} + HO\cdot \qquad (15)$$

$$SO_4^{\overline{\cdot}} + R\text{-}CH_2\text{-}OH \longrightarrow HSO_4^{-} + R\text{-}\dot{C}H\text{-}OH \qquad (16)$$

The hydroxyl radicals may abstract hydrogen from the substrate (which initiates grafting) (16) or add to a vinyl monomer (which gives homopolymer). The decomposition rate of persulfate ions is enhanced by the presence of a low-molecular weight alcohol. A high-molecular weight alcohol like cellulose can react in the same way.

The decomposition of hydrogen peroxide by ferrous ions (Fe^{2+}) is a typical redox reaction used for initiation of grafting[12]:

$$HO\text{-}OH + Fe^{2+} \longrightarrow Fe^{3+} + OH^{-} + HO\cdot \qquad (17)$$

The hydroxyl radicals formed may abstract hydrogen from the cellulose fiber substrate which gives grafting sites and subsequently grafted polymer with monomer present. The HO· radicals may also initiate homopolymerization. This means that reaction (17) is not specific for initiation of grafting. Another disadvantage is that the Fe^{3+} ions formed - if not carefully removed - may cause discoloration of the resulting product.

Grafting can be favoured by a proper choice of reaction conditions. When H_2O_2 and Fe^{2+} diffuse into the cellulose fibers at high initial concentration, a large number of grafting sites (macroradicals) are formed. A high concentration of a reactive monomer favours grafting. The presence of Fe^{3+} ions, formed in the redox process (17) or by oxidation of Fe^{2+} by air, is detrimental to grafting. Fe^{3+} ions can react with the macroradicals of the substrate which reduces the number of active sites for grafting[13]. Another possibility to enhance grafting is by periodic addition of H_2O_2 to the reaction medium at the early stages of the grafting reaction which decreases side reactions.

The detrimental effect of Fe^{3+} ions on initiation and polymerization in grafting by oxidation and termination, respectively, is a difficult problem. In attempts to decrease the reactivity of Fe^{3+} by complex formation, addition of potassium fluoride decreased the percentage of grafting somewhat without increased homopolymer formation[14]. Reactive monomers (like ethyl acrylate) are in these systems much more efficient for grafting than less reactive monomers (like vinyl acetate)[15]. A more successful approach to the Fe^{3+} problem is the use of reducing agents for conversion of Fe^{3+} to Fe^{2+} ions[16]. Addition of Fe^{2+} - sulphoxylate is claimed to increase the rate of grafting, the yield of grafted polymer and the conversion of monomer to polymer[17].

Organic Peroxides and Peroxyacids as Initiators

Although organic peroxides are commonly used as initiators for vinyl polymerization, there are only few reports of graft copolymerizations onto cellulosic materials using these initiators. Experiments with benzoyl peroxide and azoisobutyronitrile (AIBN) has given low yields of grafting of vinyl acetate and methyl acrylate onto cellulose compared with other grafting initiators[18]. These free radical initiators are assumed to effect grafting by a chain transfer mechanism, and the results obtained support these views[19].

An interesting redox system for grafting, containing cumene hydroperoxide and Fe^{2+} ions, has been developed by Simionescu et al.[20]. They grafted acrylonitrile and isoprene to rayon fibers at low temperatures (20 to 40°C) and long reaction times (up to

100 h) and found an optimum of grafting at pH 3 to 7. The radical character of the grafting process was confirmed by electron spin resonance measurements.

Peracetic acid, formed in situ from acetic acid and hydrogen peroxide in acid aqueous solution, was found to initiate graft copolymerization of vinyl monomers, e.g. methyl methacrylate and 4-vinyl pyridine, onto wood and cellulosic fibers at a good rate (1-3 h time) at 60°C[21]. The initiation reaction is not specific for grafting and considerable amounts of homopolymer is formed. An advantage is that peracetic acid can be applied as an initiator for grafting also onto lignin-containing substrates, e.g. ground wood pulp. Peracetic acid is effective also as a bleaching agent, which apparently is oxidizing phenolic groups in lignin, which otherwise may retard or inhibit radical chain reactions. The peracetic initiator has been successfully applied also for grafting onto rayon fibers and cellophane[22]. The peracetic initiator has been further developed by Hatakeyama[23] who combined the peracetic acid with Fe^{2+} or Fe^{3+} ions to a redox system found to be an effective initiator for grafting to pulps and isolated lignin. The reactions

$$CH_3\text{-}COOH + HO\text{-}OH \longrightarrow CH_3\text{-}(CO)\text{-}OOH + H_2O \qquad (18)$$

$$CH_3\text{-}(CO)\text{-}OOH \longrightarrow CH_3\text{-}(CO)\text{-}O\cdot + HO\cdot \qquad (19)$$

$$CH_3\text{-}(CO)\text{-}OOH + Fe^{2+} \longrightarrow CH_3\text{-}(CO)\text{-}O\cdot + Fe^{3+} + OH^- \qquad (20)$$

$$CH_3\text{-}(CO)\text{-}O\cdot + PH \longrightarrow CH_3\text{-}COOH + P\cdot \qquad (21)$$

involved are the formation of peracetic acid (18) and of radicals by splitting (19): the acetoxy and hydroxyl radicals may abstract hydrogen from the substrate (PH) and initiate grafting. The hydroxyl radicals may also initiate homopolymerization by addition to monomer. In the redox system (20) the hydroxyl radicals are reduced to hydroxyl ions leaving the acetoxy radicals which are more specific for grafting (21).

The peracetic acid redox systems with Fe^{2+} and Fe^{3+} salts as initiators for grafting have been further studied by Hirose and Hatakeyama[23]. From grafting experiments with methyl methacrylate, acrylamide and acrylic acid, they found that there is an optimum concentration of initiator to reach high conversion of monomer to polymer and high yield of grafted polymer. Higher temperatures than the 60°C previously used increases the rate of graft copolymerization (up to 75°C). The addition of a complexing agent (e.g. ethylenediamine-tetraacetate, EDTA) for the ferrous ions makes it possible to operate the grafting process at higher pH (complexing prevents precipitation of iron hydroxides).

A disadvantage of most peroxide and redox systems described and used for initiation of grafting onto cellulose is the bimolecular type of reaction which produces two kinds of free radicals, one of the substrate, e.g. a cellulose or lignin macroradical, and one low molecular weight radical. These two radicals initiate formation of graft copolymer and homopolymer, respectively. The reaction mechanism of grafting are further discussed for the alkali cellulose/O_2 system[24], cellulose xanthate/H_2O_2[25] and cellulose $NaClO_2$ or $NaIO_4$[26], which all initiate graft copolymerization to various degrees.

Initiation by Metal Ion Oxidation

Metal ions of transition and other elements of variable valency, e.g. Ce, Co, Fe, V, Mn, etc., are known to oxidize polysaccharides rather selectively, producing macroradicals as intermediates which are capable of adding vinyl monomers and form graft copolymers. These initiators are redox systems which differ from those previously described by not producing free radicals of low molecular weight. Only macroradicals on the substrate are formed in the redox reaction. Some homopolymer may still be formed in the process, e.g. due to oxidation of monomer or other side reactions.

Ce^{4+} Ion Initiation. Initiation of grafting with Ce^{4+} ions was introduced by Mino et al.[27]. The process has been widely studied and even applied to industrial production of cellulose and starch graft copolymers. A mechanism is derived from model experiments with low molecular weight vicinal diols in acid aqueous solution. The first step (22) is

$$Ce^{4+} + R_2\text{-}\underset{}{\overset{OH}{\overset{|}{C}}}\text{-}\overset{OH}{\overset{|}{C}}\text{-}R_2 \longrightarrow R_2\text{-}C{=}O + R_2\text{-}\dot{C}\text{-}OH + H_2 + Ce^{3+} \quad (22)$$

$$Ce^{4+} + R_2\text{-}\dot{C}\text{-}OH \longrightarrow R_2\text{-}C{=}O + H^+ + Ce^{3+} \quad (23)$$

an oxidative cleavage of the C-C bond. The second step (23) is an oxidation of the intermediate radical. With pinacol as diol acetone is the end product. It is assumed - without much evidence - that glucose units in cellulose and starch react as shown in (22) and (23) with cleavage between the carbon atoms C_2 and C_3 in the Pyranose ring. The Ce^{4+} grafting technique is operable in acid aqueous media at moderate temperatures. Only small amounts of homopolymer are formed.

The Ce^{4+} initiation of grafting and the products obtained from polysaccharides has been reviewed repeatedly[5,6]. Only some studies of the optimal grafting conditions for grafting of cellulose substrates will be treated here[28,29,30]. The reactants should be added to the cellulose suspension in the order Ce^{4+} initiator and then vinyl monomer to increase the yield of grafted polymer. The optimum conditions are not only related to the pH of the reaction medium but to the relations between the Ce^{4+} concentration and the reactivity of the monomer used. The type of Ce^{4+} salt and the acid used also influence the grafting. Sulfate ions in Ce^{4+} ammonium sulfate and sulfuric acid do not affect the reaction. When Ce^{4+} ammonium nitrate and nitric acid are used, the nitrate ion causes inhibition and termination of the grafting due to the oxidative power of the ion. The presence of occluded air in the substrate is detrimental to the grafting reaction. Organic solvents added to the aqueous reaction medium usually decreases the grafting yield and changes the order of reactivity for the different monomers used. The physical state of the cellulose substrate, e.g. crystallinity, dispersion by grinding, molecular weight, etc. highly influences the grafting yield. Grinding of the cellulose in a Wiley mill decreases the grafting yield while ball milling nearly inhibits the grafting reaction. Drying the cellulose at 105°C decreases the grafting yield while chemical depolymerization to low molecular weight causes increased grafting. The physical accessibility of the cellulose is of great importance for the grafting reaction with Ce^{4+} ions. The Ce^{4+} initiator has been applied extensively for grafting to partially substituted cellulose derivatives, e.g. carboxymethyl cellulose, cellulose acetate, cyanoethyl cellulose and allylated cellulose.

Metal Ion Initiation on Cellulose Xanthate. Vinyl monomers can be grafted to cellulose at high rates by reaction with cellulose xanthate using V^{5+}, Fe^{3+} and Cr^{6+} as redox initiators[31]. With monomers soluble in the aqueous reaction media and V^{5+} and Fe^{3+} as initiators, high yields and high grafting efficiencies (96-99%) can be obtained (only a few per cent homopolymer is formed). By this method, grafting can be achieved "on stream" during the regeneration of cellulose fibers in the viscose process. The V^{5+} ions give the highest grafting rate, but Fe^{3+} ions are more readily available and more economical to use. Grafting with Cr^{6+} ions gives lower yield and large amounts of homopolymer trapped in the cellulose framework[32].

Mn^{3+} Ions as Grafting Initiator. Phosphate complexes of Mn^{3+} ions were first described in 1961[33] for the grafting of vinyl monomers onto cellulose fibers (bleached wood pulp) as well as esters and ethers of cellulose and oxidized cellulose. Of the different phosphates studied, pyrophosphate complexes of Mn^{3+} ions were found to be most efficient as grafting initiator. This early patent specification has been overlooked in the literature of grafting initiators.

Independently the Mn^{3+} pyrophosphate complexes were developed as efficient initiators at very low concentrations (1 to 3 mmole/l) for grafting of vinyl monomers to starch[34]. High yields of polymer (over 90%) and very high grafting efficiencies (98-99%) were obtained by reactions for 1 to 3 h at room temperature (~30°C) in moderately acidic aqueous media (pH = 1.5 to 2.0). Corresponding grafting experiments have later been carried out on cellulose fibers as substrate with the same high yields and grafting efficiencies as previously obtained for starch[35]. The Mn^{3+} complexed ions appear to be the most selective and efficient initiator system for grafting to polysaccharides so far described. The mechanism of the initiation reaction has been studied in our laboratories by model experiments and ESR spectroscopy. There are two possible reactions indicated. Bond cleavage of a vicinal diol according to reaction (22) is one possibility. Another and faster reaction with Mn^{3+} giving radicals appears to be oxidation of aldehyde groups (24) to alkoxy radicals:

$$R\text{-}CHO + Mn^{3+}(Complex) \longrightarrow R\text{-}\dot{C}O + Mn^{2+} + H^{+} \qquad (24)$$

The formation R-$\dot{C}$O radicals from glucose is verified by spin-trapping experiments with tert-butyl-nitroxide and recorded ESR spectra for the resulting compound. Ring-closed vicinal diols indicate fast bond cleavage by Mn^{3+} oxidation and efficient initiation of vinyl polymerization.

It is likely that Ce^{4+} and Mn^{3+} ions initiate graft copolymerization by about the same mechanisms. Mn^{3+} ions have a lower redox potential (~1.5 V) than Ce^{4+} ions (~1.7 V), and they appear to cause less side-reactions besides the radical formation. Industrial development of the Mn^{3+} initiator for grafting is under way.

ABSTRACT

Graft copolymerization reactions - mainly based on free radicals - applied to cellulosic fibers during the last 10-15 years are described and the basic mechanisms discussed. Initiations of grafting by redox processes and chain transfer reactions give grafted polymer and homopolymer in different amounts. Several well studied initiation reactions are described in detail, e.g. grafting of cellulose xanthates, initiation by persulfate ions, hydrogen peroxide/Fe^{2+} ions, organic peroxides and peroxiacids, and various metal ions (Ce^{4+}, V^{5+}, Fe^{3+}, Cr^{6+} and Mn^{3+}). Grafting with Ce^{4+} initiation is extensively studied for cellulose and starch. More recently grafting with Mn^{3+} ions as initiators has been developed for cellulose and starch in our laboratories and shown to be the most efficient and selective initiator so far described: conversion of monomer to polymer ~90% and grafting efficiency 98-99% (only 1 to 2% homopolymer). Initiation with peroxides and peroxiacids can be applied successfully to cellulosic fibers containing lignin and extractives. Ce^{4+} and Mn^{3+} can be used only for pure polysaccharides, e.g. isolated starch and bleached wood pulps. Initiation of grafting Mn^{4+} has been studied extensively (36,37) for polysaccharides and other substrates.

LITERATURE CITED

1. Ushakov, S.N. Fiz. -Mat. Nauk. 1943, 1, 43.

2. Mark, H.F. Textile Res. J. 1953, 23, 294.

3. Immergut, E.H. and Mark, H.F. Makromol. Chem. 1956, 18/19, 322. See further Immergut, E.H. Encycl. Polymer Sci. & Technol. (Mark, H.F.; Gaylord, N.G.; Bikales, N.M., Eds.); Interscience, N.Y., 1965; Vol. 3, p. 242.

4. Burlant, W.J. and Hoffman, A.S., Eds. "Block and Graft Polymers"; Reinhold, New York, 1961.

5. Battaerd, H.A.J. and Tregear, G.W., Eds. "Graft Copolymers"; Interscience, New York, 1967.

6. Ceresa, R.J., Ed. "Block and Graft Copolymers"; Wiley; New York, 1973, Vol. 1 and 1976, Vol 2.

7. Philips, R.B.; Kobayashi, A.J.; Brown, W.; Stannett, V.T., Eds. "Pollution Abatement by More Effective Lignin Utilization. Grafting to Lignin and Lignin-Containing Pulps"; North Carolina State University, Raleigh, New York, 1970.

8. Chaudburi, D.K.R. and Hermans, J.J. J. Polymer Sci. 1960, 48, 159 and J. Polymer Sci. 1961, 51, 373.

9. Faessinger. R.W. and Conté, J.S. US Pat. 3, 330, 787 and Brit. Pat. 1, 059, 641; assigned to The Scott Paper Co., Philadelphia, Pa, USA.

10. Kraessig, H. Svensk Papperstidning 1974, 15, 417.

11. Oprea, S.; Dumitriu, S.; Bulacovski, V. Cellulose Chem. & Technol. 1979, 13, 3.

12. Gulina, A.A.; Livshits, R.M.; Rogovin, Z.A. Polymer Sci. USSR. 1965, 7:7, 1247. See further Kubota, H. and Ogiwara, Y. J. Appl. Polymer Sci. 1969, 13, 1569.

13. Morin, B.P. and Rogovin, Z.A. Polymer Sci. USSR. 1976, A18:10, 2451.

14. Misra, B.N.; Dogra, R.; Kaur, J.; Jassal, J.K. J Polymer Sci., Chem. Ed. 1979, 17, 1861.

15. Misra, B.N.; Dogra, R.; Metha, J.K. J. Polymer Sci., Chem. Ed. 1980, 18, 749.

16. Morin, B.P. and Rogovin, Z.A. Polymer Sci. USSR. 1976, A:18, 2451.

17. Morin, B.P.; Breusova, I.P.; Stanchenko, G.I.; Rogovin, Z.A.; Bereza, M.P.; Koptel'nikova, T.N. Russ. Pat. 401675, 1973.

18. Misra, B.N.; Jassal, J.K.; Paude, C.S. J. Polymer Sci., Chem. Ed. 1978, 16, 295.

19. Misra, B.N.; Jassal, J.K.; Paude, C.S. Indian J. Chem., Sect. A, 1978, 16A (12), 1033.

20. Bu, K.J.; Dimitriu, S.; Simionescu, C.I. Cellulose Chem. & Technol. 1977, 11, 455.

21. Hatakeyama, H. and Rånby, B. Cellulose Chem. & Technol. 1975, 9, 583.

22. Rånby, B. "New Methods for Graft Copolymerization onto Cellulose and Starch". (Rowell, R.M. and Young. R.A. Eds.); Acad. Press, New York, 1978, p. 171.

23. Hirose, S. and Hatakeyama, H.178th ACS Natl. Meeting, 1979, p. 199 and private communication, 1980.Jpn. Pat., S.50-119091

24. Schurz, J. and Rebek, M. Papier 1964, 18, 437.

25. Dimov, K. and Pavlov, P. J. Polymer Sci.1969, A-1, 7, 2775.

26. Morin, B.P. and Rogovin, Z.A. Vysokomol. Soedin. 1976, Ser. A 18(10), 2147.

27. Mino, G. and Kaizerman, S. J. Polymer Sci. 1958, 31, 242 and J. Am. Chem. Soc. 1959, 81, 1494.

28. Mansour, O.Y. and Schurz, J. Svensk Papperstidning 1973, 76, 258, 288, 415.

29. Mansour, O.Y. and Nagaty, A. J. Polymer Sci., Chem. Ed. 1974, 12, 141, 1887.

30. Mansour, O.Y. and Nagaty, A. J. Appl. Polymer Sci. 1979, 23, 2425.

31. Morin, B.P. and Rogovin, Z.A. Polymer Sci. USSR. 1976, A:18, 2451.

32. Morin, B.P.; Stanchenko, G.I.; Samojlov, W.J.; Rogovin, Z.A. Faserforsch. und Textiltechn. 1975, 26, 361.

33. US Patent Nr. 155, 493, 1961, assigned to Rayonier Inc.

34. Mehrotra, R. and Rånby. B. J. Appl. Polymer Sci. 1977, 21, 1647, 3407; J. Appl. Polymer Sci. 1978, 22, 2991, 3003.

35. Rånby, B.; Rodehed, C.; Stone-Elander, S.; Gädda, L.; Sundström, H. Unpublished work in 1978-1980. Cf. also ref. 22.

36. Uhlig, E. and Teichmann, R. Faserforsch. u. Textiltech. 19, 239 (1968); 20, 451 (1969); 22, 469, 609 (1971) and 23, 147 (1972).

37. Teichmann, R. Acta Polym. 1979, 30, 60.

RECEIVED February 5, 1981.

18

Fifty Years of Polystyrene

HEINZ G. POHLEMANN and ADOLF ECHTE

BASF Aktiengesellschaft, Ludwigshafen, West Germany

Just as Herman Mark was an important initiator of the scientific development of polymer chemistry and physics in our century and has left his decisive stamp on it, so did he have a particular influence on the early phase of the scientific and industrial development of styrene monomer and polystyrene. In 1980, polystyrene can look back on 50 years of industrial production, which began at the end of 1930 at Badische Anilin- & Soda-Fabrik (now BASF) in Ludwigshafen.

It is surprising that even on its golden jubilee polystyrene is known only to the specialists, although it has become one of the most commonly used plastics and has had its place in our daily life for many decades. We come across it in household appliances and toys, in furniture and electrical articles, in vehicles, in building and in mechanical engineering, in refrigeration and in packaging, at the workplace and in leisure activities - and scarcely anybody knows it by name.

Roots (from the 18th Century to 1930)

Actually polystyrene or its starting material styrene monomer is much older than 50. As early as the 18th century a chemist named Neuman obtained the styrene monomer as an essential oil when he subjected "storax" to steam distillation (1). Its original name "styrol", which is still used in German, was coined by E. Simon, a Berlin pharmacist, who discovered the correct elementary composition and observed that "styrol" solidified into a gel after standing for some time (2). J. Blyth and A.W. Hofmann correctly interpreted this in 1845 as a reaction without any change in the composition and called the solid reaction product "metastyrol" (3). Subsequently, styrene was described by a number of workers, including Berthelot, who obtained it by passing a mixture of benzene and ethylene through red-hot tubes (4). Berthelot also found that the polymerization of styrene can be accelerated by catalysts (5). Closer investigations of the polymerization reaction

0097-6156/81/0175-0265$5.50/0

were made by Kronstein (6) and Stobbe and Posnjak (7). The first practice-orientated work dates back to 1911 when polystyrene was described in two patents of Matthews (8) for the first time as an engineering material and "vehicle for paints and varnishes" without there being any practical consequences. The second effort was made in the twenties by Naugatuck Chemicals in the U.S.A., which tried to manufacture polystyrene on the basis of patents of Ostromislensky, but the technical and economic difficulties proved too great (9).

However, what polystyrene actually was, its constitution as a high molecular weight substance with a chainlike structure, the mechanism of its formation from the monomer in a chain reaction, - these facts had come to light in university and industrial laboratories since 1920. H. Staudinger, K.H. Meyer and H. Mark are the most important names on this path. They are the true founders of polymer chemistry as an independent discipline.

In his famous work of 1920 Hermann Staudinger first described the correct structure of polystyrene (10). It was Staudinger, too, who gave polystyrene its name and elucidated the mechanism of its formation (11). The polymerization of styrene provided access to a big class of substances and made a significant contribution to the understanding of natural polymers and to the synthesis of industrial plastics. A whole new branch of the chemical industry is based on the key substance polystyrene.

Three further conditions, however, had to be fulfilled before polystyrene could be manufactured on a large scale as an engineering plastic: the monomer had to be accessible at a reasonable price and in sufficient quantities; polymerization on an industrial scale had to be mastered; and there had to be an efficient method of fabrication available.

The first of these obstacles was surmounted by H. Mark and C. Wulff in 1929 at BASF Ludwigshafen. They succeeded in dehydrogenating ethylbenzene to styrene monomer over metal oxides (12). This is the process which is still used today. The second condition was also fulfilled in Ludwigshafen: C. Wulff and E. Dorrer (13) developed a continuous polymerization process that was ready for operation in 1930. It was a happy coincidence that at the same time it became possible, in collaboration with Dynamit AG, Troisdorf, to transfer the injection molding process for metals to plastics processing. This is described in a patent of Deutsche Legrit-Gesellschaft, Berlin (14).

Thus the stage was set for the production of polystyrene.

The First Steps (1930-1945)

Polystyrene as an Industrial Product. The beginnings were modest. The first Ludwigshafen polystyrene plant produced

6 tons per month. Up to the beginning of 1932, however, only 62 tons of polystyrene ("Polystyrene I") had been manufactured. The finished articles produced from it were brittle and developed cracks because they contained too much residual monomer and solvent. The removal of these constituents with steam at subatmospheric pressure ("Polystyrene II") and a higher molecular weight ("Polystyrene III") remedied this. New grades were added with the advent of emulsion polymerization which provided polystyrenes with a substantially higher molecular weight (polystyrene E and polystyrene EF).

Output in Germany soon increased appreciably. In 1936, 845 tons of ethylbenzene, 714 tons of styrene and about 550 tons of polystyrene were produced. By 1942 output had risen to almost 5000 t. At this time the first copolymers were being produced at Ludwigshafen by emulsion polymerization: polystyrene EN, a copolymer of 3 parts of styrene and 1 part of acrylonitrile as a chemical-resistant brand and polystyrene EH (2 parts of styrene, 1 part of acrylonitrile and 1 part of vinylcarbazole) as a boiling-water-resistant brand. Polystyrene EN became a precursor of the present SAN copolymers. Industrial production of styrene monomer in the U.S.A. did not begin until 1935, polystyrene following in 1938. In Great Britain polystyrene was produced from 1941 onward; other countries did not follow until after 1945.

Foamed polystyrene, too, dates from the thirties. The method of adding blowing agents to polystyrene in an extruder so that the extrudate expands was developed by Dow (15). It is still used today on a large scale.

Whereas, at the beginning of the thirties, polystyrene had been the driving force in the styrene monomer and polystyrene fields, this development was soon reversed. Under Germany's efforts to become self-sufficient there was a much bigger demand for styrene monomer for the manufacture of synthetic rubber than for polystyrene. As early as 1938 approximately 2500 t of styrene monomer was produced in Ludwigshafen for the newly commissioned rubber plants. When the Allies were cut off from their Asian rubber plantations in the Second World War, the U.S.A. followed suit with large styrene monomer capacities for the manufacture of rubber. Thus there were big capacities for styrene monomer available by 1945 for other uses.

Manufacturing Processes. By 1945 three processes were competing with each other for the further expansion of polystyrene capacities:

a) the continuous mass polymerization process
b) the batch suspension process, and
c) the emulsion polymerization process.

All three were developed at BASF Ludwigshafen and operated in various I.G. Farben plants.

The BASF continuous mass polymerization process employed a tower reactor with an upstream continuous stirred tank reactor (16) (Figure 1).

The conversion was 30-35% in the prepolymerization tank reactor and about 98% at the discharge end of the tower. The polymer was then fed by means of an extruder to a perforated die, extruded as strands and granulated. This principle is still being used today by many companies, even if in modified form.

The suspension process, developed in the U.S.A. (17) and also tested during the war in Ludwigshafen for polymerizing styrene (18), did not fully come into its own until after 1945.

The emulsion process, however, competed strongly in the initial phase with the continuous mass polymerization process, one reason being the easier heat removal but the main reason being that high molecular weights were obtained in a simple manner. The process first appeared in the patent literature (19, 20) in 1927 and was further improved by H. Fikentscher (21), finding wide application in the whole field of polymer chemistry.

Processing and Uses. Polystyrene is an outstandingly good injection molding material. In this process the material is melted and forced through dies into a mold where it cools and solidifies. The process is very adaptable and is particularly suitable for the mass-production of small articles. Components for the electrical industry, objects for everyday use (buttons, combs, tube tops, fountain pens, etc.) and also articles for the military field (components for detonators in grenades, parts of gas-masks etc.) were manufactured in this way at that time.

Beside the injection molding process, extruder processing was introduced. Extruded, biaxially stretched blown film of polystyrene was used as insulating material for deep-sea cables and as condensor film. It was here that particularly the high molecular weight emulsion polymers were used. Polystyrene EN containing acrylonitrile was used for making printing type and boiling-water-resistant polystyrene EH plays a role in high-frequency technology.

Research. We have already pointed out the contributions made to macromolecular chemistry by H. Staudinger, K.H. Meyer and H. Mark. It was not until 1930 that a basic consensus of opinion was reached on the molecular structure of high polymers. 1932 saw the publication of Staudinger's monograph on "High molecular weight organic compounds", in which he was the first to treat polymerization as a free-radical chain reaction (11). For his fundamental work he was awarded the Nobel Prize in 1953. Work was now carried out on a broad front in this field. Staudinger demonstrated the relationship between viscosity and molecular weight; in 1932 Fikentscher published his

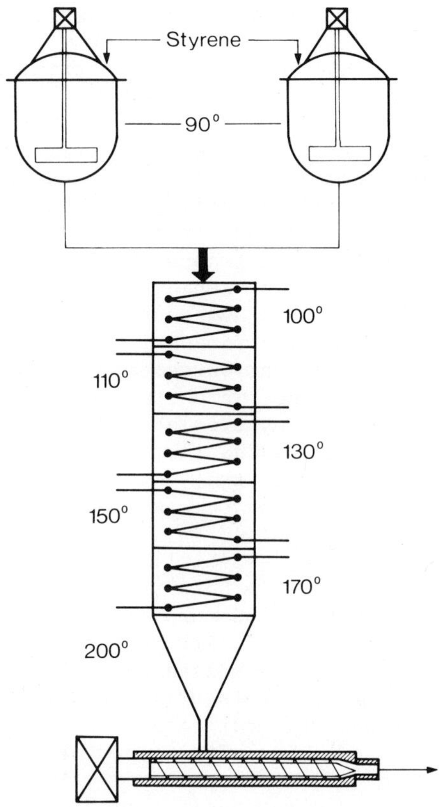

Figure 1. Ohlinger's continuous tower process of 1936 (Ludwigshafen, Germany) (16).

famous viscosity/concentration function with its constant which is still used today as "K value" for estimating the molecular weight (22). P. Flory (23) and G.V. Schulz (24) succeeded, independently of each other, in deducing an expression for molecular weight distribution based on kinetic considerations. The transfer reaction was first described by Mayo (25), the theory of emulsion polymerization by Smith and Ewart following a little later (26, 27) after preliminary work by H. Fikentscher (21). During this period it also proved possible to describe copolymerization using kinetic models (28).

Polystyrene's Development to a Major Industrial Product (1945-73)

The Polystyrene Family. After 1945 there was a drastic drop in the demand for styrene monomer because the need for synthetic rubber suddenly declined. On the other hand, a big demand existed for goods in the civilian sector and this provided a strong impetus to the expansion of polystyrene. In many cases, however, the mechanical strength of polystyrene was inadequate and this initiated numerous research efforts, especially in the immediate postwar period.

One of the most important outcomes of these efforts was impact-resistant polystyrene, which was obtained by modifying the brittle material with rubber. The first products were blends of polystyrene and synthetic rubbers; recourse was soon made, however, to a principle that Ostromislensky (29) had suggested as early as 1927: styrene monomer was polymerized in the presence of rubber dissolved in it.

Copolymers of styrene, especially with acrylonitrile, also attained increasing importance both in the unmodified form (30) and modified with rubber as ABS copolymers. The first products of this kind were blends of nitrile rubber and SAN (31). However, these only had mediocre mechanical properties because the interfacial compatibility was insufficient. The breakthrough came when nitrile rubber was replaced by a polybutadiene rubber which was grafted in emulsion with styrene and acrylonitrile (32).

It was possible to cover additional applications with these new types. In 1951 BASF found it was possible to impregnate polystyrene direct with expanding agents in the suspension process. This opened up great new possibilities for polystyrene foam (STYROPOR(R), BASF, 33), for example in the packaging and building fields.

Manufacturing Processes. The three manufacturing processes already mentioned (continuous mass polymerization, batch suspension and emulsion polymerization) continued to compete with each other after 1945. Whereas the third one gradually decreased in importance, the other two were given preference in further development. Up to this time, the production of polystyrene had been mainly carried out in the various works of I.G. Farben, but now the leading role in process development was taken over by the U.S.A.

Union Carbide (34) and in particular Dow adopted the continuous mass polymerization process. Credit goes to Dow (35) for improving the old BASF process in such a way that good quality impact-resistant polystyrenes became accessible. The result was that impact-resistant polystyrene outstripped unmodified crystal polystyrene. Today, some 60% of polystyrene is of the impact-resistant type. The technical improvement involved numerous details; it was necessary to learn how to handle highly viscous polymer melts, how to construct reactors for optimum removal of the reaction heat, how to remove residual monomer and solvents, and how to convey and meter melts and mix them with auxiliaries (antioxidants, antistatics, mold-release agents and colorants). All this was necessary to obtain not only an efficiently operating process but also uniform quality products differentiated to meet the requirements of various fields of application. In the meantime this process has attained technical maturity; over the years it has been modified a number of times (Shell in 1966 (36), BASF in 1968 (37), Granada Plastics in 1970 (38) and Monsanto in 1975 (39)) but the basic concept has been retained.

The continuous mass process is divided into 4 steps: rubber solution in styrene monomer, polymerization, devolatilization and compounding. In 1970 N. Platzer (40) drew up a survey of the state of the art. Polymerization is divided into prepolymerization and main polymerization; for both steps reactor designs other than the tower reactors shown in Figure 2 have been proposed. Main polymerization is taken to a conversion of 75 to 85%; residual monomer and any solvent are separated under vacuum. The copolymer then passes to granulating equipment, frequently through one or more intermediate extruders in which colorant and other auxiliaries are added.

The second large-scale process was the batch mass suspension process. Monsanto did the pioneer work on this (41). In this process, prepolymerization is carried out in bulk and main polymerization in suspension; the latter is taken to conversions of over 99%. In contrast to the continuous mass process, peroxide starters are used in order to achieve a high conversion at tolerable reaction times. Figure 3 shows a basic flow diagram of such a plant. A detailed discussion of advantages and disadvantages of the two processes can be found in R. Bishop's monograph published in 1971 (42), and it is continued in a paper by Simon and Chappelear in 1979 (43). It was a decisive factor for the economic success of impact polystyrene that these processes had been completely developed and mastered in theory and practice.

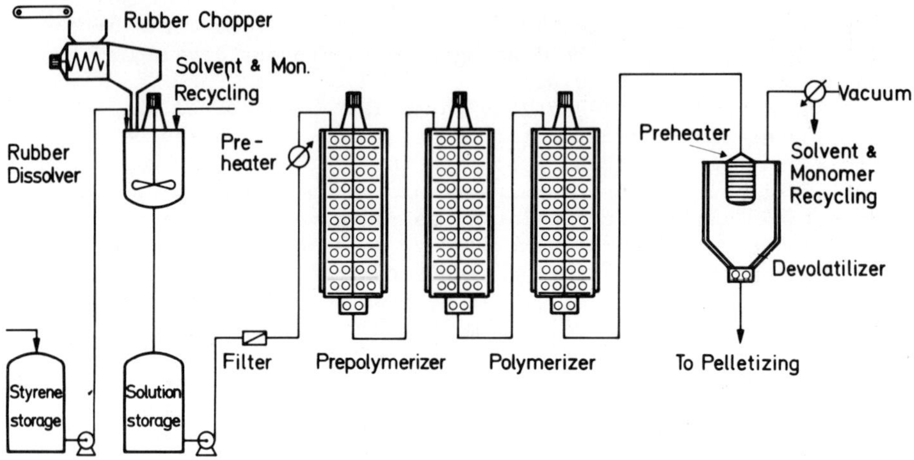

Figure 2. Continuous mass process (40) for crystal and high impact polystyrene.

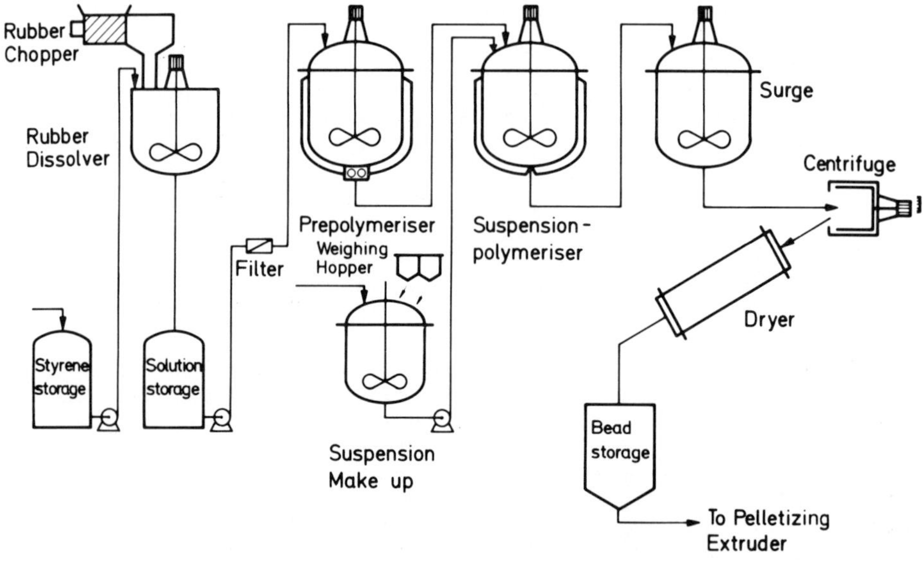

Figure 3. High impact polystyrene by mass-suspension polymerization (40).

Economic Development. In terms of output, polystyrene experienced a dramatic development in the postwar years. Whereas in 1940 world production was under 10,000 tons per year, by 1950 consumption in the western world had risen to 138,000 tons. The U.S.A. alone accounted for 85% of this. In 1960 550,000 tons and in 1970 2.1 million tons were used. Capacities increased parallel with the increase in consumption up to 1970 (Figure 4) after which they forged ahead of consumption. Petrochemistry on the one hand and plastics fabrication machinery on the other made enormous contributions to this rapid expansion.

For example, as the result of a great effort in Europe in the fifties, there was a switch from coal to oil as the feedstock. Only the petrochemical industry was able to supply the increasingly large amounts of raw materials for plastics. This is particularly true of benzene and ethylene, the two starting materials for styrene monomer and polystyrene. Ever-growing capacities and plants supplied these chemicals up to 1973 at ever lower prices which brought down the price of polystyrene and opened up further fields of applications.

Processing and Application. Polystyrene and the other thermoplastics would not have been able to enter on their triumphal march if the manufacturers of processing equipment had not been able to provide efficient fabrication methods. A decisive factor was that between 1950 and 1960 the injection molding machine with a screw preplasticizer was developed and extrusion through slit dies to give sheeting or film followed by thermoforming was evolved.

The screw injection molding machine was given a very considerable impetus by H. Beck's design (44); this was followed by the well-known Windsor machine with a double-screw preplasticizer (45), which found practical application earlier than the single-screw system. Injection molding techniques then progressed to greater and greater shot weights and locking pressures until it finally became possible to make the inner boxes for refrigerators from easy-flow high-impact polystyrene by injection molding. Machines for shot weights of over 10 kg and locking pressures of up to 3000 tons were needed for this (46).

Extrusion techniques for sheet and film were brought to a high degree of perfection, particularly in the early postwar years (47). They formed the basis for the use of impact polystyrene in the manufacture of large parts (refrigerator inner boxes and door liners) and later for the mass production of packaging articles (especially drinking cups). It has retained its important position as a fabrication method up to the present day; about 25 - 30% of impact polystyrene has been and still is going into extrusion.

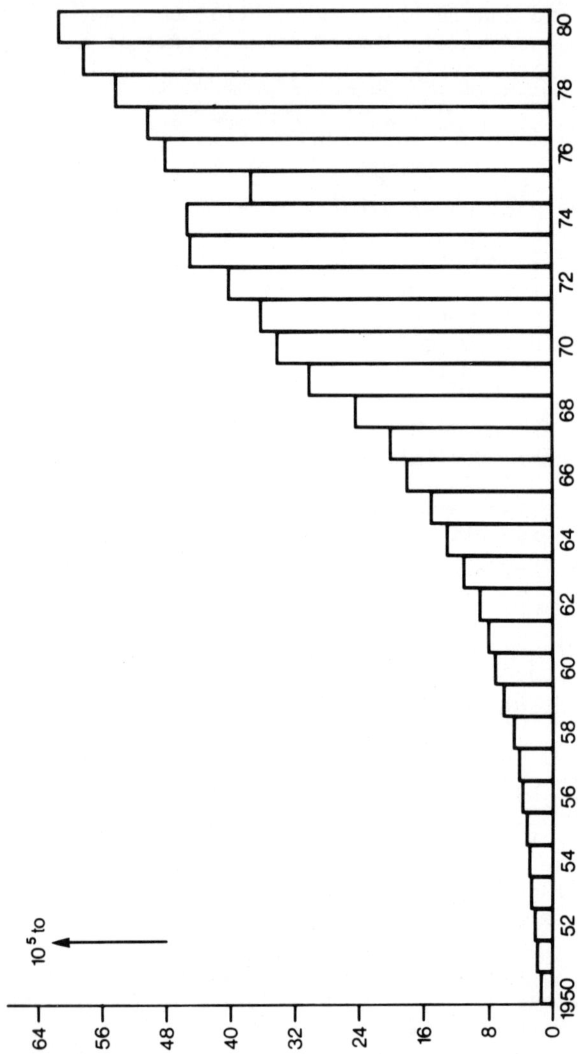

Figure 4. Production of styrene polymers in the western world from 1950–80.

More recently Thermoplastic Foam Molding (48) has been attaining increasing importance. This involves mixing impact polystyrene with chemical blowing agents and injection molding it. Molded articles of expanded polystyrene are formed with a relatively high density and a compact surface. This process is preferably used for furniture parts.

Polystyrene has a wide range of applications. Because it can be processed simply and efficiently to a great variety of articles, has an attractive combination of stiffness and toughness, is an outstanding electrical insulator, can be colored as desired and - above all - because the price was right, it has since 1950 conquered many more fields of application, ousted traditional materials and created many new uses for itself. A comparison of the applications in the U.S.A. in 1955 with those in Western Europe in 1979 conveys an idea of how the emphasis has shifted.

Table

Consumption of polystyrene in various fields (in %)

	U.S.A. 1955	Western Europe 1979
Packaging	15	43
Refrigeration	22	8
Toys	10	5
Housewares	11	15
Decorative articles	14	1
Electrical appliances	8	9
Automotive industry	-	2
Furniture	-	5
Miscellaneous	13	13

Even if the different consumer habits in the U.S.A. and Western Europe are taken into account, the comparison shows clearly that the various applications have undergone different developments. There has been a sharp drop - in relative terms - in the use of impact polystyrene for refrigeration - initially its main application - and for toys. In Europe, decorative articles play hardly any role. The packaging sector has made particularly big advances. Housewares, too, are on the way up.

Test Methods. The vigorous expansion of plastics necessitated selecting and standardizing suitable test methods. This took place very early in the industrial countries through the adoption of standards, for example ASTM in the U.S.A. and DIN in Germany. Tensile tests, impact and notched impact strengths, melt index and Vicat softening point were of particular importance for styrene polymers. These more general tests were complemented by some specific methods which provided insight into the morphological structure of the products; this was of decisive significance for the development of custom-tailored impact polyyrene grades. They include dynamic mechanical analysis with the torsion pendulum according to Schmieder and Wolf (49), which is an elegant way of determining both the shear modulus and the glass transition temperatures of the two constituents of the system (50). Another important milestone in this direction was the discovery of a method of analysis in BASF's laboratories in 1963 (51) to render the distribution and structure of the rubber particles in the polymer matrix visible by electron microscopy. For this purpose ultrathin sections are stained by osmium tetroxide, which reacts preferentially with the polybutadiene phase and thus makes it visible under the electron microscope. This pioneering method was not published, in order to keep it from competitors. It was, however, developed independently a short time afterward by K. Kato (52). The elegant method permits a clear insight into the structure of two-phase systems; it was the prerequisite for later intensive studies on the relationship between structure and properties, and is used today all over the world. Particularly the two-phase nature of impact polystyrene has sparked off numerous investigations and stimulated the extension of methods in polymer analysis, molecular weight determination, particle size determination, fracture mechanism, rheology, long-term studies, etc. A summary of the test methods for polymers may be found for example in the monograph entitled "Polymeranalytik" by Hoffmann, Krämer and Kuhn (53).

Progress in Polystyrene Research. As so often happens, practical applications have a stimulating effect on investigations of the underlying fundamentals. This is true of polystyrene, too. Styrene is a particularly versatile monomer, as it is copolymerizable with numerous other monomers, may be copolymerized by any known method, and is also easy to handle.

For instance, the mechanism of free-radical chain polymerization and free-radical copolymerization has been investigated, mainly using styrene as the model (cf., e.g., 54). The emulsion polymerization of styrene has also been studied by numerous authors (55). However, styrene is also polymerizable with Ziegler catalysts; an isotactic, crystalline polystyrene is obtained which has a high heat distortion temperature, but is very brittle (56). Styrene may also be anionically polymerized to give "living" polymers (57), and copolymerization of styrene with other monomers such as, for example, butadiene, gives block copolymers made up of segments of both polymers. Because of their incompatibility, there is phase separation in microscopic ranges, and the domains of each polymer block form ordered structures having a crystal-like arrangement (58, 59). Their macroscopic properties combine the features of both constituents. They possess good physical properties without vulcanization and can be processed like thermoplastic materials. For this reason they are often called thermoplastic elastomers. They have played a big part in the recent development of new high-impact polystyrenes. Styrene may also be polymerized cationically, but at ambient temperature this only gives low molecular weights.

Polystyrene also occupies an important position in polymer physics. A topic of intensive investigations and extensive controversy was in particular the question of the energy dissipation mechanism in two-phase materials. The observation of crazes in plastic parts under stress and how they differ from cracks was made in polystyrene (60, 61) and their role in fractures in polystyrene and other thermoplastics has since been investigated by many authors. It is now uncontested that crazing is the main energy-absorbing fracture mechanism in polystyrene and that the rubber particles act as nuclei for craze formation and as terminating sites for growing crazes (62, 63). In ABS polymers and rubber-modified PVC the fracture mechanisms can be different (64).

For clarifying the colloidal processes during polymerization in a two-phase system (oil-in-oil emulsion) (65-69) a solution of polybutadiene in styrene monomer was the most investigated model (70-73). In addition to forming a homopolymer, polystyrene is also grafted onto polybutadiene, and the graft copolymers formed stabilize the dispersed particles of the two-phase system and thus determine the properties of the end product. It has further been shown that phase/volume ratio, particle size, degree of crosslinking and molecular weight of the matrix contribute in a complex manner to the combination of properties (74, 75, 76).

The general principles of the structure of glassy polymers and their rubber-modified derivatives, and of the molecular and morphological parameters have been thoroughly investigated. This knowledge ensures that a high standard of quality of the commercial products is maintained and that polystyrene can continue to adapt to changing economic situations and new applications.

New Developments and Trends (1973-1980).

The Oil Crisis. Up to 1973, polystyrene consumption grew unchecked. A seemingly secure supply of inexpensive materials continuously opened up new applications and growing markets for this versatile and inexpensive material. However, the situation changed in 1973, when it was found that crude oil supplies were not limitless and subject to political manipulations.

Whereas crystal polystyrene cost about DM3 per kg in Western Europe in the middle of the fifties, the price had dropped to DM1 per kg at the beginning of the seventies. This development was abruptly stopped by the first oil crisis in 1974, and a second vigorous price hike was experienced in 1979, when the price of crude oil, and, in its train, naphtha, benzene and ethylene, again rose dramatically. Particularly the relatively heavy increase in the price of aromatics - boosted by anti-lead legislation for fuels - compared with aliphatics has influenced the competitiveness of polystyrene. Dwindling growth rates from 1974 onward and consumption which has at times stagnated - due in part to increasing use of the less expensive polypropylene - speak for themselves (Figure 5).

Polystyrene will in the coming years have to show that it can master this situation by further process rationalization and the development of even more sophisticated products. In spite of unfavorable cost trends, polystyrene has up to now continued its march. Production and sales have risen steadily - except for a brief decline in 1975 - and for 1980 we are expecting, in spite of the second price explosion in 1979, a doubling of 1970 sales in the western world to 4.2 million tons. If the estimated 1980 consumption of ABS/SAN polymers and EPS is added to this figure, an amount of 6.1 million tons of styrene polymers is obtained.

New Developments. Research and development work over the last decade has brought interesting innovations, which depart from the broad path usually trodden by products after their maturing phase and show the attempts being made to open up new outlets for the old polystyrene. The specialization which is now taking place will create new possibilities for polystyrene and help to ensure its future.

The first move in this direction was to improve the weatherability of impact-resistant polystyrene. Because polybutadiene, the most widely used rubber in impact-resistant polystyrene, is unsaturated, it is sensitive to photooxidation, and impact-resistant polystyrene is therefore not suitable for outdoor applications. A saturated rubber might be able to help here. In the ABS sector this has been successfully tried out with acrylate rubber (77) and EPDM (78, 79), and the latter has also been used in impact-resistant polystyrene (80, 81). This development has elicited satisfactory responses only in certain areas and more work still has to be done. For instance, attempts have been made to improve resistance to weathering by using silicone rubber (82). This approach is effective, but economic factors still stand in its way. Further impetus may also be expected from stabilizer research. Hindered secondary amines (83), to which considerable attention has recently been paid, are a first step in this direction.

Another successful development was based on morphological studies of traditional impact-resistant polystyrenes. Products with unusually big particles and a certain combination of composition and properties are particularly resistant to stress cracking (84). They have achieved considerable success on the

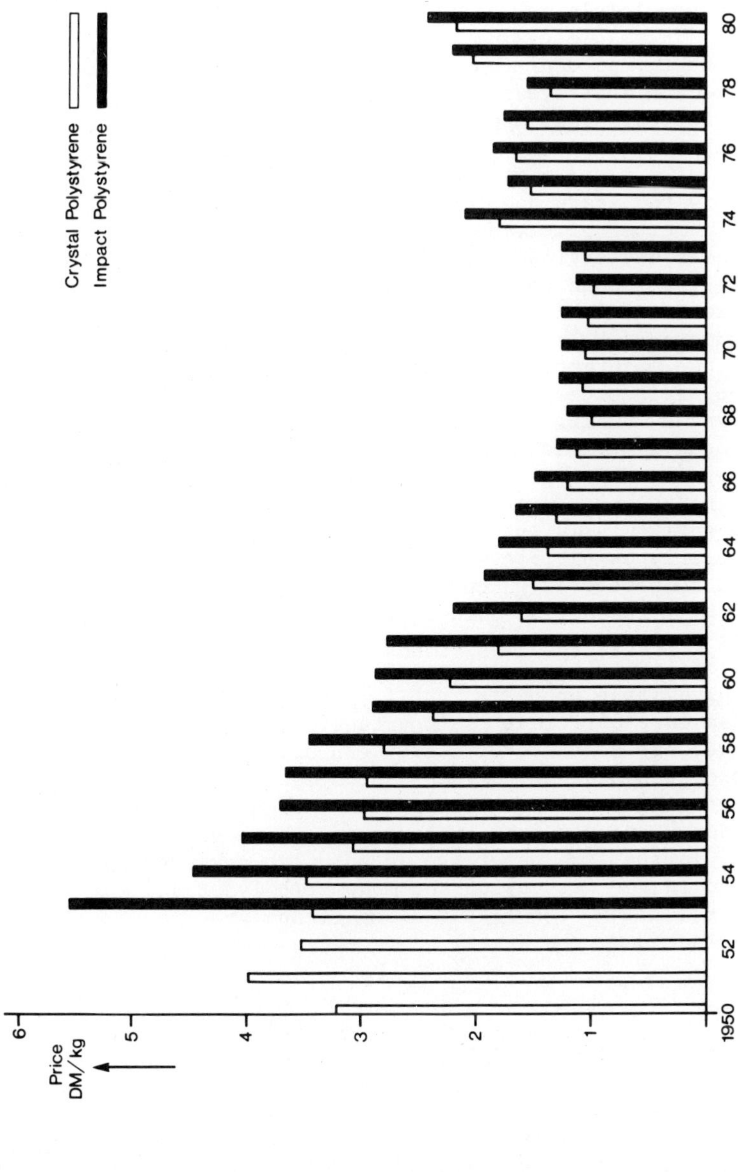

Figure 5. Polystyrene price development from 1950–80 in West Germany.

European market in refrigerators and freezers, and are also used for packaging fats. Here, too, development work is still in progress.

Morphological investigations have also been the basis for improvements in transparency. The use of block copolymers, and improvements in grafting during polymerization are opening up the way to completely new particle structures (85, 86, 87) (Figure 6). A systematic evaluation of these investigations has only just begun; products having capsule particle morphology are translucent and have excellent gloss.

Even further advantages in transparency are obtained with special styrene-butadiene block copolymers consisting predominantly of styrene units (88, 89). Such products have been in use for a few years now for transparent packagings and other purposes; numerous other applications are feasible.

The heat distortion temperature of styrene polymers is insufficient for some applications, but can be improved by copolymerization with monomers such as α-methylstyrene or maleic anhydride. Maleic anhydride copolymers are excellently suited to the manufacture of foamed articles. The advantages of glass-fiber reinforcement are greater in such copolymers than in polystyrene itself.

The heat distortion temperature of impact-resistant polystyrene may also be improved by polymer blends. Those of impact-resistant polystyrene with poly-2,5-dimethylphenylene-1,4-oxide (PPO) are particularly interesting (90). Polystyrene and PPO are molecularly compatible and mixtures of them have glass transition temperatures which vary virtually linearly with composition. A further advantage of these compositions which should not be under-estimated is their better flame resistance.

In addition to the development of new products with previously unavailable property combinations, the task of making the process more efficient is important, particularly in this day and age. The cost factor energy can still be reduced if, for example, the heat of polymerization can be better utilized. It has been suggested that heat pumps be used for this purpose and the energy recovered be employed for the devolatilization step (38). In the same paper the author also suggests the integration in one factory of the monomer/polymer and end product fabrication, the latter since the polymer is already available in the molten state.

Milestones

If the history of polystyrene up to the present day is reviewed, the following milestones emerge.

1929 Technical-scale dehydrogenation of ethylbenzene to styrene by H. Mark.

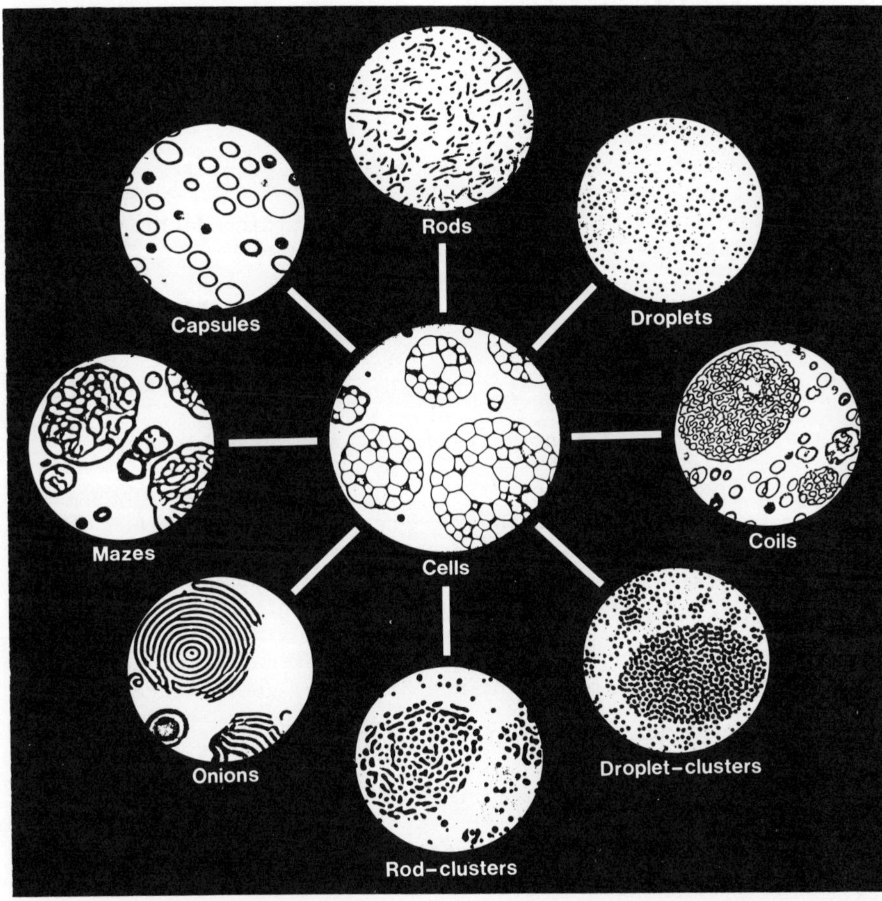

Figure 6. Different particle structures in high impact polystyrene (73) (85).

1930	First commercial-scale production of crystal polystyrene in Ludwigshafen by BASF's continuous tower process; processing of polystyrene in ram injection molding machines.
1936	Improvement of tower process by adding stirred prepolymerizers. First industrial-scale manufacture of SAN polymers in Ludwigshafen (emulsion process).
1938	First production in the U.S.A. of crystal polystyrene (Dow).
1943	Manufacture of polystyrene foam in Great Britain and the U.S.A. Introduction of screw injection molding machine (BASF).
1948	Introduction in the U.S.A. of ABS polymers (U.S. Rubber Co.).
1949	Processing of polystyrene by extrusion and thermoforming. Impact-resistant polystyrene from latex blends.
1951	STYROPOR(R), BASF - polymerization of styrene monomer in suspension in the presence of pentane as the blowing agent. Manufacture of new insulating packaging materials, etc. Suspension polymerization of crystal polystyrene in Ludwigshafen.
1952	Continuous mass process for impact-resistant polystyrene from styrene and SBR (Dow) goes onstream.
1953	Nobel Prize awarded to Staudinger.
1956	Anionic polymerization of styrene.
1957	ABS by graft polymerization.
1960	Processing of impact-resistant polystyrene to large injection moldings (weighing up to 10 kg). Block copolymers of styrene and butadiene.
1961	SBR in impact-resistant polystyrene ousted by polybutadiene.
1962	Blends of impact polystyrene with SBS block copolymers give super-high-impact polystyrene.

1963 Easy-flow, impact-resistant grades for large injection moldings.

1966 Low residual monomer products for flavor-neutral food packaging (BASF).

1966 Weather-resistant ASA polymers containing acrylic rubber in an SAN matrix (LURAN S(R), BASF).

1968 Polyblends of impact-resistant polystyrene and PPO (NORYL(R), General Electric).

1970 Transparent block copolymers of styrene and butadiene, having polystyrene character (Phillips). Weather-resistant, impact-resistant polystyrene with EPDM rubber (Mitsui Toatsu, Hoechst).

1972 Novel laminates produced by coextrusion using bonding agents.

1975 Stress-crack-resistant HIPS for refrigeration equipment and packaging of fats (BASF). Translucent HIPS for clear-on-contact packagings (BASF).

1978 Further reduction in the residual monomer level for food packaging applications (Dow, BASF, etc.).

Prospects

The development of polystyrene is by no means finished. On the contrary! The opinion is often put that the heyday of thermoplastics - and thus of polystyrene - is over and gone, because the raw material from which they are made, crude oil, continues to increase in price and is in short supply. However, this view is superficial, because what is overlooked is the fact that traditional materials (metals, glass, porcelain, ceramics, wood, paper, wool, cotton, etc.) cannot be produced without energy. If the total energy balances for different materials and their processing are compared, it will be seen that plastics, including polystyrene, come off better than their competitors. Plastics, compared with these other materials, require very little energy for fabrication (91, 92, 93) and they do not take up valuable land, that can be better used for agriculture or forestry.

Therefore, the use of plastics will increase and not diminish. One need only think of the importance attaching to heat insulation in the building industry for saving valuable energy, or of the automotive industry, where plastics can help to trim vehicle weight and thus save energy. There are so many things in life today which are made much easier, or indeed possible, by plastics, for instance supplying a continuously growing population in conurbations with day-to-day goods by means of plastic packagings. We employ an army of electrotechnical and electronic slaves for all sorts of jobs in production and the service business which free us from daily drudgery and toil - thanks to plastics. There are so many articles in daily use which only mass production keeps reasonably priced - household equipment, toys, writing utensils, medical equipment, furniture parts and decorative articles, dishes, large containers, and parts for boat and airplane construction - to mention only a few examples from the range of molded plastics of which polystyrene is one. They can be colored and printed, galvanized and flameproofed, combined and laminated, reinforced with glass fibers and filled with minerals. And this does not even take into account the significance of plastics as raw materials for textiles, paints and other surface coatings, paper and adhesives! Plastics also help to keep the environment inhabitable, because by replacing wood and fibrous materials they protect us against over-exploitation of our woods and crops.

However, they do more than merely replace. They are lighter than metals, less brittle than glass, tougher than wood, less permeable than paper, more resistant than ceramics, and more decorative than cardboard. Plastics, and polystyrene as one of them, have secured their own firm place in the modern world and it is hard to imagine life without them.

The future will bring many tasks and challenges. Polystyrene and the other plastics will have to be improved so that they can be used in new outlets. In the case of polystyrene this means developing grades which are even higher-impact, even stiffer, even more weather-resistant, even more resistant to heat distortion, and even more transparent. It will be made even easier to process. More than has hitherto been the case, it will be filled with cheap, organic or inorganic natural materials, and by no means have all modifications, for example those obtained by the introduction of functional groups, by copolymerization, or by morphological variation, been exhausted.

For polystyrene and its family, we therefore look optimistically into the future. The path which, in 1930, Herman Mark was one of the first to tread, will lead on, to the benefit of mankind.

Literature Cited

1 A Dictionary of Practical and Theoretical Chemistry, London, 1786, cited by A.J. Warner in Boundy-Boyer, "Styrene", New York, 1952; p. 2
2 Simon, E. Ann. 1839, 31, 265-277
3 Blyth, J.; Hofmann, A.W. Ann. 1845, 53, 292-329
4 Berthelot, J. Compt. Rend. 1866, 63, 788-793
5 Berthelot, J. Bull. Soc. Chim. (2) 1886, 6, 294
6 Kronstein, A. Ber. 1902, 35, 4153
7 Stobbe, H.; Posnjak, G. Ann. 1909, 371, 270
8 Matthews, F.E.; Brit. Pat. 16,277 and 16,278, (1911)
9 Ostromislensky, I.I.; U.S. Pat. 1,683,401-5 (Naugatuck) (1928)
10 Staudinger, H. Ber. 1920, 53, 1073
11 Staudinger, H. "Die hochmolekularen organischen Verbindungen"; Berlin, 1932
12 Mark, H.; Wulff, C.; German Pat. 550,055 (I.G. Farben) (1929)
13 Wulff, C; Dorrer, E.; German Pat. 634,278 (I.G. Farben) (1930)
14 Posner, F.; German Pat. 540,552 (Deutsche Legrit-Gesellschaft) (1929)
15 McIntire, O.R.; U.S. Pat. 2,450,436 (Dow) (1948)
16 Ohlinger, H. "Polystyrol", Berlin, 1955
17 Crawford, J.W.C.; McGrath, J.; U.S. Pat. 2,108,044 (ICI) (1938)
18 Hölscher, F.; BASF Company Archives Series, Ludwigshafen, 1972, p. 48
19 Luther, M.; Heuck, C.; German Pat. 558,890 (I.G. Farben) (1927)
20 Dinsmore, R.P.; U.S. Pat. 1,732,795 (Goodyear Tire & Rubber Co.) (1927)
21 Fikentscher, H. Angew. Chem. 1938, 51, 433
22 Fikentscher, H. Cellulosechemie 1932, 13, 58
23 Flory, P. J. Am. Chem. Soc. 1936, 58, 1877
24 Schulz, G.V. Z. Phys. Chem. 1939, B 43, 25
25 Mayo, F.R. J. Am. Chem. Soc. 1943, 65, 2324
26 Smith, W.V.; Ewart, R.H. J. Chem. Phys. 1948, 16, 592
27 Hohenstein, W.P.; Mark, H. J. Polym. Sci. 1946, 1, 127
28 Alfrey, T.; Mayo, F.R.; Wall, F.T. J. Polym. Sci. 1946, 1, 581
29 Ostromislensky, I.I.; U.S. Pat. 1,613,673 (Naugatuck), (1927
30 Brit. Pat. 371,396 (I.G. Farben) (1932)
31 Daly, L.E.; U.S. Pat. 2,439,202 (U.S. Rubber Comp.) (1946)
32 Hayes, R.A.; U.S. Pat. 2,802,808 (Firestone) (1953)
33 Stastny, F.; Gaeth, R.; German Pat. 914,322 (BASF) (1949)

34 Allen, I. jr.; Marshall, W.R.; Wightman, G.E.; U.S. Pat. 2,496,653 (UCC) (1943)

35 Amos, J.L.; McCurdy, J.L.; McIntire, O.R.; U.S. Pat. 2,694,692 (Dow) (1950)

36 Brit. Pat. 1,063,603 (Shell) (1966)

37 Bronstert, K.; Buchholz, K.; Echte, A.; Hofmann, J.; Germ. Publ. Appl. 1,770,392 (BASF) (1968)

38 McCurdy, J.L. Mod. Plast. Int. 1979, 9, 118

39 Carter, D.E.; Simon, R.H.M.; U.S. Pat. 3,903,202 (Monsanto) (1975)

40 Platzer, N. Ind. Eng. Chem. 1970 (1), 62, 6-20

41 Stein, A.; Walter, R.L.; U.S. Pat. 2,862,906 (Monsanto) (1957)

42 Bishop, R.B. "Practical Polymerization for Polystyrene"; Boston, 1971

43 Simon, R.H.; Chappelear, D.C. "Technology of Styrenic Polymerization Reactors and Processes"; Am. Chem. Soc. Symp. Ser. 104, Washington D.C., 1979, p. 71

44 Beck, H.; German Pat. 858,310 (BASF) (1943)

45 Brit. Pat. 760,480 (R.H. Windsor Ltd.) (1953)

46 Hofherr, F.; Wild, H.O. "Kunststoffhandbuch"; Vol. V. "Polystyrol", Vieweg/Daumiller: Munich, 1969; p. 604-606

47 Dow Digest, cited in Kunststoffe 1954 (1), 44, 36-40

48 Eckart, H. Kunststoffe 1980 (3), 70, 122-127

49 Schmieder, K.; Wolf, K.A. Koll. Z. 1952, 127, 65

50 Cigna, G. J. Appl. Polym. Sci. 1970, 14, 1781

51 Stabenow, J.; internal BASF report, 1963, unpublished

52 Kato, K. J. Electron. Microsc. 1965, 14, 219

53 Hoffmann, M.; Krömer, H.; Kuhn, R. "Polymeranalytik" Thieme: Stuttgart, 1977

54 Henrici-Olivé, G.; Olivé, S. Fortschr. Hochpolym. Forsch. 1961, 2, 496

55 Gerrens, H. Fortschr. Hochpolym. Forsch. 1959, 1, 234

56 Natta, G. Angew. Chem. 1956, 68, 393

57 Milkovich, R.; Levy, M.; Szwarc, M. J. Am. Soc. 1956, 78, 2656

58 Vanzo, E. J. Polym. Sci 1966, A 1, 4, 1727

59 Hendus, H.; Illers, K.H.; Ropte, E. Kolloid-Z.Z. Polym. 1967, 216/127, 110

60 Sauer, J.A.; Marin, J.; Hsiao, C.C. J. Appl. Phys. 1949, 20, 507

61 Hsiao, C.C.; Sauer, J.A. J.Appl. Phys. 1950, 21, 1071

62 Bucknall, C.B. "Toughened Plastics"; London, 1977

63 Retting, W. Die Angew. Makrom. Chem. 1977, 58/59, 133

64 Breuer, H.; Haaf, F.; Stabenow, J. J. Macrom. Sci.-Phys. 1977 B 14 3, 387

65 Molau, G.E. J. Polym. Sci. 1965, A, 3, 1267

66 Molau, G.E. J. Polym. Sci. 1965, A, 3, 4235

67 Molau, G.E.; Keskkula, H. J. Polym Sci. 1966, A 1, 4, 1595

68 Bender, B.W. J. Appl. Polym. Sci. 1965, 9, 2887
69 Manaresi, P.; Passalacqua, V.; Pilati, F. Polymer 1975, 16, 520
70 Willersinn, H. Die Makrom Chem. 1967, 101, 296
71 Freeguard, G.F. Brit. Polym J. 1974, 6, 205
72 Rosen, S.L. J. Appl. Polym. Sci. 1973, 17, 1805
73 Echte, A. Die Angew. Makrom. Chem. 1977, 58/59, 175
74 Wagner, E.R.; Robeson, L.M. Rubb. Chem. Technol. 1970, 43, 1129
75 Cigna, G; Matarrese, S.; Biglione, G.F. J. Appl. Polym. Sci. 1976, 20, 2285
76 Stein, D.J.; Fahrbach, G.; Adler, H. Die angew. Makrom. Chem. 1974, 38, 67
77 Zahn, E.; Otto, H.W. Kunststoffe 1967, 57, 12, p. 921
78 Meredith, C.L.; Barrett, R.E.; Bishop, W.A.; U.S. Pat. 3,538,191 (Copolymer Rubber & Chemical Corp.) (1967)
79 Bishop, W.A.; U.S. Pat. 3,538,192 (Copolymer Rubber & Chemical Corp.) (1967)
80 Tourata, M.O.; Ito, S.; Otsuka, I.; German Laid-Open Appl. 2,048,557 (Mitsui Toatsu Chem. Inc.) (1970)
81 Freund, G.; Lederer, M.; Stobel, W. Die Angew. Makrom. Chem. 1977, 58/59, 199
82 Saam, J.C.; Lindsay, S.E.; Duncan, A.W.; Falender, J.R. SPE Journal (April 1973) 29, 58
83 Heller, H.J.; Blattmann, H.R. Pure & Applied Chem. 1973, 36, 141
84 Mittnacht, H.; Echte, A.; Jenne, H.; Lieb, M.; Bronstert, K.; Adler, H.J.; U.S. Pat. 4,144,204 (BASF) (1975)
85 Echte, A.; Gausepohl, H..; Lütje, H. Die Angew. Makrom Chem., 1980, in press
86 Riess, G.; Schlienger, M.; Marti, S. J. Macrom. Sci.-Phy. 1980, B17 (2), 355
87 Walter, M.; Stein, D.J.; Fahrbach, G.; Jung, R.H.; Echte, A.; German Laid-Open Appl. 2,613,352 (BASF) (1976)
88 Kitchen, A.G.; Szalla, F.J.; German Publ. Appl. 1,959,922 (Phillips) (1969)
89 Fahrbach, G.; Gerberding, K.; Seiler, E.; Stein, D.J.; German Publ. Appl. 2,550,226-7 (BASF) (1975)
90 Cizek, E.P.; U.S. Pat. 3,383,435 (General Electric) (1968)
91 Kindler, H.; Nikles, A. Chem. Ing. Techn. 1979, 51, 1
92 Brighton, C.A.; Pritchard, G.; Skinner, G.A. "Styrene Polymers: Technology and Environmental Aspects"; London, 1979, p. 9
93 Axtell, W.A. Polym. Plast. Techn. Eng. 1980, 14, (1), 1

(R) = registered trademark

RECEIVED February 5, 1981.

19

Morphology of Polymers: Some Industrial Aspects

J. BRANDRUP, K. H. BURG, and H. CHERDRON

HOECHST AG, Kinststoff–Forschung, 6230 Frankfurt/Main 80, West Germany

One of the most important differences between low molecular weight compounds and macromolecules is to be seen in the many possibilities of influencing the morphology of polymers, and by this, in the possibilities of changing their physical properties in a controlled manner. Pronounced effects of the morphology exist on the mechanical and optical properties and in some cases also on the rheology of the polymer melt. The broad term morphology includes not only the shape and arrangement of amorphous and crystalline polymers or polymer segments, but also the crystallite structure of macromolecules. Some of the most important ways of influencing the polymer morphology are:

- Biaxial or multiaxial orientation
- Formation of multiphase polymer systems by mixing of two or more different polymers
- Influencing the crystallization process.

From the industrial point of view the change of polymer properties by influencing the morphology is of great importance for several reasons:

- instead of changing the chemical structure or composition of a polymer similar or better results may be achieved by mixing two or more polymers which is in many cases the more economical way (e.g. high impact copolymers versus high impact polymer blends).
- In several cases the formation of a special morphology imparts mechanical properties to "simple" macromolecules which hitherto have only be obtained with "exotic" polymers (e.g. high strength polyethylene fibers).
- Last not least, in some cases changing the morphology of a polymer is the <u>only</u> way to achieve certain properties (e.g. polymer fibrids).

0097-6156/81/0175-0289$5.00/0

1. Amorphous Polymer-Blends

The mixing of two or several polymers generally yields a two- or multiphase system although some homogeneous blends are known. This behavior is due to the low entropy of mixing two polymers. Thus even small positive changes in enthalpy of mixing yield a positive free energy of mixing which then causes incompatibility.

The distribution in space of the two phases depends on the mode of fabrication, on physical and chemical factors and certainly on the concentration range considered.

The field of polymer blends has been extensively investigated both from a theoretical and a practical point of view during the last decade and several reviews are published (1, 2, 3, 4).

Theoretically there exist four possibilities to combine hard and soft polymers with each other. They are listed in Table 1 together with some examples which have already found commercial interest.

In this report we will only consider type 3, i.e. mixtures of a rigid amorphous thermoplast with small amounts of an elastomer which is the underlying principle for all rubber toughened plastics to improve impact behavior.

In solid polymers, the energy of impact is dissipated by the formation of crazes or shear bands in the matrix[5]. Both mechanisms are enhanced by the addition of a second elastomeric phase to the rigid polymer, but not altered in principle.

Chemical structure, concentration, size and distribution of the rubber phase as well as adhesion to the matrix determine processing and final properties, specifically the impact strength. In the case of a powder blend, like PVC/rubber-systems, the characteristic powder data have also to be taken into account.

The type of rubber chosen depends on the desired mechanical properties and on the end use. If, for example, excellent outdoor weatherability is required, the use if saturated rubbers as elastomer phase is possible.

The preferred morphology of these rubber modified amorphous thermoplastics is the distribution of distinct rubber particles unfilled or filled in an isotropic matrix of the basic polymer. This was shown to be the case for rubber modified polystyrene and for ABS-type polymers.

Since these systems have been extensively described in the literature we want to concentrate on another industrially important amorphous polymer mixture, namely the PVC/rubber-blends.

Unexpectedly, PVC/rubber blends show under certain circumstances quite a different morphology. This was shown experimentally for chlorinated polyethylene and ethylene-vinylacetate copolymers as rubber phase.

Figure 1. Morphology of rubber modified high impact polystyrene.

TABLE I: Polymer Mixtures (Polymer Blends)

Possible Combinations	Combinations of Practicle Interest
Type 1: Soft matrix + soft dispersed phase	Elastomer-Blends
Type 2: Soft matrix + hard dispersed phase	Polymer filled Elastomers
Type 3: Hard matrix + soft dispersed phase	High impact thermoplastics • High impact polystyrene and ABS • High impact PVC
Type 4: Hard matrix + hard dispersed phase	• PPO/PS-Blends • PC /ABS-Blends • PVC/ABS-Blends

Prerequisite for these observations was the development of proper analytical procedures to visualize the morphology of these blends by electron microscopy.[6,7]

The morphology of the PVC/rubber-blends depends primarily on the maximum temperature reached during manufacturing and to a limited extend on the history of the sample (powder blending or graft polymerization).

A network of rubber around distinct PVC-particles is formed up to temperatures of 170 - 190°C, while very fine rubber particles are distributed in a homogenous matrix above this temperature (Fig. 2). This process is irreversible, a sample heated to high temperatures will not form a network again when the temperature is decreased.[8]

In accordance with this dynamic mechanical measurements show that the rubber partially forms the outer phase below 170 - 190°C.[9]

Prerequisite for the formation of a network structure is the existence of distinct PVC globules of ca. 1 µm diameter generated in the polymerization directly as was shown by Berens and Folt.[10]

The breakdown of these particles apparently caused by melting of the small crystalline segments in PVC causes the change of morphology observed.

Rheological data are in agreement with this. While a particle flow mechanism was shown to apply below this temperature range, viscous flow occured above it.[9]

In Fig. 3 is shown that the molecular flow depends on particle size of the PVC particles at low temperatures while it depends on molecular weight at higher temperature. It depends on molecular weight only if the product was dissolved and represcipitated before measurement.

If the torque as measured in a Brabender is plotted in an Arrhenius plot against 1/T characteristic linear dependencies are observed (Fig. 4). While powder flow is observed up to the softening point of PVC a distinct linear relationship between shear rate and reciprocal temperature exists between 150 - 180°C indicating one mechanism of flow. The deviation above ca. 180°C indicates a change in flow mechanism.

The modulus of blends can be calculated by methods developed by Takayanagi, Nielsen and Kerner. Different values are obtained for interpenetrating networks and for particle-matrix morphologies.[11]

a

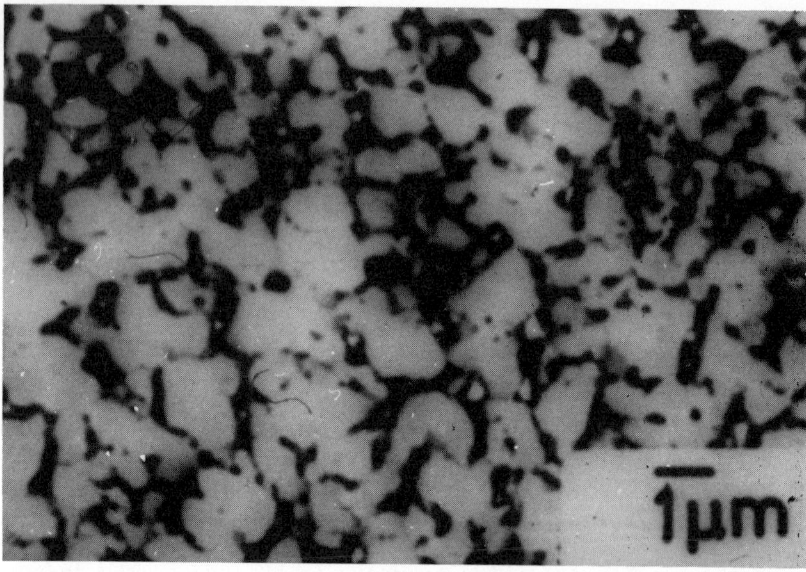

b

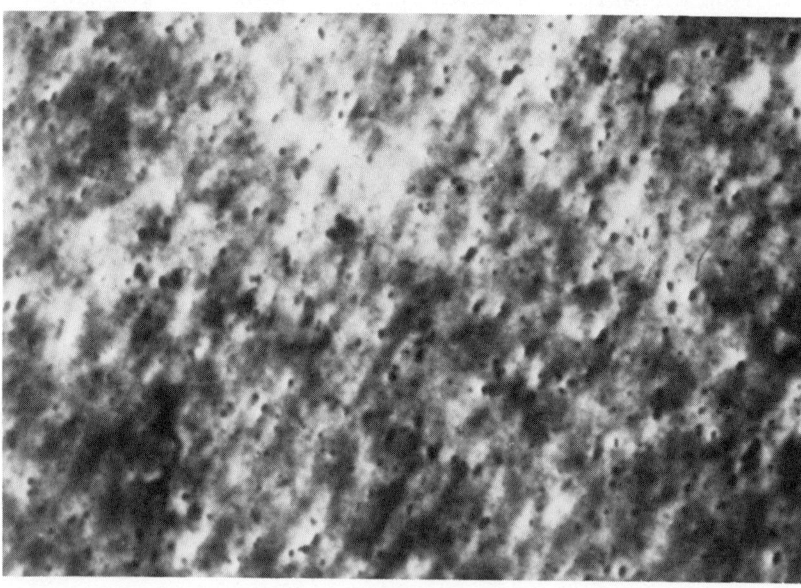

Figure 2. Morphology of PVC/CPE (10%)-blends after processing; (a) *below 170°C,* (b) *above 180°C. Scale identical in both figures.*

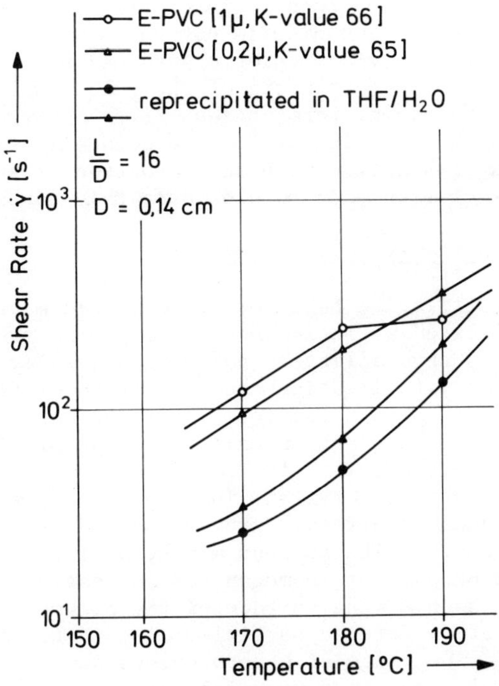

Figure 3. Shear rate dependence of PVC melts on temperature.

In accordance with these models, modul calculations for low temperature blends fall on the line of a network morphology, while all calculations for high temperature blends fall on the line of a particle/matrix morphology (Fig. 5).[9]

Careful interpretation of data in the literature indicates that these observations are valid for several other PVC/rubber blends (Acrylate rubber, MBS-rubber).

Comparison of mechanical and processing properties with morphology shows that the formation of a network structure is the desired morphology for high impact strength, low viscosity and low shrinkage as indicated in Figure 6 for the impact behavior.[12]

The mechanism how a rubber distributed in a network influences the rupture mechanism is not quite well understood yet. It is known that poly(vinyl chloride) forms shear bands when stress is applied and that parts of the rubber which are located in these shear bands may form crazes.[13] It might well be that a network structure is efficient for the delocalization of stress energy only in combination with the formation of shear bands. Experimental work is needed to elucidate this further.

2. Spherulites in Crystalline Polymers

The crystallization behavior of polymeric materials is by far more complicated as it is known from low molecular weight compounds. Thus crystallizable polymers generally exhibit only a partial crystallization and one has to differentiate between the growth of spherulites or similar morphological structures and a subsequent annealing process within the boundaries of the spherulites.[14,15,16] Spherulites are composed of radiating aggregates of crystals in spherical arrays. These fibrillar or lamellar crystals grow by secondary nucleation from a primary nucleus.[17] The nucleus may be a single crystal from the crystallizing polymer (homogeneous nucleation) or another heterogeneity, such as the surface of the polymer containing vessel or any other foreign particle. The latter possibility has led to the wide range of heterogeneous nucleation, which is being extensively used in commercial formulations of semicrystalline polymers. Spherulites grow outward from the central nucleus until they ultimately reach the next growing entity. They therefore finally appear as polyhedra. Because of their radial growth the fibrillar or lamellar crystals have only little space-filling ability as they move away from the central nucleus. Typically for spherulitic structures, an irregular noncrystallographic branching usually at small angles can be observed.

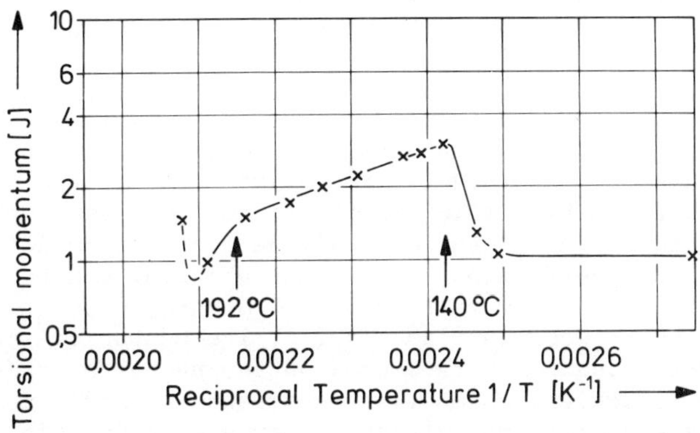

Figure 4. Arrhenius plot of PVC melt viscosity as measured in a Brabender.

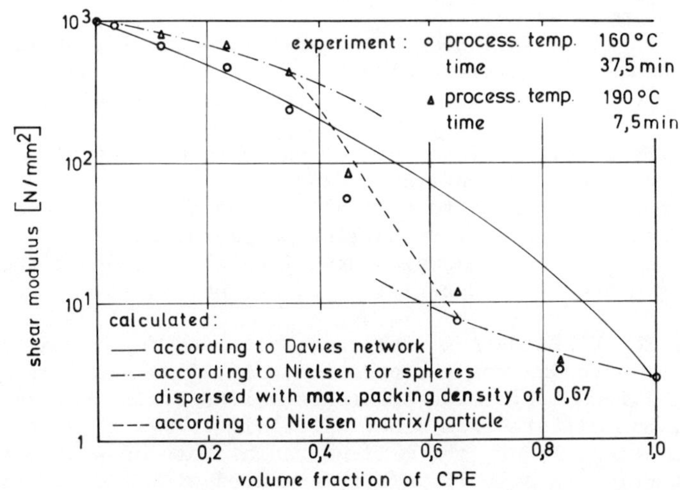

Figure 5. Dependence of compression modulus on processing conditions; experimental and calculated data.

The rate of spherulitic growth is extremely temperature sensitive and seems to be independent of the nucleating agent. It is of course important to note that the overall rate of crystallization is not only determined by the growth rate of the spherulites, but also by the amount of nuclei being present in the system. This possibility is used as an effective method to influence the total crystallization rate of commercial polymeric materials in a controlled manner and to influence the size of spherulites and thus the physical properties of finished articles made from semicrystalline polymers.

During the spherulitic growth of high polymers the predominant mode of packing in the crystalline state appears to be folding of long polymer chains. Since these chains can be simultaneously involved in more than one adjacently growing crystal, links develop between the lamellae.[18] The mechanical properties of semicrystalline polymers depend on the number of these so-called tie molecules, which must be present in small and large spherulites and, in some instances, may also bridge interspherulitic regions.[19] Tie molecules aid in dissipating the absorbed energy.

Among the phenomena mentioned above important features of spherulitic crystallization of polymers are:

a. the lamellar or fibrous growth habit starting from primary nuclei
b. the irregular noncrystallographic branching of lamellae usually at small angles
c. the existence of tie molecules between lamellae and spherulites
d. the possibility to control the size of spherulites via heterogeneous nucleation and
e. the possibility to control the overall rate of crystallization of semicrystalline polymers.

In the following, examples are given where some of these features are used in industrial applications.

<u>POLYETHYLENE TEREPHTHALATE (PET)</u> has been used for a long time only as raw material for the production of films and fibers. PET crystallizes by forming spherulites like other semicrystalline polymers, but differs from other materials in its rate of crystallization. Thus, the maximum radial rate of spherulite growth of PET[20] is only 10 μm/min, compared with 400 μm/min for polyoxymethylene[21] and approximately 5000 μm/min for high density polyethylene[22]. Since PET mouldings with maximal stiffness and optimal crystallization are important for technical applications, suitable measures must be taken to accelerate the rate of crystallization. As already mentioned, the overall rate of crystallization can be influenced by the temperature and by the addition of nucleating agents. Using sodium montanate[23,24,25] the half life period of crystallization can be reduced by a

factor of three to ten depending on the amount of nucleant being present in the material. Thus having a polymer on hand with an optimal number of seed crystals the injection moulding processor can additionally increase the mould temperature up to 140°C. This ensures economically reasonable cycle times together with optimal physical properties due to the high heat capacity of PET, although this mould temperature does not meet the maximal rate of crystallization for PET which was determined to be at 170°C.

ISOTACTIC POLYPROPYLENE (PP) recently has gained interest as an especially suitable raw material for the manufacture of thin-walled one-way containers for food. Up to now polystyrene (PS) and polyvinylchloride (PVC) have been the preferred materials for this field of application. Important advantages of PP over PS and PVC are the lower density, the lower water uptake, the lower permeability for water vapor and the higher heat deflection temperature which allows to fill the containers at temperatures up to 130°C. Of course, there were also disadvantages of PP during the industrial manufacture of said containers which had to be overcome. The main problems were a too low stiffness during mould release in the injection moulding process, too low cycle times and especially an unsufficient transparancy. By proper selection of kind and amount of a suitable nucleant, the cycle time has been lowered by 20% and the transparancy was improved up to a level that these PP-grades can be used in the field of one-way packaging for food. The effect of nucleation on the transparancy of PP is illustrated in Fig. 7. A summary of the most important nucleants for PP is given in.[26]
In the past POLYOXYMETHYLENES (POM) have gained considerable importance in the field of engineering applications because of their good mechanical properties, their high heat deflection temperature, their excellent electrical properties and their favorable gliding and wear behavior. For these materials also, different possibilities for nucleation have been found[27,28,29,30] and the influence of these measures on the properties of POM has been investigated. Figure 8 shows microphotographs taken from microtome cuts of the outer zone of 4 mm thick injection moulded plates from nucleated and unnucleated POM. As can be seen the plate from nucleated material exhibits a microcrystalline, essentially more uniform structure as compared to the sample from unnucleated material. As a consequence of these differences in the morphological structure an increase in the heat deflection temperature, in hardness, stiffness, tensile- and impact properties has been

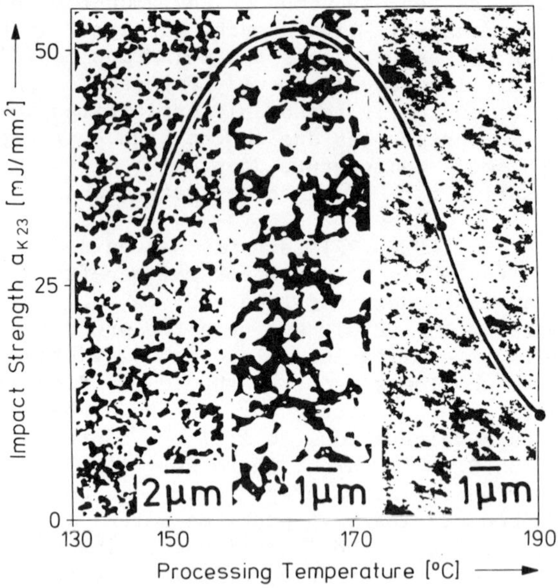

Figure 6. Dependence of impact strength of PVC/CPE (10%)-blends of processing temperature and on morphology.

Figure 7. Nucleation of isotactic polypropylene; right unnucleated sample.

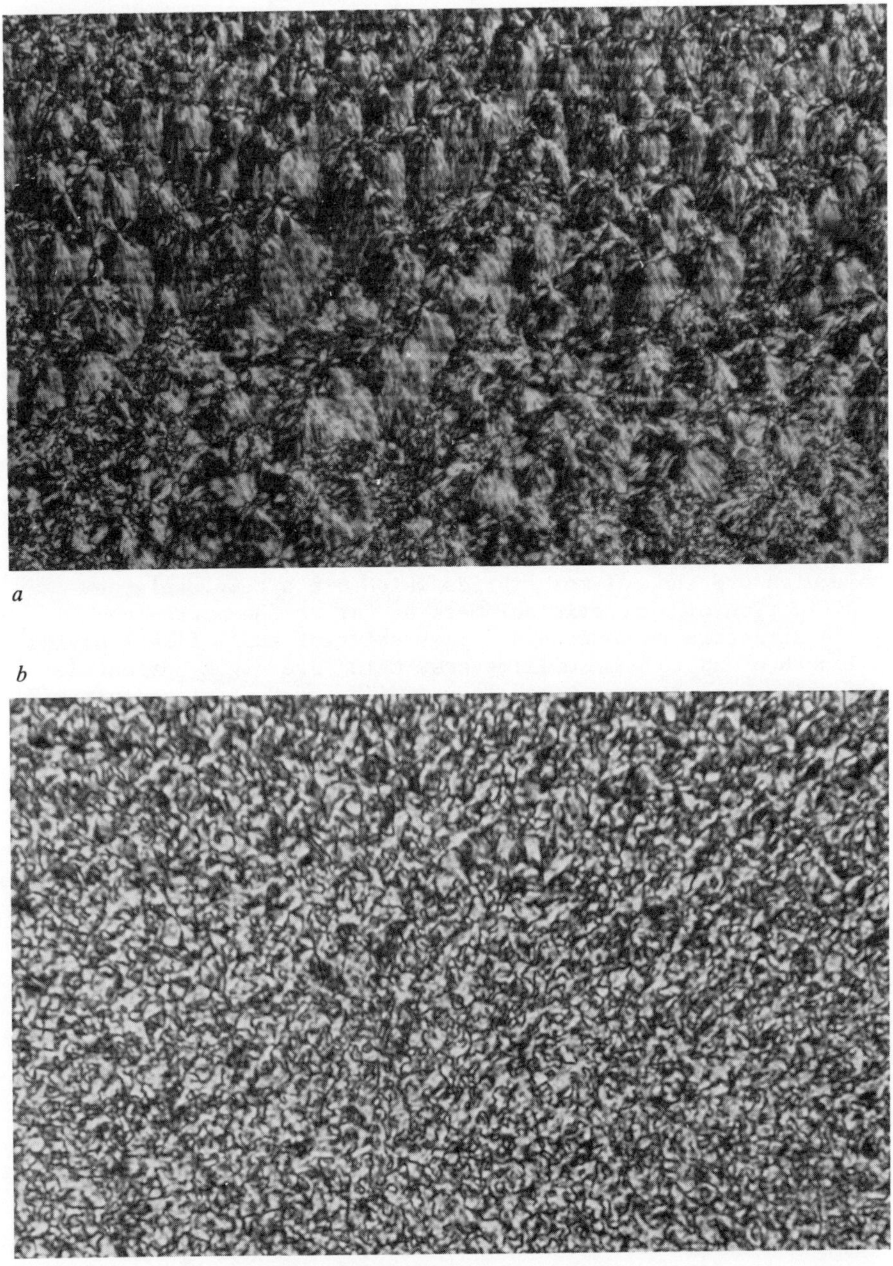

a

b

Figure 8. Nucleation of polyoxymethylene; (a) unnucleated; (b) nucleated.

observed when POM was nucleated.[31] Further advantages of nucleated POM over unnucleated POM are shorter cycle times, improved dimensional stability and less warpage of injection molded articles.

Experiments with dyestuffs being dissolved in molten POM have demonstrated that during spherulitic crystallization low molecular weight substances migrate to the interfaces of lamellae and spherulites.[32] Taking into account that nucleation leads to an increase of the number of spherulites per unit volume, it can be concluded that nucleation creates an increase in the total interface between the spherulites. During spherulitic crystallization the separating low molecular weight substances are distributed over a larger area and thus form thinner layers between lamellae and spherulites. So, tie molecules can develop easier and thus ensure a better cohesion in the semicrystalline structure. As a consequence, better mechanical properties are observed.

3. Polymer Fibrids

The latest results of a controlled crystallization of macromolecules are the polymer fibrids which are a completely new modification of synthetic polymers as far as the micro- and macro-structure is concerned. They exist of small fibers having a length of up to some millimeters, which are highly oriented, and which have a macro-morphology similar to that of cellulose pulp.

Usually, synthetic polymers crystallize[14,15] from a melt or a solution in form of folded lamellae. Under specific circumstances it is sometimes also possible to obtain extended chain crystals which is the preferred arrangement in the crystallites of many natural polymers (cellulose, silk). Recently it has been found[33,34] that in some cases another crystalline modification can be obtained, the so-called shish-kebabs, which are a sort of hybrid between folded lamellae and extended chain crystals. These shish-kebabs are obtained by shear-induced crystallization, a process in which the polymer crystallizes from solution under the influence of an elongated flow.

By special processes, like polymerization under shear conditions or flash-spinning of polymer solutions, it is furthermore possible to obtain materials which do not only have this new shish-kebab micro-morphology, but also a new macro-morphology, namely the polymer fibrids.[35]

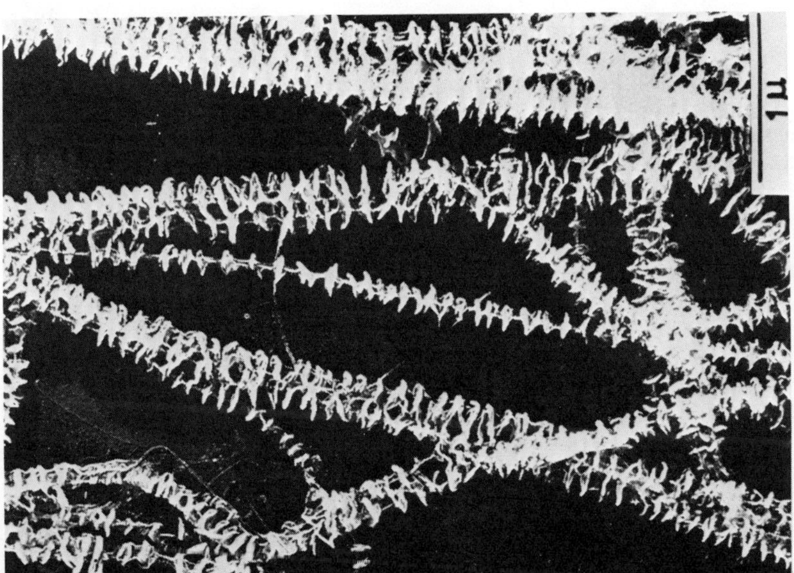

Figure 9. Electromicrograph of a high density polyethylene shish-kebab.

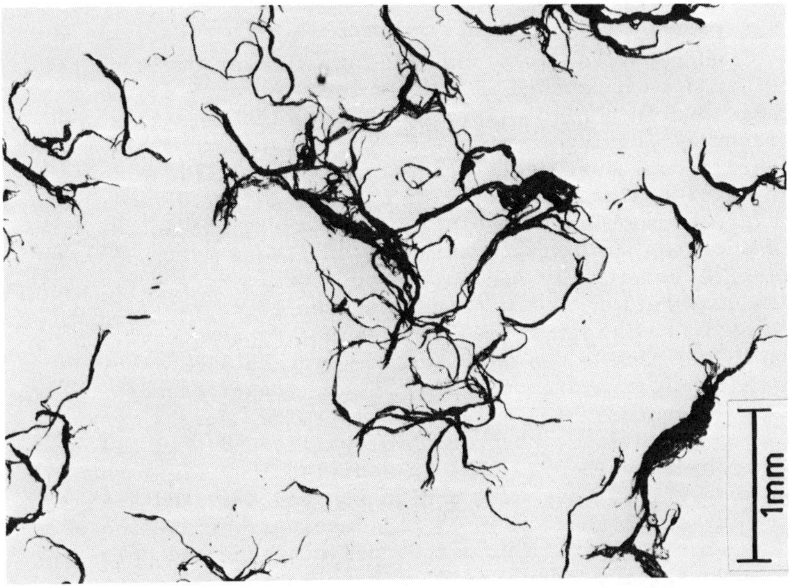

Figure 10. Light micrograph of polyethylene fibrids.

In the case of polyethylene the process is as follows: A solution of polyethylene in a low boiling hydrocarbon is heated up to 150°C and then pressed with high speed through a specifically designed nozzle into a vessel which is held at normal pressure. During the rapid flow through the nozzle the coiled macromolecules elongate and partially parallelize. This conformation of the solution is also maintained in the solid state because as soon as the solution leaves the nozzle the solvent evaporates explosively which causes the polyethylene to crystallize immediately and in fibrous form. Moreover, the sudden evaporation of the solvent causes the crystallized polyethylene to split along the fiber axis. By a special treatment these fibrids can also be made hydrophilic. They are then able toform hydrogen bonds. Thus, these polymer fibrids very much resemble cellulose pulp: They have a very similar macro-morphology, they are hydrophilic and they are able to form hydrogen bonds. But in addition they still retain lyophilic properties, they have a high surface area and are meltable. It is for the reason of this unusual combination of properties, that polymer fibrids from polyethylene and polypropylene have already found application[35] in very different areas like for example in paper and cardboard, as thixotropant in the building industry, as bonding fiber in paper and nonwovens and even as component in chromatography and enzyme immobilization.[36]

Outlook

The search for new or improved polymer properties meant for a long time primarily the chemical modification of the macromolecules for example by copolymerization. In recent years influencing the morphology has more and more become a very important way to alter the polymer properties either alone or in combination with chemical modifications. And there are profound indications that this will even more be the case in the future. Activities in this field will include the optimization of already existing products, like polymer-blends, block copolymers and segmented polymers, multicomponent films and fibers or polymers with specific crystallite structure, as well as new products like liquid crystalline polymers and the wide field of synthetic membranes. All in all, polymer morphology is not only a challenge to the polymer scientist, but also an area with very promising industrial aspects.

Acknowledgement

This review is to a large extent based on the experimental and theoretical work of our colleagues Drs. K. D. Asmus, E. Fischer, D. Fleischer, M. Fleibner, W. Gordon, G. Heufer, F. Kloos, H. J. Leugering, H. Schaum and E. Wolters which we want to greatfully acknowledge.

LITERATURE CITED

1. L. Bohn, Kolloid-Z. 1966, 213, 55, Rubber Chem. Tech. 1968, 41, 495; see also "Polymer Handbook" ed. J. Brandrup, E. Immergut, Wiley III 1974, 211.
2. Schmitt, B. J., Angew. Chem. 1979, 91, 286.
3. Paul, D. R. and Newman, S., "Polymer Blends" Academic Press New York 1978.
4. Paul, D. R. and Barlow, J. W., Macromol, J., Sci. Rev. Macromol. Chem. C 18 (1).
5. Bucknall, C. B. and Clayton, D., Nature Phys. Sci. 1971, 231, 107.
6. Fleischer, D., Fischer, E. and Brandrup, J., J. Macromol. Sci. Phys. 1977, 14, 17.
7. Bassewitz, K. V. and Menzel, G., Angew. Macromol. Chem. 1975, 47, 201.
8. Fleischer, D., Scherer, H. and Brandrup, J., Angew. Macromol. Chem. 1977, 58/59, 121.
9. Fleischer, D., Kloos, F. and Brandrup, J., Angew. Macromol. Chem. 1977, 62, 69.
10. Berens, A. R. and Folt, V. C., Trans. Soc. Rheolog. 1967, 11, 99; Polymer Eng. Sci. 1968, 8, 5; 1969, 9, 27.
11. Nielsen, L. E., Rheolog. Acta 1974, 13, 86; Kerner, E. H. Proc. Phys. Soc. (London) Sect. B. 1965, 69, 803.
12. Fleischer, D., Brandrup, J. and Heinzmann, O., Kunststoffe 1977, 67, 3.
13. Haaf, F., Breuer, H. and Stabenow, J., Angew. Makromol. Chem. 1977, 58/59, 95.
14. Geil, P. H., Polymer Single Crystals, Interscience Publishers, New York 1963.
15. Mandelkern, Crystallization of Polymers, McGraw-Hill, New York 1964.
16. Magill, J. H., in: J. M. Schultz, Properties of Solid Polymeric Materials, Academic Press, New York 1977.
17. Price, F. P., J. Polym. Sci. 1959, 37, 71.
18. Keith, H. D., Padden, F. J. and Vadimisky, R. G., J. Polymer Sci. 1966, Part A-2 4, 267.
19. Magill, J. H., Polymer 1961, 2, 221.
20. Keller, A., Lester, G. R. and Morgan, L. B., Phil. Trans. R. Soc. 1954, A 247, 1.
21. Inoue, M. and Takayanagi, T., J. Polymer Sci. 1960, 47, 498.
22. Price, F. P., J. Phys. Chem., 1960, Ithaca 64, 169.
23. Asmus, K. D. and Fleibner, M., Br. Polymer J. 1970, 2, 295.
24. US-Patent 3.619.267 (Hoechst AG, 1971).
25. US-Patent 3.632.402 (Hoechst AG, 1971).
26. Beck, H. N., J. Appl. Polym. Sci. 1967, 11, 673.
27. US-Patent 3.467.625 (Hoechst AG, 1969).
28. US-Patent 3.704.275 (Hoechst AG, 1972).
29. US-Patent 3.767.610 (Hoechst AG, 1973).
30. US-Patent 4.181.685 (Hoechst AG, 1980).

31. Wolters, E. and Gleisberg, P., Plastverarbeiter 1976, 26, 201.
32. Leugering, H. J., Burg, K. and Cherdron, H., Makromol. Chem., Suppl. 1975, 1, 621.
33. Pennings, A. J., IUPAC-Symposium on Macromolecules, Prague 1965, Preprint 216.
34. Pennings, A. J., J.M.A.A. van der Mark and H. C. Booij, Kolloid-Z. 1969, 236, 99.
35. Gordon, W., Leugering, H. J. and Cherdron, H., Angew. Chem. 1978, 90, 833; Angew. Chem. (International Edition) 1978, 17, 820.
36. Manecke, G. and Vogt, H. G., Angew. Makromol. Chem. 1980, 85, 41; 1980, 88, 37.

RECEIVED April 1, 1981.

20

Block Copolymers: Maturation of an Old Concept into a Versatile Modern Tool

P. TEYSSIE

University of Liege, Laboratory of Macromolecular Chemistry and Organic Catalysis, Sart Tilman, 4000 Liege, Belgium

As one tries to write down an analysis of the developments in the block polymers area, one realizes very soon that it is going to consist of a series of variations on a theme : a theme which is the increasingly stronger reality, in our everyday scientific life, of what can be now really called "the molecular engineering of polymers properties", i.e. the possibility to control, through precise (although sometimes small) modifications of molecular structures, the final bulk properties and macroscopic behaviour of polymeric materials. In other words, one deals there with a very active version of the studies on structure-properties interrelationships, a question which by the way has always been a focal point among the many diversified interests of Professor Mark.

More specifically, the use of well-designed block copolymers can lead to a fine-tuning of polymeric materials morphology, obviously a key-point in devising their final (bulk) physical properties. Among the spectacular consequences of such a capability, suffice it to recall :

- the use of these block copolymers as high-performance engineering products, particularly in the field of thermoplastic elastomers (1) (Kraton-type);
- their ability to stabilize emulsions, and in a very important extension of that concept, to stabilize fine dispersions of the corresponding homopolymers, bridging the "compatibility gap" existing between these individualistic molecules (2), and promoting the development of a new "Plasturgy";
- their "organization" into mesomorphic phases (3), giving rise to highly original, often anisotropic, properties.

0097-6156/81/0175-0307$5.00/0

Of course, all these achievements imply to meet a strict and difficult requirement, i.e. to be able to control precisely and simultaneously, both the structural and molecular characteristics of the block polymers involved : it is the role (maybe too often underestimated) of synthetic polymer chemistry, to meet these challenges.

The approaches analyzed hereafter are meant to illustrate these views, mainly on the basis of current work performed in our laboratory, and so to show the direct and powerful connection between the synthetic performances, the resulting physical properties, and the potential applications based upon them.

I. The older route

Until quite recently, the best known answer to the challenge of combining different properties in an additive manner in a single polymeric product, has been the use of so-called "living" initiators to produce block copolymers, in a 2- or 3-steps process depending wether a di- or a tri-blockwas needed. (see scheme I).

This very efficient, time-honoured pathway suffers however a very severe limitation, i.e. the relative reactivity of the living centre C* must be adapted to the structure and reactivity of monomer M_2, a requirement which is not very often met.
Fortunately this approach, illustrated mainly in anionic copolymerization, has been extended significantly along two different lines of achievements.(4)

First, new "living" initiators have been discovered (although not always as efficient), which respond to other mechanisms, i.e. cationic (5) or even radical ones (6), and can accordingly accomodate other types of monomers. A recent typical example is the coordination polymerization of butadiene by bis (η^3-allyl-trifluoroacetato-nickel) to yield a "living" pure 1.4 cis-polybutadienyl-nickel, able to initiate in turn the polymerization of monomers like isoprene or styrene (7).

Second, the conversion of one of the blocks into another type of structure by a suitable quantitative chemical reaction, allows a broad diversification of the properties of the available block copolymers. The best example of such an opportunity is probably the hydrogenation of poly(butadiene-b-styrene) copolymers, which yields a product close to a low density poly(ethylene-b-styrene) when starting from an anionically prepared diblock (including a certain amount, ca. 10 %, of 1.2 units), while a high density poly(ethylene-b-styrene) is obtained (crystalline m.p. ca. 125°C) when hydrogenating the polymer obtained from the η^3-allyl-nickel system mentioned above (7).

An attractive application offered by these last diblock copolymers is their use as blending aids (or "emulgators") for mixtures of homopolystyrene (PS) and polyethylene (PE, both low and high density). In that capacity they perform indeed two important functions (8) :

- first, they control the overall morphology of the blend by reducing, thanks to their interfacial properties, the size of the domains well under 1 μm, ensuring accordingly homogeneous physico-mechanical properties. Ofcourse, the keeping of these domains is a must, since only an heterophasic situation will allow a cumulative addition of properties in the final blended material;

- on the other hand, if the molecular weights of the two blocks are at least close to those of the homopolymers to be blended, interactions take place (f.i. through entanglements and Van der Waals type forces) which anchor the phases into each other (stronger types of interactions might furthermore overshadow that M.W. requirement). As a result, good physico-mechanical properties are obtained over the whole range of blending compositions, instead of having a monotonous variation, or worse (as often observed) a depressed situation (fig.1). Still more interestingly."synergistic" behaviours appear in some cases, where the level of certain properties can be higher than the one of the best performing component, but also where rather antagonistic properties like strain and tensile strength can be improved simultaneously.

II. A newer scheme

Despites all of its virtues, the classical approach described hereabove suffers too many constraints to be fully satisfactory, and active efforts have been devoted to the successfull designing of a more general and versatile route to these desirable block structures. Since quite a few years now, the pathway described in scheme 2 has been illustrated by an increasing number of very interesting achievements. (9). It implies again the formation of a "living" polymer chain P_1, the active endgroup of which (C_1^*) is then converted into a functional group able to promote the formation of a new "living" initiator, which will then ensure the polymerization of the second monomer. Triblocks can be obtained by using difunctional starting polymers, or by further coupling of the final products.

Scheme 1.

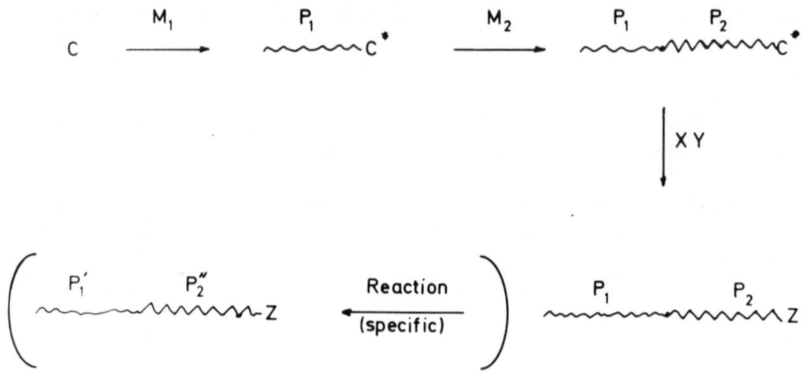

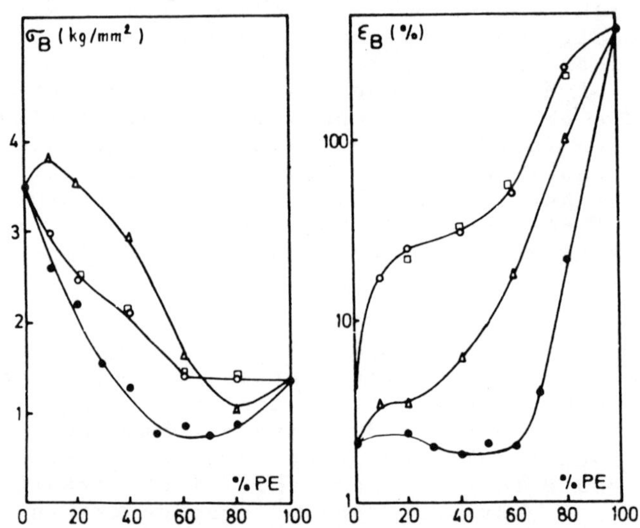

Figure 1. Ultimate tensile strength (σ_B) and elongation at break (ϵ_B) of 1.d. PE ($\overline{M}n = 40{,}000$)/PS) $\overline{M}n = 10^5$) blends. Key: o, *without copolymer;* △, *with 9% of a polystyrene-b-hydrogenated butadiene),* $\overline{M}n$ *total* = 58,000; ○, *with 9% of a polystyrene-b-hydrogenated butadiene,* $\overline{M}n$ *total* = 155,000.

Although the basic principles underlying this second method are not that different from the first one, it enjoys obviously a much broader efficiency and versatility, if one is able to design properly C_2^* to fit the structural and kinetic requirements of monomer M_2. Two very different illustrations of this concept will be presented hereafter, somewhat showing the diversity of consequences implicit in that kind of approach.

A. The amazing block copolymers of ε-caprolactone (CL) and vinyl-type monomers.

It has been shown recently (10) that such block structures could be tailored precisely by the general method summarized hereabove. It is indeed possible to convert the hydroxyl end-group of a vinyl polymer PA (f.i. polystyrene, or polybutadiene obtained by anionic polymerization terminated with ethylene oxide), into an aluminum alcoholate structure; since it is well known that CL polymerizes in a perfectly "living" manner by ring-opening insertion into the Al-O bond (11), the following reaction sequence provides a direct access to the desired copolymers, with an accurate control of the molecular parameters of the two blocks :

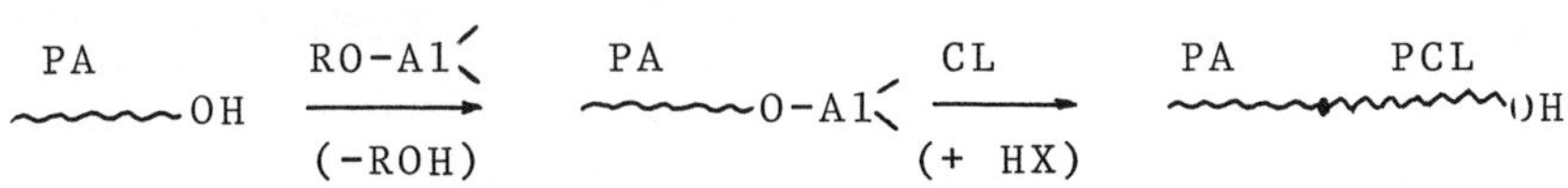

Thanks to the amazing properties of PCL itself, these block copolymers enjoy several interesting characteristics, leading to new types of applications.

First of all, these materials are attractive "per se". In particular, poly(styrene-b-caprolactone) is a semi-crystalline product displaying an amazing resistance to cold-fracture, and also macroscopically biodegradable at least when PCL represents the continuous phase. It can be "organized" into single crystals, wherein lamellae of PSt and PCL alternate with a periodicity of about 80 Å.

But they are also very helpful as blending aids again, thanks to the miscibility (if not true compatibility) of PCL with several other important polymers, including PVC and acrylonitrile copolymers (i.e. SAN). Indeed, the use of a PA-PCL diblock will now allow to stabilize fine dispersions of homo-PA and any PB miscible with PCL. Along these lines, good blends of PVC with PS or even with PE have been obtained (i.e. through hot-milling), which display a fine morphology (< 1 μm) and good mechanical properties, again as long as the corresponding molecular parameters have been properly optimized (12).

In other words, this example stresses clearly the very broad applicability of the "emulgator" concept reminded above, to practical problems involving large-scale polymers.

B. The challenge of synthesizing high-performance thermoplastic elastomers (TEP).

It is another well-known problem that classical triblock TEP's, i.e. those of the Kraton-type, are often confined to applications under rather mild conditions, due to the relatively low Tg of the glassy phase, and/or the medium thermal stability of the rubbery phase. In principle at least, good answers to that challenge can be offered by the type of synthesis discussed in this section.

Such an example has been recently worked out, combining a very hard block i.e. polypivalolactone (PPVL), with a thermostable elastomer segment i.e. polydimethylsiloxane (PDMS), using the reactions sequence described in scheme 3. (One should note that the starting PDMS has been obtained through a previously described procedure (14), avoiding any unstable Si-O-C bond in the final product).

As expected, the resulting polymers are indeed heterophasic materials, exhibiting (besides that one of PDMS blocks at -42°C)acrystalline melting point at 225°C, very close to the one of pure PPVL. Optical micrographs under polarized light demonstrate an overall formation of very large spherulites, while transmission electron micrographs show very small, randomly distributed PDMS domains (ca. 200 A), an ideal situation for interesting macroscopic physico-mechanical properties.

III. The last general answer

Thanks to its versatility, the general approach described hereabove gives a good grip of the synthesis of many different families of block polymers. However, in several cases where the formation mechanisms of the two types of chains are not amenable to this quantitative active end-group conversion technique, a still more general answer is needed.
In principle, it should be found in the coupling of 2 different polymers, PA-x and PB-y, through interaction of their reactive end-groups x and y, to yield a copo-

Scheme 2.

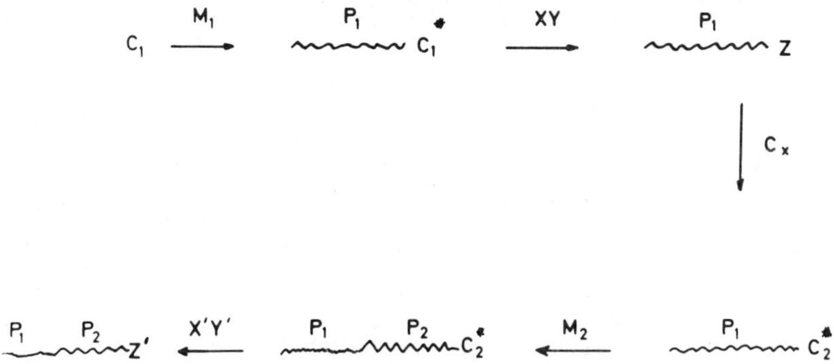

Scheme 3.

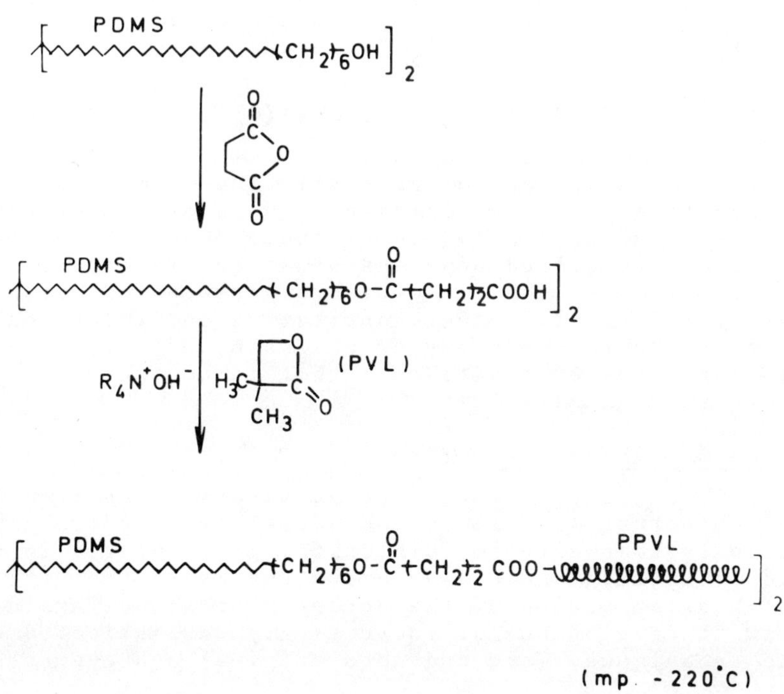

lymer PA-z-PB. It is known however, that this simple idea is usually wishful-thinking when it comes to combine two different polymers of high molecular weight, and so for two obvious reasons. A kinetic one, which is the very low concentration of reactive end-groups; but also a thermodynamic one : i.e. the repulsion of the two uncompatible polymeric coils of different nature, as soon as their interaction parameters differ even by a small value, what is usually the case for the desirable pairs.

An answer to that important challenge can now be provided, thanks to the use of a two-phases liquid system. In that system, the two different polymers are dissolved each in a good solvent of theirs, the two solvents used being immiscible. The critical point is to provide one of the two partners with a reactive end-group (i.e. x on PA in solvent A') having a preferential compatibility with the solvent (i.e. B') of the other partner (i.e. PB-y), and vice-versa.
In this manner, it becomes possible to bypass both the kinetic limitation (since there is an increased concentration of active groups at the interface), and the thermodynamic one (each polymer being in a favorable solvent).

The efficiency of this new concept has been demonstrated (15) by the facile synthesis, in 90 % yield in 2 hours at 50°C, of the multiblock poly(α methylstyrene-b-ethylene oxide)$_{\bar{n}}$ ($\overline{Mn}$ ca. 70.000), starting from $Br(CH_2)_5O[CH_2-CH_2-O]_n(CH_2)_5-Br$ in water and

$[(CH_3)_2N(CH_2)_6N^+(Br^-)(CH_2)_6(C(CH_3)(C_6H_5)-CH_2)_n]_2$ in toluene ($\overline{M}_n$ ca. 10.000 each).
If one is able to control precisely the formation of end-functionalized polymers carrying groups of the proper reactivity and philicity, this scheme should open efficient and diversified new ways of access to interesting block copolymers. The basic problem remains accordingly a catalytic one, i.e. the quantitative end-functionalization of growing homopolymer chains by efficient termination or transfer reactions : that is fortunately a rapidly improving field.

IV. The multiblock structures

The interest of that type of material has been largely demonstrated, f.i. by the Hytrel-type of products (i.e. polyether-ester multiblocks), particularly in the field of thermoplastic elastomers thanks to the efficient cross-linking action of the harder blocks. Due again to the swift development of increasingly sophisticated catalytic techniques, more and more original achievements are regularly reported. Suffice it to cite (16) the "coding" 1.4 polymerization of butadiene by "tailored" coordination catalysts (namely η^3-allyl-nickel-X complexes), yielding a multistereoblock polymer, poly (cis 1.4-b-trans $1.4)_n$-butadiene, the first example of a thermoplastic elastomer (semi-crystalline m.p. ca. 135°C) obtained in one step from one single monomer.

Another research avenue has been recently opened in that prospect, based on the idea that it should be possible to replace the harder blocks by single groupings, pro vided their structure promote very strong mutual interactions in the medium wherein they are used.
A typical example along these lines is the synthesis of so-called "halato- (or salt-) telechelic polymers", displaying the general structure $[\sim\!\sim\text{PA}\sim\!\sim COO^- M^{2+}\,{}^-OOC]_n$ and obtained from the complete neutralization of any α,ω-dicarboxylato-polymer (PA) by a divalent metal derivative (17). When PA is a lipophilic chain, the corresponding products will still be soluble in hydrocarbons (at least if the charge density is not too high, i.e. if PA molecular weight is large enough), a kind of medium wherein the ion-pairs associate in clusters of variable sizes (depending on the conditions).The same type of association will obviously occur in the neat product for the same reasons.
These associations generate a whole set of interesting properties, i.e. :
- the formation of strong elastic gels, exhibiting resonance phenomena, at very low critical concentrations (ca. 1-2g/dl). This gelification is totally reversible both upon dilution or increase of temperature;
- the formation of periodical structures in concentrated solutions or in the neat state, structures which are probably built-up of equidistant small "plates", the periodicity being ca.70 Å for an essentially 1.4 polybutadiene, quite irrespective of ion nature and presence of solvent. In fact, that periodicity seems to be largely controlled by the mean polymer chain lenght, and the use of a vinyl-type polyisoprene reduces it to 55 Å;
- unexpected rheological responses, including anti-thixotropic and "bouncing-solutions" behaviours.

Obviously enough, that kind of approach represents a very flexible method for the molecular tailoring of broadly diversified materials, by changing the nature (and dimensions) of the polymer and the ions involved, as well as that of the medium. The main problem to be met here is the necessity, like in any step-wise type of polymerization, to ensure a very high conversion (ca. 99 %) to reach a high degree of polymerization, a key to the generation of the most interesting properties; that goal has been successfully achieved by the use of very reactive metal intermediates.

In conclusion, the examples which illustrate the analysis presented hereabove, certainly support an optimistic view of the future of block polymers, particularly along the three main lines of application reminded in the introduction. After a sometimes discouraging maturation period (considering the huge research effort involved in their development), their come of age has been witnessed by several brilliant achievements, including very successful industrial developments.and will undoubtedly lead to further, and hopefully unexpected, realizations.
In particular, they have contributed, more than any other type of product, to translate that general concept of molecular engineering of bulk materials physical properties into a very actual and vivid reality.

Stil more important, now that we master their synthesis quite satisfactorily, they give us new keys to improve and to diversify broadly (even used in small amounts) classical polymeric materials of large diffusion : that is certainly a very worthwhile goal in our today's technology and economy.

o

o o

Acknowledgements : the author is deeply indebted to his coworkers for their devoted and enthusiastic collaboration (particularly to Dr R. Jérôme, and also to Drs G. Broze, R. Fayt, P. Hadjiandreou, J. Heuschen and P. Lefebvre), as well as to the Services de la Programmation de la Politique Scientifique (Brussels) for their very efficient support.

o

o o

References

1. G. Holden, E.T. Bischop and N.R. Legge, J. Polym. Sci. Part C (1969), 36, 37.
2. S. Marti, J. Nervo and G. Riess, Progr. Colloid. Polymer Sci., 58, 114 (1975); D.R. Paul, "Polymer Blends", Vol.2, D.R. Paul and S. Newman Edit., Academic Press, (1978), p.168.
3. C. Sadron and B. Gallot, Makromol. Chem., 164, (1973) 301; B. Gallot, "Liquid crystalline order in polymers", A. Blumstein Edit., Academic Press, (1978), p.191.
4. M. Szwarc, "Carbanions, living polymers and electron-transfer processes", Wiley-Interscience, 1968.
5. G. Berger, M. Levy and D. Vofsi, J. Polym. Sci., B4, 183 (1966); Y. Yamashita and K. Chiba, Polymer J, 4 (1973), 200.
6. R.B. Seymour, G.A. Stahl, D.R. Owen, and H. Wood, "Copolymers, Polyblends and Composites", N.A. Platzer Edit., Advances in Chemistry Series 142, (1975) p.309.
7. P. Hadjiandreou, Ph. D. Thesis, University of Liège, 1980; Ph. Teyssié et al., Symposium on Coordination Polymerization, A.C.S., 181th Meeting, Atlanta, 1981.
8. R. Fayt, R. Jérôme and Ph. Teyssié, I, Polymer Letters accepted for publication; II, J. Polym. Sci., part B, in press.
9. D.H. Richards, Symposium on Anionic Polymerization, J. McGrath Edit., A.C.S. 179th Meeting, Houston,1980.
10. J. Heuschen, Ph.D. Thesis, University of Liège, 1977; J. Heuschen, R. Jérôme and Ph. Teyssié, Macromolecules, in press.
11. T. Ouhadi, A. Hamitou, R. Jérôme and Ph. Teyssié, Macromolecules, 9, 927 (1976); J. Polym. Sci.,A15 , 865 (1977).
12. J. Heuschen,R. Jérôme and Ph. Teyssié, French Patent 2.383.208.
13. P. Lefebvre, Ph. D. Thesis, University of Liège, 1978; R. Jérôme, P. Lefebvre and Ph. Teyssié, J. Polym. Sci., in preparation.
14. P. Lefebvre, R. Jérôme and Ph. Teyssié, Macromolecules, 10, 871 (1977).
15. G. Broze, R. Jérôme and Ph. Teyssié, U.S. Pat. application pending; submitted for publication in Macromolecules.
16. Ph. Teyssié et al., in "Preparation and Properties of Stereoregular Polymers", p.144, R.W. Lenz and F. Ciardelli Edit., D. Reidel Publ.C°, 1980.
17. A. Moudden, A. M. Levelut and M. Pineri, J. Polym.Sci. B15, 1707 (1977); G. Broze, R. Jérôme and Ph. Teyssié, Macromolecules,in press; J. Polymer Sci., Part A, in press; J. Macromol. Sci., in press.

RECEIVED May 19, 1981.

21

Current Perceptions of Sino–U.S. Trade

JULIAN M. SOBIN

Sobin Chemicals Incorporated, 1900 Prudential Tower, Boston, MA 02199

In the People's Republic of China, such expressions as "Dida", "wubo" and "renduo" are used frequently.

Founded in 1949, the People's Republic of China has a total land area of 3.7 million square miles, including Tibet and Siankian Province, representing seven percent of the world's entire land surface. The southern and northern portions of China extend into the tropical and subfrigid zones to conform perfectly with the expression "dida", which means vastness of land.

"Wubo" means an abundance of untapped subterranean resource deposits, potential sources of electric power and farm products. It is estimated that China's coal deposits amount to 1600 billion tons, iron ore deposits 120 million tons, and petroleum several billion tons. China's hydroelectric power alone is estimated to total 35 billion kilowatts.

In addition, the estimated deposits of nonferrous metals, including tin, antimony and tungsten, are said to be among the world's largest. Agricultural production, when modernized, will certainly increase greatly. The expression "wubo" has come to be widely used in China.

"Renduo" is an expression meaning large population, and is frequently used to refer to China's approximately one billion people, who account for about a quarter of the world's total. This is "renduo" indeed.

The rich deposits of natural resources are important to China's economic development. Therefore, an effective integration of these factors is of the utmost necessity for the development of her national economy. How else can China think of achieving her goal of "The Four Modernizations": Science and Technology, Industry, Agriculture and Military Security?

The founding father of China, Sun Yat-sen, once stated that China had been subjected to semicolonial treatment and continued oppression by Britain and other European powers for many years. Under such circumstances--and with the outbreak of revolutionary

0097-6156/81/0175-0319$5.75/0

wars and foreign invasions following the collapse of the Ching Dynasty in 1911--China witnessed constant warfare being staged throughout the country. In addition, the exploitation that took place under a system of feudal landownership, complete with fiefs and landowners, made it utterly impossible for China to build a national economy.

Consequently, the country had little accumulated capital and her industrial development was limited to light industries. There had been no foundation for modern industrial development. As a result, the technology needed for industrialization of the country did not progress at all, and the country had to depend upon primitive agriculture, with its extremely low productivity. This prevented the accumulation of capital necessary to initiate production on a progressive scale.

Since the founding of the People's Republic of China 31 years ago, however, the country has made great steps forward toward modernization. This has been done through reorganization of its social structure, which used to be the fundamental defect in the national economy, in addition to investments and the accumulation of capital.

Following the successful formation of the People's Republic of China, the government of China carried out the first, second and third five-year programs and achieved considerable success. Through the Great Cultural Revolution, which began in 1966, and the Ninth National Convention of the Chinese Communist Party, the nation's economy was steered onto a path toward stabilization.

The country then started its fourth five-year program in socialization of the national economy and, in particular, the development of heavy industries. Success in heavy industrial development contributed much to establishing the bases for the development of modern industries. On the other hand, modernization of agriculture through the formation of People's Communes was carried out extensively. Since mid-1968, in particular, China's economy has made marked progress, to reach an all-time high in 1979-and well in 1980.

Trade With China

The key objectives of China's foreign trade were the establishment of an independent trade system through elimination of undeveloped trade setups and unequal semicolonial conditions that existed prior to the founding of the People's Republic of China. Meanwhile, a program was carried out to build trade to meet the demands of developing a socialist economy. In 1949, China declared that one of its general principles would be to restore and develop commerce and trade relations with foreign governments, with equality and reciprocity being the most important principles.

As a result, the Chinese government is directly responsible for foreign trade and the establishment of trade systems in response to the demands of an independent and planned economy. For this particular purpose, the foreign trade of China is controlled entirely by the state, with trade corporations and the Bank of China, which are government organizations, in charge of actual business transactions. Consequently, the country has closely coordinated foreign trade with domestic economic development programs, thereby making the most of import trade in developing her economy.

The actual operation of China's trade policy, therefore, is mainly carried out in accordance with governmental trade agreements. After these agreements are concluded, arrangements are made based on long-range treaties and the policy of maintaining a balance between exports and imports.

While the United States now has a governmental trade agreement with China, trade is still mainly carried out with China in accordance with private trade agreements. US trade with China therefore has been done through private, "friendly" transactions.

Sino-US trade has grown favorably since trade relations between the two countries were reopened in 1971. From $90 million beginning in 1972, it grew to more than $900 million by 1974. The US ranked second among Chinese trade partners then, but slipped to $450 million in 1975, a poor third.

Foreign trade in 1976 was weakened by the death of China's three leaders, Chairman Mao Tse-tung, Premier Chou En-lai and Chu De, father of the People's Liberation Army, and continuing low foreign trade statistics reflected the miseries of the Chinese people, who at that time brought about, and suffered from, serious production and transportation difficulties. These were incited, as the Chinese like to say, by the oppressions of the infamous "Gang of Four".

In 1976, the Gang of Four was crushed, when it was perceived that Chairman Mao's last years were being orchestrated by the oppressive and exploitive Gang, carrying on the repressions of the Cultural Revolution for their own purposes.

The year 1977, with a Sino-US trade volume of $374 million, was characterized by a liberalization program encouraged by the collective leadership that took over power, Chairman Hua Guofeng, wearing the Mao mantle, along with Vice Premier Deng Xiaoping, carrying out the promises of Chou En-lai to make China into a powerful and modern industrialized nation by the year 2000.

An important upsurge in Sino-US trade, totaling $1.2 billion, occurred in 1978, when by the year's end, we had normalized our relationship with the new China.

And 1979 advanced our continuing "private" trade relationship to $2.3 billion, but we must remember that fully fifty percent of that figure reflects U.S. agricultural sales to China.

China's Place in the International Community

People's Republic of China is not a major trading nation by international standards. The total trade of the United States is about 15 times that of China, and it was only in 1973 that the level of the mainland part of China's foreign trade surpassed that of Taiwan. Also, China's total international trade is a relatively low percentage of its gross national product--about six percent. Yet this is not surprising for a major country that has a massive population, a large domestic market and low per capita income.

Although China plays a minor role in world international commerce, foreign trade plays a major role in stabilizing and developing China's domestic economy. China depends on agricultural imports to help feed and clothe its huge population. China relies on imports of machinery and equipment, metals and advanced technology to support the growth and modernization of the agricultural and industrial sectors of its economy.

On the other hand, the nature of these transactions--licensing, planning, supply and installation of complete plants, particularly in the fields of chemicals and steel--indicates that China is not wholly abandoning self-reliance. Mainly, these transactions are used to enhance "confidence in one's own resources".

But it is already known that trade between China and the United States has continued to increase this year. Already, the number of contracts signed is markedly higher than in any recent year.

China is inviting a growing number of industrial missions. Many of these missions are from individual companies and are not planned to negotiate or close the contracts. Rather, they are designed to discuss possible future business, and companies often accept these invitations in order to inform the Chinese of developments in our Western technology.

American traders who have been visiting China since the Chinese opened the door to the US continue to meet their share of frustrations. The Chinese are only now allowing foreigners to deal with end-users of the products, but trade continues to be conducted "Chinese style"--that is, the Chinese reluctantly enter into long-term contracts--but there is still an absence of a conventional legal system in which to conduct trade.

Nevertheless, if maximum calm and patience is exercised in dealing with the Chinese, the experience usually proves quite satisfying.

No small part of the success of Sobin Chemicals Inc. in China is due to the fact that we learned during our first visit in early 1972 that, in China, business, politics, psychology, sociology and human relations are interwoven. Throughout the negotiation process, the interplay of these elements often times caught us off guard. From the Chinese vantage point, these factors can never be divorced from each other. And, to the Chinese, foreign trade serves as a vital extension of Chinese foreign diplomacy.

The means of utilizing trade for accomplishment of that purpose is distinctly different than that to which we are accustomed. However, in the past few years, since the "third coming" of Deng Xiaoping, China has exhibited a more practical approach toward trade. Politics is taking a back seat to hard economic facts and realities.

Negotiating patterns have reflected this reordering of priorities. In the last few years, major technical sales contracts to China, which are being negotiated in Peking, have increased in frequency and breadth. As China gears up to modernize fully her industrial base by the year 2000, she casts a sharper eye on more Fortune 500 companies in order to meet her formidable development needs. On the other hand, while these top companies more frequently are being asked to make presentations to end-users and to foreign trade corporations, such as Sinochem, Machimpex and Techimport, the resulting scenarios--played out in Peking, Shanghai and Canton against a backdrop of Chinese turf and a Chinese set of rules--really have not changed since our trade and exchanges reopened in early 1972. What has changed is the posture China is now adopting, the magnitude and scope of the contracts and the perspective with which China's foreign trade corporations approach the bargaining table.

It is, of course, in the contract that the exchange of goods finds its legal expression. Nothing dampens the Chinese process of carrying out the terms of a contract from beginning to end in accordance with the principle of "honoring the contract and keeping one's word." This, the Chinese assert is the strongest way to help the development of their foreign trade relations.

One must view so-called traditional business--the delivery of semifinished and finished products--as being increasingly supplemented by the sales of large capital equipment and whole plants. Whereas the former has usually been carried out through standard-form contracts of the respective foreign trade corporations, the latter transactions are negotiated in detail.

Negotiating with the Chinese

China is an increasingly competitive market. Out of China's total economic activity, only about six percent is devoted to foreign trade. To become part of that small percentage requires the utmost patience, tact and skill.

The Chinese are hard, protocol-minded businessmen who prefer to negotiate with our high-level executive officers and with those people who can make not only day-to-day but also long-range buying and selling decisions. The Chinese are formidable negotiators who are specialists and whose lives are devoted to developing negotiating skills.

Certainly they "role-play", anticipating our new ploys and counterploys. They are almost always well-versed on the American company with which they are dealing, as well as with world market prices. They are also thorough in familiarizing themselves with changing market conditions and are resilient and flexible enough to be able to respond on a moment's notice. No one is more expedient than the Chinese, and all manner of negotiating techniques are satisfactory to them in the pursuit of a principle upon which they stand resolutely.

The contract is just such a principle, and its contents are cast in concrete--and you had better not leave anything out. No one is more a comparison shopper than the Chinese, and no one knows better how to play off competitive buyers and sellers with each other.

All of the above may be in praise of Chinese negotiating style, because we American traders surely meet our peers in Peking, Shanghai and Canton. On the other hand, a cautionary word is that Chinese negotiating methods cannot be stereotyped. One has to appreciate the flexibility of Chinese negotiators who function in the interest of their end-users, but one must not overlook the strong "tributary" tendencies. For example, the foreigner with a strong desire to do business must respond to the traditional Chinese concept of international relations. The seller must almost always travel to China to negotiate personally; the seller must adjust himself to the procedure of the Chinese, who establish the agenda, change the topics and pace the negotiations; the seller must stay in China as long as the Chinese consider it adequate.

Contract drafts brought along by the seller are largely ignored by the Chinese.

The Chinese are not bashful about asking for commercial and technical presentations in Peking. Oftentimes, these days, the Chinese side may extend an unsolicited invitation to an American supplicant firm. If such visits result in contract negotiations, they are divided into technical and commercial parts. Only after the technical questions have been made clear, do the commercial negotiations begin. Such negotiations are in both parts arduous and difficult, but still quite agreeable and educational.

Among the Chinese negotiating team, there is clearly leadership at both the technical and commercial levels, with the commercial leader appearing to be superior. Sometimes the leader is introduced as such or his position only becomes apparent in the course of negotiations. Oftentimes, this leadership position is occupied by a woman. The leader possesses both professional and ideological competence. On the other hand, the ideological element is sometimes apparent in the person of a so-called "people's engineer". Occasionally the ideological element cannot be identified with a particular negotiating member but is represented somewhere in the group.

As far as the Chinese technical side is concerned, it must be acknowledged as including specialists coming partly from the negotiating foreign trade corporation, partly from the end-users and partly from scientific advisory institutes. It is easier to see the difference in their competencies than it is to discern the allocation of functions on the Chinese commercial side. If there are banking or shipping people present, they are rarely recognizable. More important, the Chinese negotiate without legal advisors, and this does lead to misunderstanding. The Chinese clearly do not share the importance we give to contractual legal questions or to the meaning and the relevance of our legal institutions.

It is frequently disconcerting while participating in important negotiations with the Chinese to witness the structural composition of the Chinese negotiating team changing, as it tends to do. Explanations must be repeated, and this may reflect the Chinese desire to use the opportunity to learn as much as possible about our negotiating techniques and practices. Some members of the Chinese team may be simply note-takers--or they are political figures, keeping the negotiations within the framework of ideology?

How the Chinese check with superior authority is not well known. Important negotiations take place in Peking so that the Chinese can refer to their leadership. While that is obvious, it seems that the competence of the Chinese negotiators to make decisions is probably quite limited in the technical sphere as well as in the commercial field.

It is obvious that the main reason for the relatively long negotiations that take place in Peking is the constantly moving vertical structure of the Chinese decision-making process. On the other hand, direct contact by our side with superior officials in Peking is impossible. The higher leadership of the relevant Chinese corporation appear only when the contract is finally ready to be signed.

In contrast to more private negotiations in Peking and Shanghai most negotiations in Canton are conducted in the open halls of the Canton Fair building. Foreign traders waiting their turn are free to eavesdrop as their competition negotiates. Despite this disadvantage, the Chinese display an uncanny sense of fair play. (For instance, I recall negotiating a contract to buy a particular chemical. After signing the contract, I was astonishingly pleased to find that the Chinese wanted to lower the price to me. When I asked why, the Chinese negotiator pointed to a newspaper report indicating that the price of the chemical had fallen subsequent to my signing of the contract.) Sometimes to our chagrin, the Chinese may reverse the process!

The Chinese negotiators are generally meticulous and astute. They prefer hard bargaining, searching for the truth of the marketplace no matter where it may be. And they have a strong distaste for Madison Avenue sales tactics. The Chinese are often both subtle and blunt about what they want. They almost insist that US company negotiators include both technical representatives and commercial personnel with authority to make final decisions.

The negotiating team will most often include US representatives who act on the US client company's behalf. It is clear, however, that the Chinese are reluctant to have American sellers or buyers represented by third-country parties. The Chinese prefer to deal with Americans representing American products and American companies. They are interested in learning about our thought processes, our marketing practices, our corporate management skills, as well as about our products, services and technology. Yet they have a sense of humor not unlike ours.

While it is possible that negotiation may take place through an exchange of communications via mail or telex--depending upon the type and size of the proposed contract, and of course upon the nature of the products and technology under discussion--negotiations generally require face-to-face contact in China.

More often than not, the Chinese are appreciative of those firms that take the trouble to visit the semiannual Canton Trade Fairs, to see the state of the Chinese product and trade art.

To be sure, negotiations of capital equipment and high technology sales contracts generally take place in Peking, and they begin with the all-important technical seminar.

Basic negotiations for such sales involve four phases:

•Phase 1 - The introductory step, which enables each side to get to know each other. It is a "feeling out" process. The American negotiating team, consisting of one or more persons knowledgeable in both technology and sales, exchanges general information on the company, its products, technology and services.

•Phase 2 - The discussion phase, which is the period during which the Chinese indicate their level of interest regarding the purchase of this equipment, technology or product. Negotiations at this stage center around technical questions, capabilities and specifications.

•Phase 3 - Dealing with contractual principles and price quotations. Matters pertaining to payment, delivery and inspection also are discussed.

•Phase 4 - The stage at which the deal is consummated. Whereas Phase 3 ironed out all major questions of the contract framework, this last stage gets down to a review of the contract, exact delivery dates and payment terms, and is concluded with the signing of the contract.

The pace of negotiations is erratic and grueling. While there may be difficulties in arriving at a final agreement, there is no deal until the understanding of both negotiating teams has been reduced to writing and signed by them. If there is interest and it is immediate, the sale will be concluded that much more quickly. But one cannot place time limits on negotiation sessions. Negotiations have been known to drag on for days, months, even years.

Someone once observed that negotiating in China is similar to a group encounter session. It is a time-consuming experience, and you cannot really perceive the outcome until the very conclusion. And, before you can reap the benefits, you must be willing to inject adequate doses of cultural understanding, cost, time and talent into your efforts.

Economics

An important wrinkle in the Chinese business environment is the magnitude of the deficit spending in 1979. Treasury Minister Wang Bingian revealed a budget deficit of $11.5 Billion in 1979, which the Chinese hope to have reduced by half in 1980. The effect of the resultant cost-consciousness will be a sharp cutback in capital investment, mainly in industrial projects. Foreign borrowing will be held to 20% of exports which should reach $18 Billion in 1981. Joint ventures will be encouraged but will be concentrated in light industries and primarily designed for export earnings.

To achieve better value for the money, and to squeeze waste out of the economy, Chinese industry will be made freer, and therefore, it is hoped, more productive. Yao Yilin, the new Chief planner (Vice Premier of the State Council and currently Chairman of the State Planning Commission), in a speech delivered on August 30, 1980 at the Fifth NPC, confirmed reports that experimental autonomy granted to selected factories during the past 18 months will be broadened to include all state enterprises. Factories are encouraged to adjust their output to market demands and purchase their necessary inputs in the open market--once state production and procurement targets are met.

Factories can and will deal directly with foreign buyers once approval is granted from the top. Furthermore, they will be given greater responsibility for their use of allocated assets, and more flexibility in making their own adjustments to changing conditions. Economic market forces, the suppression of which in the past led to dislocation of supply, production, and distribution in Mao's China, will be restored to prominence. Starting next year, the State will charge a fee for the money it provides; where possible, state investment is to be replaced by interest-bearing bank loans--a move designed to establish financial accountability at the factory level. Plants must become "genuinely independent units of cost-accounting responsible for their own profit and losses".

Another important goal is to reduce the profligate waste which is endemic to centrally-planned economies. A high priority will be given to energy-saving. Although China will become a major producer of energy in time, its energy output in 1981 is projected to increase by only 0.9%. China plans to save 7-8 million tons of coal and 2.5 million tons of oil per year by shifting investment from high-to-low energy-using industries and by converting oil-fired power stations back to coal. Reflecting the importance of energy, a new energy commission was created at the 1980 Summer NPC to oversee this vital sector. YuQuili will be the Commission's first administrator. The creation of this Commission comes at a time when Petroleum Minister Song Zhenming resigned as a result of a scandal surrounding an offshore oil rig accident ("Bohai 2") in 1979. It was headlined in People's Daily as a classic case of "politics overruling technical expertise", and denounced as symptomatic of much that ails China today. This is not to be in the order of things in Deng's new China.

There is more to the resignation of Petroleum Minister Song Zhenming (and the demerit meted out to Vice Premier Kang Shen) than meets the eye. The explicit and implicit message in these unprecedented reprimands will have a profound and enduring impact on the Chinese bureaucracy. The purpose of this aposematic public purge is twofold: First, to remove another loyal Maoist who owed his rise to Mao's blessings, and strike a blow against the Maoist dictum (politics overruling science). This completes the dismantling of twin Maoist pillars: "In Agriculture, learn from Tachai (Dazhai); in Industry, learn from Taching Daqin." Ex-Vice Premier Chen Yungui of the now discredited Dazhai production brigade in Shanxi Province, Mao's former egalitarian paradgm, and Petroleum Minister Song Zhenming of Taching (Daqin) have both exited from the scene. From now on, strict adherence to scientific methods and procedure ("seeking truth from facts" or "shih-shih ch'ui shih") is to the rule of the land. To emphasize the point, Chinese scientists have been invited to give lectures and seminars in the Chinese leadership compound of Chungnanhai. The necessity of having the right leaders at every level is the main reason for the campaign to study the criteria of truth. Second, to serve warning to unreconstructed Maoist and bureaucratic deadwood that cooperation and support will be demanded of them in the modernization drive. Removal and early retirement are distinct possibilities for intransigents. As an editorial in the People's Daily puts it, ".....with the settlement of the idealogical line, we must settle the organization line. Whether a correct political line is carried out by people who support it or oppose it, produces very different results. The most fundamental thing in an organizational line is to select and cultivate successors."

Another important if implicit message is to foreign businessmen that they no longer have a friend in the Petroleum Ministry. Private sources indicate that for the wealthiest ministries (hard-currency earners) of China's State Council (i.e., Petroleum and Metals & Minerals), frequent ministerial changes will be the rule rather than the exception. This is to guard against entrenchment and to reduce the chances of over-fraternizing with foreign businessmen. While China is neither ready nor able to pursue a systematic crackdown on incompetence, it nevertheless demonstrated its intention to begin with the most flagrantly negligent example as epitomized by the incident of the barge, "Bohai #2".

This series of unprecedented high-level shake-ups demonstrates the determination of the present leadership to perform the necessary surgery to streamline industry--particularly the petroleum industry. The Chinese leaders have reasoned (undoubtedly accurately) that improved performance of the Chinese Petroleum Ministry and its attendant petrochemical industries is crucial, if they are to play a vital role in modernizing China and improve the living standard of its one billion citizens.

How will China conduct business, especially with foreign partners? The two income-tax laws passed in China in September by the NPC should partially fill the legal vacuum in which would-be foreign partners have been operating.

Resident foreigners will do their bit to help China fill its coffers by paying income tax rates of up to 45%, retroactive to January of 1980. (It is a progressive levy that ranges from 5% and starts with earnings of $540/month for persons who are resident in China for a year.) Foreign companies involved in joint ventures will be taxed at 33% (How China will tax joint-venture profits: 30% to national government and 3% to municipal and provincial governments, plus 10%--if profits were repatriated in the first ten years of a venture.)--except in special export zones--where the rate will be 15%. Another 10% will be levied on remitted profits, but there will be reductions and exemptions for the first three years so as not to discourage potential investors.

The Chinese tax structure lacks strong incentives for a foreign company to move in, especially in view of the fact that the Beijing government has yet to unveil such basic guarantees as a labor law, a law on contracts, insurance or a civil code.

In spite of these problems, the present may turn out to be the optimal time to contract with China--before any new host of rules reduces negotiating flexibility. CITIC officials have offered assurances in private that potential investors can write into their contracts the protection normally afforded by typical Western commercial legislation. The contractual arrangements will be immune from adverse changes (i.e., quantum leaps in raw material and labor costs) emanating from subsequent legislation. Indeed, the early investors would acquire "pioneer" status, and the attendant prestige would produce advantages not available through any legislation. Finally, the Chinese leaders place a premium on the success of the early joint ventures to create good will and a favorable climate for investing in China.

As a further stimulus to international business, China has announced its willingness to consider and/or accept such unprecedented arrangements as:

a) long-term credit and the financing of capital imports through foreign bank deposits of the Bank of China;
b) barter and compensation deals for plant purchases;
c) the importation of materials to reprocess for export;
d) designation of certain factories to produce goods for specific foreign markets, establishment of special export zones (SEZ):
e) fuller compliance with international standards on patents, trademarks and copyrights;
f) direct government-to-government loans;
g) joint ventures with foreign firms;
h) private foreign investments (even 100% foreign ownership is under consideration);
i) seeking loans from both the IMF (International Monetary Fund) and the World Bank.

Wang Bingqian, Minister of Finance, in a speech given at the Washington, D.C. IMF meeting in October, 1980, stated that new Chinese laws were being drafted "to guarantee the legitimate rights and interests of foreign investors and collaborators." It is a clear signal that China intends to move into the "establishment" which manages world economic and politcal affairs.

Science and Technology

The history of science in China is long and impressive, to which Joseph Needham's multi-volume treatise on the subject will attest. The Chinese, having acquired immense knowledge in Chemistry through practice, as far back as 5,000 years ago, made wine and sugar, and 2,000 years ago, made pesticides such as sulfur dioxide and arsenic oxide by indigenous methods. The history of modern science in China goes back only to the 1800's, and organized modern research did not take root in China until 1914.

Scientific progress in the PRC is subject to the buffeting effects of political vicissitudes. The Chinese Academy of Sciences was established after the founding of the PRC in 1949. The new government was strongly commited to science and technology and it set about establishing a significantly expanded network of scientific and technological institutions.

By 1965, Chinese science and technology had registered impressive growth and development despite major disruptions stemming from political turmoil during the Great Leap Forward (1958-1960). The period of stability following the GLF was ephemeral and ended with the advent of the Great Cultural Revolution in 1966. The capricious decisions of one man, Mao Zedong, coupled with the new brand of oriental obscurantism championed by the Gang of Four during the Great Cultural Revolution, disrupted, and in extreme cases, as no catastrophe, natural or man-made could do, shattered the warp and woof of Chinese society. Animosities built up during that tumultuous period continue to exact a toll on economic growth and scientific progress, even today.

The most serious effect of the Cultural Revolution on science was the devastating impact it had on higher education. Thus, China now enters its drive for comprehensive modernization with a missing age stratum of scientists resulting from ten years of omitted higher education and advanced training. (The number of graduates in Engineering & Sciences dropped from about 100,000 in 1965 to a low of 5,000 during the height of the Cultrual Revolution in 1967. The number of yearly graduates hovered around the 10,000 mark until 1970. In 1979, China's colleges and universities were able to accommodate only 270,000 freshmen. China, with a population of 1 billion and a labor force of 430 million, is now seriously handicapped by the severe shortage of skilled labor that can be pressed into service in its modernization drive.)

With the death of Mao, removal of the Gang of Four and the subsequent rise of Deng Xiaoping, scientific and technological policy and organization have been under intense national review and discussion. The consensus is unequivocal. The development of science and technology and science education must be given top priority for "without modern science and technology, there can be no modern agriculture, industry, and national defense." Science and technology will serve as "the key" to China's "Four Modernizations."

China is now ready to make a national political commitment to rapid scientific and technological development. As Kuo Mojo, late president of the Chinese Academy of Science, proclaimed, "Springtime for Science--the most splendid in the annals of the Chinese civilization--has come and is here to stay....the time when the reactionaries can trample on science has indeed gone forever!"

It will be sometime before reality fully catches up with political slogans. The current crop of Chinese leaders is bent on "seize the day and seize the hour, defy all hardships, and reach the pinnacles of world science at the earliest possible date." In fact, political and social expedients demand it, but political and social considerations also demand that "Springtime for Science" will not simultaneously affect negatively the status quo nor alter the structural integrity of its socialist society.

Caveat notwithstanding, one may rightfully expect significant scientific advancements to be forthcoming from the extraordinary people who have invented--among other things--the compass, gunpowder, the use of oil as a fuel, papermaking, printing and porcelain, and was scientifically leagues ahead of Europe until the 13th century.

Scientifically, China is characterized by extremes. It has been described as a vast underdeveloped nation with islands of high technology and science. Chinese scientists are confident that China can do well in any technological or scientific field, within the state of the art, by giving it top political priority and allocating sufficient funds to see it through. It is in the area of development, production, and especially transferring technology from the laboratory to the plant level (where it has to survive the rigors of a host of economical and structural limitations), where the Chinese readily concede their insufficiency.

It remains to be seen whether China can create its own innovation chain from R&D to commercial implementation. Given the prerequisites of political stability, committed leadership, adequate allocation of resources and a functioning infrastructure of research and training institutions, the future can be very promising.

There is a vast reservoir of residual goodwill toward the U.S. stemming from the generation of Chinese scientists educated in the United States in the 1930's and 1940's. This aged group of Chinese scientific elite will soon pass on the reins of scientific leadership to a generation of scientists who were trained in the U.S.S.R. in the 1950's. The correct aim for the U.S. is to make a vigorous bid to establish strong and ongoing bilateral links with the Chinese scientific community (especially those trained in the U.S.S.R.), because scientific and technical ties are invaluable in promoting closer political cooperation.

The establishment of intensive links with the scientific community of an emerging superpower will certainly be advantageous in the long run. In the short run, the attendant boost to U.S. sales of technology, hardware, and management are some additional inducements for closer ties. Someday, perhaps, Chinese scientists may even reciprocate the initial one-way flow by helping to advance U.S. research in some areas.

Chemical Industry In China

China's chemical industry began with a 40,000 Ton/per year Caustic Soda Plant in 1914. A few plants producing basic industrial chemicals were built in the 1930's. By 1949, what the Sino-Japanese War (WW II) did not destroy of China's chemical industry, the Soviets managed to dismantle and remove.

In 1948, three small oil fields produced 876,000 barrels per year, and in 1956, with the discovery of oil in Xinjiang, China's oil production rose to 22 million b/year. China was a net importer of oil until 1963. Her Chemical Industry did not really really take off until 1973, when oil from the fields in North China started to flow. China's Chemical Industry is still in the embryonic states. Given the present production level of 2.3 million barrels/day, China has finally entered the petrochemical era. She has developed a credible, if still thin, infrastructure, research backup, and downstream integration for its petrochemical industry.

They have the raw materials for most resins and are producing most of the basic feedstock. Therefore, they are certainly more interested in acquiring foreign manufacturing technology for resins rather than finished plastic goods.

The long-term strategy will be, first, to concentrate on adapting and digesting foreign technology and on short-term research, and later advance to longer-range and more basic research.

After going through what Deng Xiaoping described as "paid our tuition and suffered a little...", the Chinese have adopted a very pragmatic approach to acquiring foreign technology. These Chinese appear to have been turned off from their proclivity to buy whole huge plants, toward agreements providing for the transfer of technology. Importing more technology and less hardware is implicit in the Chinese aphorism of "buying hens to lay eggs", instead of buying "eggs" themselves.

The fundamental strategy of the PRC's leaders will be to use petroleum and petrochemicals as a launch pad for a Chinese version of "better living through chemistry". Petroleum and petrochemicals will provide much needed raw materials and capital necessary to finance China's ambitious modernization drive. China is at the point where it can make the now quite attainable breakthrough to a decent consumer society.

Chemicals will play a central role in China's economic masterplan. They represent 5% of the PRC's yearly exports and 17% of imports. Upgrading of diet and clothing is contingent upon the expansion of chemical manufacturing and process chemical exploitation of resources of various kinds.

All petrochemical projects, either planned or under construction, are large-scale plants of international proportions. When completed, the Chinese hope to develop the capacity to manufacture a wide variety of general-use plastics, synthetic rubber, synthetic textiles and other petrochemical-derived products.

In the cavernous halls of the Shanghai Industrial Exhibition, one can see a cornucopia of consumer goods (clothes of polyester, polyacrylic, and polyvinyl alcohol fiber; shoes and sandals of polyvinyl chloride; suitcases and television set frames of acrylonitrile-butadiene-styrene plastic; toys and containers of polyethylene, and many other plastic products (China produced approximately 800,000 tons of plastics in 1980) of convenience we take for granted in the West) that the Chinese government will try to deliver, in quantity, to its citizens in the years to come.

With all the imported technology and plants, China's petrochemical industry will undoubtedly increase production by leaps and bounds, in due time. Factors limiting its future expansion may be China's future production of energy resources, particularly petroleum, and China's allocation of such resources to its petrochemical industry.

Estimated Production of Chinese Crude Oil by Oilfield

(Unit: Million tons)

Oilfield	1979	1980	1981	1982	1983	1984	1985
Daqing	53	55	55	54	50	49	48
Shengli	20	22	24	26	28	30	30
Huabei (Dagang, Renqiu)	11	12	14	15	15	15	15
Others Onshore (Incl. Shale Oil)	24	25	26	28	29	29	30
Bohai Bay	0	0	2	5	10	20	30
Others Offshore	0	0	0	0	5	10	20
TOTAL	108	114	121	128	137	153	173

A recent report by Robert Silin of BA ASIA LTD. reported that by 1985, China will allocate an estimated 10%--or 17 million tons/year (350,000 barrels per day)--of its total oil production of 170 million tons/year to its expanding petrochemical industry. Also in 1985, the Chinese expect that 70% of the materials used in light industry will come from its own petrochemical industry, up from the present 5-10 percent.

The current state of Chinese petrochemical technology is quite low. The Chinese plant managers and engineers have not reached the level of sophistication where they can fully visualize the plant as a continuous process. This deficiency is caused by two interrelated factors--one technical, the other structural.

Technically, the Chinese are quite new to the game and are inexperienced. But it is with this level of inexperience that they have to operate 300,000 ton capacity ethylene plants employing state-of-the-art technology. Even if they do finally master all the intricacies of running such giant plants, Chinese managers and engineers are handicapped by the lack of industrial infrastructures (factors beyond their control) that are essential to the smooth functioning of the plants. A streamlined plant requires a systematic, continuous flow of feedstock, utilities and finished product shipment--a complex process whose successful execution is contingent upon a level of managerial sophistication and industrial capabilities (i.e., adequate transportation systems) that China has not yet acquired.

Structurally, the weakness of the centralized management system typical of socialist countries, becomes clear. Managing the economy by the apparatus of government, making enterprises its mere appendages, blocked horizontal linkages between enterprise and enterprise; plant and plant. The vertical organization structure, bureaucratism, and technical inexperience present formidable bottlenecks to the rapid absorption of imported technology.

Chinese leaders are learning that a competent manager cadre class is not an automatic by-product of turnkey contracts. (Modernization plans cannot proceed without participation of professional administrators; modern technology and modern management are but two sides of the same coin, and it is only natural that one has to merge the latter with the former.) Some of Beijing's grandiose investment projects will not work because they cannot be sustained on China's "widget" industrial base. The real need is to import more varied foreign technologies, in a much cheaper and cost-conscious way. That is the way Japan developed in the '50s and '60s. The eventual big successes were Japanese modifications of rather unexpected foreign technologies that worked.

The Chinese state planners are stepping up the effort to maintain and expand the coal-based energy supply and to minimize the use of oil for primary energy. Alternate sources of energy, such as methane gas produced from agricultural and human wastes, are also being promoted. This system of energy and resource allocation is designed to save petroleum (which, in Chinese eyes, is almost too valuable to be just burned) and divert as much petroleum as feedstock to increase the flow of petrochemical products to the people, one of the major goals of the modernization. China, with the third largest coal reserves in the world, is pursuing the development of coal-based technology to produce additional feedstock.

In 1979, China produced approximately 106 million metric tons of oil (ranked ninth in the world) equivalent to approximately two million barrels a day. Of the total oil produced, China exported about 10 million metric tons. Assuming that China consumes 10% of its naphtha and kerosene domestically as feedstock, she should have been able to produce a theoretical maximum of 2.5 million tons of ethylene for 1979. Given the forecasted production figure of 170 million tons of oil in 1985, and if the allocation of oil for feedstock is increased to 15%, China should be able to produce a maximum of 3.6 million tons of ethylene.

By way of comparison, in 1979, Japan--with only one-tenth of the PRC's population--consumed three times the total oil output of China, or some 300 million tons. In 1978, Japan's petroleum industry produced eight times more ethylene than China, 400 times more LDPE, nine times more HDPE, five times more polypropylene (PP) and 13 times more PVC. Even if China manages to get all the projects onstream, it will still have only one-half the total ethylene capacity of Japan.

Chemically, the simplest class of commercial polymers in the polyhydrocarbons. The most important commercial hydrocarbon polymer is polyethylene. The consumption of monomers from which polyethylene is derived can be used as an index for comparing the relative living standard of nations.

Table I: Consumption in 1978, Kg per Capita

	U.S.	Western Europe	China
Ethylene	55	40	0.3
Plastics	65	50	0.6

The level of U.S. consumption is about 110 times that of the Chinese level for plastics and 180 times for ethylene. Given the absolute shortage of resources relative to its population, China will not be able to attain the level of petrocivilization enjoyed by the U.S., nor is China likely to try. If China's economic planners can quadruple the consumption levels of ethylene (to 1.2 kg/cap) and plastics (2.4 kg/cap) by 1985 (implying an annual growth rate of 32%), they will have done well.

As can be deduced from plant purchases, the PRC is still at the formative stage where emphasis is on producing only the most basic petrochemicals. No plants were purchased for producing dibasic acids (phthalic and maleic anhydrides, etc.) and flurocarbon or tetrafluoro ethylene or some of the advanced engineering plastics like ABS polyacetals, polycarbonates, polyimides or any other unsaturated polyesters. Another important area of low Chinese activity is thermoplastics for space and defense applications.

The gaps in China's petrochemicals industry and research will present major outlets for U.S. plants and technologies in the latter half of the 1980's. China's strong commitment to applied chemistry and her perception of the U.S.A. as preeminent in both basic and applied research will translate into important competitive advantages for the U.S. when China is in the market for the necessary technology and hardware.

While China is borrowing from foreign technology the innovations necessary for its economic development, the leadership is also strongly committed to having indigenous R&D capability for effectively absorbing foreign technology and for the very nationalistic reason of becoming eventually "self-reliant" in science and industry.

With China's oil reserves, the growth of her petrochemical industry and the related growth of a polymer industry are high on China's priorities. With the growing need in China for cheap materials and reduced dependence on agriculture as a material source (it currently supplies 40% of materials used in light industries), expansion of polymer research and production is strongly on the agenda.

Chemical research in general and polymer research in particular are highly application-oriented, concerned with quotidian problem-solving, and the generation of production processes leading to the improvement of products. In the polymer area, there is not much theoretical work, and there is no basic research in the sense we know it. Synthetic polymer work in China is highly applied, and includes effort to synthesize polymers by the usual radical and ionic routes, to make block polymers by the living polymer technique, and to make condensation polymers of polyester, polyamides, and polyimides. The strongest efforts are to find new polymerization catalysts, using rare earths.

The Chinese have decided to emphasize research to maximize short-term developments of their production capacity. As China progresses in its industrialization program, the emphasis will include long-range research.

Agriculture and Industries

Yu Qiuli, former Chairman of the State Planning Commission, seems to have shared in the opprobrium resulting from the same "Bohai #2" incident. Yu Qiuli was made head of a new State Energy Commission, but his new position is in fact a demotion from his previous job. Another veteran administrator, Yao Yilin, replaced the tarnished Yu as the Chairman of the State Planning Commission.

"The Party and the Government are Making Tremendous Efforts to Increase People's Income and to Improve Their Life in Both Urban and Rural Areas."

Yao Yilin's Economic Report at 30 August, 1980,
National People's Congress Session

Yu Qiuli is acknowledging the past imbalance of neglect of improving the salary structure because of over-accumulation of capital for industry. China's work force has had its wages frozen for 20 years (1957-1977) and this has seriously undermined worker and peasant incentives. The Chinese government granted salary increases to 40% of its work force in order to enlist worker support in the modernization drive. Now the government must provide a modest increase in consumer durables.

For the first time since the founding of the PRC, the improvement of the quality of life for her citizens has been given such high priority. It even edged out defense spending (a perennial favorite) which suffered a cut of $2 Billion, or 13 percent.

Long-term trend (implicit in this Economic Report) is the Chinese policy to gradually decrease the present dependence on imports for food, fibers and industrial materials and eventually lead to greater use of export earnings for the purchase of high-technology equipment and processes. The strategy for the '80s and '90s seems aimed at initially heavy exploitation of petroleum and petroleum derivatives to create the conditions and environments for domestically-driven "self-reliant" economic growth.

To implement the betterment-of-life plan, four major areas are targeted for improvement. The are 衣 (clothing), 食 (food), 住 (housing) and 行 (transportation), and--not coincidentally--these four areas are directly dependent on petroleum and petroleum derivatives.

Chinese planners are counting heavily on petrochemicals and polymer science to provide the diverse goods needed by the Chinese consumer.

Clothing 衣

China's textile raw material needs by 1985 will amount to at least 4.26 million M.T. In synthetics, China is committed to developing polyester fibers as its primary manufacture, with output for 1985 targeted at 1.5 MMT (equivalent to the 1979 Japanese level of synthetic production. The present Chinese output is only 326,000 M.T.) Limited quantities of nylon, polypropylene, vinylon, acrylic and polyester fibers are already produced.

There will be import substitution by increased production in China of polyesters, other synthetic textiles, plastics, synthetic rubbers and many chemicals. Beneficial ripple effects of the increased production of indigenous synthetics will be to:

1) Reduce the imports of synthetic fiber, synthetic thread, rayon thread, and other textile goods, the total import value of which exceeded $300 Million in 1979.
2) Expansion of synthetic fiber production is crucial if the textile industry is to supply more and better clothing to consumers. Synthetics will substitute for natural homegrown cotton, wool or silk in the production of textiles and thus release land previously devoted to cotton, for food production. Cotton cloth production at the 1980 level of 12 billion square meters is insufficient to meet domestic demand, and is still rationed.

Food 食

The critical role of agriculture cannot be overemphasized as an obvious axiom of the policy of self-reliance, which is the ability to feed its own population. Political slogans such as "Agriculture as Foundation" and "Industry Must Aid and Serve Agriculture" are just some indications of the primary importance accorded to agriculture. (A separate Ministry of Food was established on June 12, 1979, with Zhao Xinchu--full member of the 11th Central Committee of the CCP--as its first Minister.) Agricultural development will continue to be viewed as the fundamental basis for economic modernization, and the point of departure for pursuit of import substitution policies.

Improving agricultural yield is still one of China's highest priorities. More indigenous production of chemical fertilizers will help toward that. It will also reduce the present large-scale imports of fertilizers, a drain on the balance of payments, as has the buying-in of foreign grain. Chemical fertilizer has been one of the fastest growing Chinese industrial sectors during this decade and should continue to grow rapidly in the immediate future. By the end of 1980, there will be 13 new Urea plants, bought from the West, completely on stream, with a production target of about 18 million tons of nitrogen fertilizer per year.

The Chinese are also counting on petrochemical feedstock to promote synthetic substitutes for the huge quantities of edible oils used industrially (i.e., in the paint industry). Edible oil supply is still deficient in China (China imported $36 million worth of soybean oil from the U.S. in 1979 to help alleviate the shortage), and is rationed. The use of petrochemical derivatives should help boost the per-capita edible oil consumption (less than one-quarter kg/month), and correct some dietary deficiencies.

The eventual goal is to shift the composition of Chinese diets away from grains toward higher intakes of vegetables, edible oils, and animal protein, that is, into a pattern akin to those associated with more affluent LDC's.

Housing 住

During the self-induced turmoil and excesses of the Cultural Revolution and its aftermath (1966-1976), the construction industry was crippled and construction of residential housing was brought to a standstill. During the same period, the population increased by 300 million, thus putting severe strains on the already stressed housing situation (where 9 square meters per person is considered spacious).

In 1979, only 120 million square meters of new residential construction was undertaken. Given a population of 1 billion, that works out to be an insignificant increase of 0.12 m^2/person. Massive building programs projected for the next decade will attempt to correct past neglects in the housing sector.

A new Ministry of Construction Materials Industry was formed on March 5, 1979, with Song Yangchu as its Minister. The mission of the new Ministry will be to supervise the production and promotion of heretofore non-existent synthetic building materials. The increased use of petrochemical derivatives in building construction will ease the demand on timber and produce a decidedly beneficial side effect: that of revitalizing the chronically depressed paper industry in a land which prides itself as the inventor of paper.

Transportation 行

Chinese planners clearly perceive that upgrading their inadequate transportation system and equipment is essential to the success of their modernization program.

China now produces less than 90,000 metric tons of natural rubber, and in 1978, she subsidized domestic needs by importing 229,000 metric tons. China produced only 70,000 metric tons of synthetic rubber in 1979; by comparison, the U.S. produced 2.5

million MT and Japan 1.1 million MT. Currently, less than 25% of China's total rubber consumption currently goes into vehicle tire production, and in 1979, China produced only 12 million units of outer rubber tires. The continuing shortfall in rubber goods has led to greater emphasis on the development of synthetic rubber capacity.

On a more pedestrian level, China's Textile Ministry is promoting the greater use of synthetics for shoes. While there is ample supply of PVC and polyethylene for informal wear ("bubble sandals" in China), the Chinese will aim for the greater use of synthetic leather for more general-purpose footwear.

World Scene

Many economically important polymers are derived from gas and oil. In terms of raw materials, the U.S. enjoys a very distinct advantage in having an abundant supply of natural gas (a resource of negligible importance in Europe and Japan). Furthermore, the U.S. imports only half of its oil needs, which means that there is a small but significant advantage in having half of its oil supply at prices slightly below that of the world market.

In contrast to a relatively self-contained U.S., Europe and Japan are subject to political turmoils and are at the mercy of political vicissitudes in the Middle East. Europe and Japan are especially vulnerable to adverse changes because they both suffer from low levels of natural resources.

OPEC nations are, of course, gearing up to be the primary producers of petrochemical products in the coming decade instead of exporting only crude petroleum. Mexico, with its vast oil reserves, galloping full speed ahead at a 24% growth rate in petroleum output (the world average is 4%), may in fact overtake OPEC nations in the shift from crude petroleum to petrochemical products.

China is likely to become a major influence in petrochemicals in the late 1980's, particularly in Asia, if the current growth schedule holds firm. That China will be a significant exporter of petrochemicals is due to two important factors. One, the need for massive offshore funds, and Two, the inability of China's downstream capacity and domestic demand to absorb this sudden surge in output. The exportable surplus of petrochemicals will include substantial quantities of benzene, propylene and perhaps C4 fractions, HDPE, PVC, ethanol, SBR/BR, and possibly polypropylene.

China's petrochemical exports are still quite limited at present. International sales are handled primarily through SINOCHEM. In the near future, we may expect the Ministry of Chemical Industry to be more directly involved in exporting through China National Chemical Industry to be more directly involved in exporting through China National Chemical Construction Corporation (CNCCC). This trend is impelled by the new emphasis placed on ministries to earn their own foreign exchange requirements. Although to date, the Ministry has had little experience in this area, CNCCC has recently established a special unit to tackle the matter. The MCI will probably emerge as the final authority on China's petrochemical industry. The MCI already controls production of petrochemical intermediates and finished products, and refineries processing feedstock.

Another development is that China National Technical Import Corporation (CNTIC), a unit under Ministry of Foreign Trade, is losing importance to CNCCC, which is under MCI. CNCC is responsible for purchasing and constructing chemical and petrochemical plants with MCI controlling the bulk of funds and the allocation of crude oil, which is used as feedstock.

Trade has been characterized as the binder that ties the loose leaves of cultures together. It has been the vital function of trade since the dawn of civilization.

The opening of China has been hailed as the single most important event in the latter half of the 20th century, and businessmen have been in the vanguard of this developing relationship. U.S.-China trade ballooned from $96 Million in 1972 to about 4 Billion in 1980. The projected figure for 1985 is around $10 Billion between the U.S. and China as compared with a forecast of $7 Billion between the U.S. and U.S.S.R. The Trade Agreement signed in 1980, and the extension of Eximbank credits both augur well for the expansion of U.S.-China trade. China's twin goals of improvements in living standards and modernization should provide the U.S. with major outlets for agriculture (will reach $2 Billion/year next year) and high-technology products.

China's reduction in whole plant purchases in favor of the importation of equipment and technology, as well as the renovation of existing plants, will further enhance the United States' competitive position. America's preeminent technical leadership in petroleum and petrochemical industries will undoubtably attract Beijing to favor U.S. companies in these high-priority ventures.

We are Sobin Chemicals have been privileged to play a small part in this historic interchange. We feel China is a market and a culture which will have an increasingly important and positive impact on our society in the years to come.

While China trade promises no economic panacea, it does offer entrepreneurial challenges found nowhere else. China trade is still in a dynamic state of flux in spite of, or because of, the host of new regulations and laws. These newly-passed regulations and laws, in my opinion, are meant to serve as general guidelines only and do not necessarily carry the full weight nor the finality of law as we know it. China trade is a market and a gave of imagination for the intellectually agile and the financially astute entrepreneur with integrity.

At this point, I would like to relate what a Chinese trade official said to a group of apprehensive and visibly agitated foreign businessmen in response to their anxious queries. He said, with a smile, "We are fated to live in interesting times, and provocative times!"

China is a country rich in talent and human energy waiting to be harnessed to the twin yokes of modern management and modern technology. I am convinced that a fruitful interchange between the U.S., the nation with the most advanced technical skills, and China, the nation with by far the largest management challenges will produce far-reaching and rewarding ramifications. Forging China into the business communities of the democratic nations through a network of bilateral ties (scientific, trade and cultural), will bolster Chinese experiments that strive to advance quietly toward economic growth and freedom, light years away from previous doctrinaire, authoritarian practices. The establishment of these ties is of paramount importance because nothing the West can do will stop China from reaching to seek one. There is no foolproof insurance against "losing" China a second time. But the longer the nascent understanding between the United States and China is allowed to mature, the better the prospect that any Chinese turn toward the U.S.S.R. will not mean a corresponding tilt away from the West.

I have presented a general socio-political overview of China. The in-depth analysis of China's political environment is both necessary and relevant. The prevailing political climate in China determines not only China's economic development path, but also its scientific and technological development path. Politics is China's "Great Helmsman". In the process, I hope I have provided sufficient background information on the Chinese psyche and society to aid in interpreting nuances when interacting with the Chinese.

RECEIVED July 27, 1981.

APPENDIX

Awards and Honors of Herman F. Mark

Honorary Degrees

Universitaet Liege	Dr.Ing.h.c.	1950
Universitaet Uppsala	Dr.Phil.h.c.	1953
Free University of Berlin	Dr.Phil.h.c.	1954
Technical University of Berlin	Prof.Chem.h.c.	1955
Lowell Technological Institute	Dr.Sci.h.c.	1956
Technical University of Munich	Dr.Eng.h.c.	1960
Johannes Gutenberg University Mainz	Dr.Sci.h.c.	1962
Karl-Franzens-University in Graz	Dr.Phil.h.c.	1964
Technische Hochschule in Vienna	Dr.Ing.h.c.	1965
Polytechnic Institute of Brooklyn	Dr.Sci.h.c.	1965
Karls University, Prague	Dr.Phil.h.c.	1965
University of Jassy, Romania	Dr.Sci.h.c.	1971
Universidad Autonoma de Madrid	Dr.Sci.h.c.	1973
Technion-Israel Institute of Tech.	Dr.Sci.h.c.	1975
Long Island University, N.Y.	Dr.Sci.h.c.	1976
Montanistische Hochschule Leoben	Dr.Mont.h.c.	1976
University of Nottingham	Dr.Sci.h.c.	1979

Medals and Awards

Heinrich Hertz Medaille	1928
Wilhelm Exner Medaille	1934
Medaille des Internationalen Chemiker und Koloristen Vereins	1937
Harrison Howe Award-American Chemical Society	1948
Franqui Medaille	1948
Honor Scroll of the American Institute of Chemists	1953
Legion d'Honneur	1953
Medal of Honor of the Polytechnic Institute of Milano	1954
Golden Medal of Honor of the University of Vienna	1955
Transenster Medal of the Association of Belgian Engineers	1955
Nichols Medal of the ACS	1960
Distinguished Service Medal-Syracuse University	1961
SPE International Award in Plastics Science and Engineering	1962
Gold Medal Award-Indian Association for the Cultivation of Science	1962
ACS Award in Polymer Chemistry	1965

Olney Medal Award	1965
Austrian Honor Cross For Arts and Sciences	1966
Elliot Cresson Medal	1966
Swinburne Medal of the Plastics Institute, England	1968
Science Award of the City of Vienna, Austria	1970
Distinguished Service Award-Polytechnic Chapter Sigma Xi	1970
Scientific Achievement Medal Award-City College Chemistry Alumni Association	1972
Chemical Pioneer Award-American Institute of Chemists	1972
Willard Gibbs Medal Award-American Chemical Society	1975
Silver Award of the Republic of Austria with the Star	1975
150th Anniversary Prize of the Aachen and Munich Vericherungsgesellschaft	1975
ACS Award in the Chemistry of Plastics. and Coatings Sponsored by Borden Foundation	1976
Harvey Prize in Science and Technology-Technion, Israel	1976
Distinguished Service Award-Polytechnic Institute of New York	1977
Humboldt-Preis	1978
Plastics "Vision" Award-Society of Plastics Engineering	1978
The Wolf Prize in Chemistry	1979
Perkin Medal	1980
The National Medal of Sciences	1980

Membership of Academies

Oesterreichische Akademie	1933
Bukarest Akademie der Wissenschaft	1934
Oesterreichische Akademie	1935
Budapest Akademie der Wissenschaft	1938
New York Academy of Science	1943
Oesterreichische Akademie der Wissenschaffen	1947
Amsterdam Akademie der Wissenscaft	1949
Indian Academy of Sciences	1950
American Academy of Arts and Sciences	1956
National Academy of Sciences	1961
Soviet Academy of Sciences	1966
International Academy of Wood Science	1968
Austrian Academy of Sciences	1970
The New York Academy of Sciences	1977

Honorary Membership

Chemical Society Madrid	1936
Chemical Society Bukarest	1937
Verein der Chemiker Koloristen	1937
Oesterreichische Gesellschaft fuer Roentgen-forschung	1937
American Institute of Physics	1944
Royal Institute of Great Britain	1947
Max Planck Gesellschaft	1947
American Leather Society	1949
Wiener Physikalisch Chemische Gesellschaft	1950
Textile Institute of Great Britain	1950
Verein der Papier Chemiker	1951
Oesterreichische Gesellschaft fuer Hozforschung	1952
National Institute of Sciences of India	1952
Weizmann Institute of Science	1953
Italian Chemical Society	1954
Phi Lambda Upsilon Honorary Chemical Society	1962
Plastics Institute of America	1965
The Fiber Society	1965
The Franklin Institute	1971
Society of Polymer Science of Japan	1972
Honorary Fellowship of the Indian Chemical Society	1973
Society of Plastics Engineers of SR Croatia	1974
Plastics Hall of Fame	1976
The Chemists Club	1977
The Chemical Society of Japan	1978
The American Institute of Chemists	1978
Society of Plastics and Rubber Engineers	1979

INDEX

INDEX

Jacket design by Carol Conway.
Production by Cynthia Hale.

Elements typeset by Service Composition Co., Baltimore, MD.
Printed and bound by The Maple Press Co., York, PA.